Home B

One woman's solo jo
continents o... ...o wheels

Steph Jeavons

stephjeavons.com

© Stephanie E Jeavons

Steph Jeavons has asserted her right to be identified as the author of this work in accordance with the Copyright, Design and Patents Act 1988.

ISBN 978-1-8381239-0-1
British Library Cataloguing in Publication Data.
A catalogue record for this book is available from the British Library.

Printed and bound by booksfactory.co.uk
Edited by Paul Blezard and John Pearce
Cover design by Dylan Chubb
Typesetting by Peter Jennings-Bates
Graphics by Paul Tomlinson, 2can design

Previous publications by the author
Embrace the Cow (How to Ride Around the World on a Budget)
Available on Amazon Kindle

For Tim, and in his memory
Thank you for believing in me
It meant the world

For my parents
Linda and Peter Jeavons
Thank you for never giving up on me
It meant everything

The greatest glory in living lies not in never falling, but in rising every
time we fall.
– Nelson Mandela

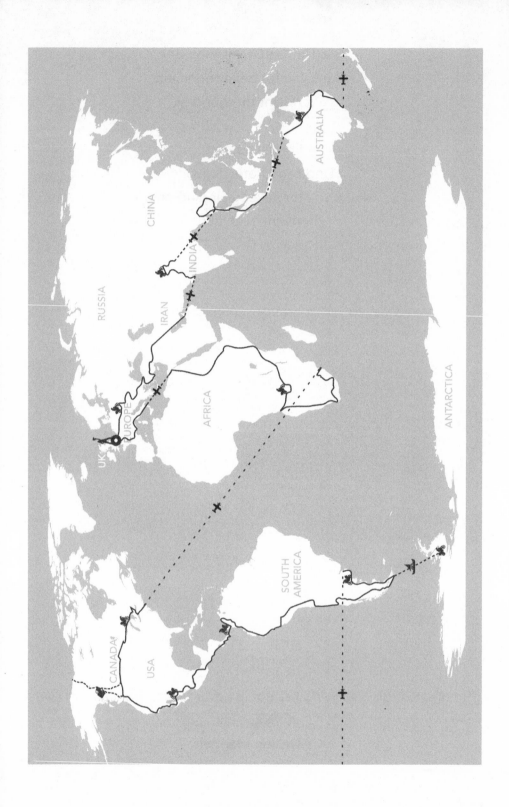

Foreword

I guess you could say that I have always lived my life behind bars of one description or another. Either motorcycle 'bars, or otherwise! The irony is that they both set me free.

Writing this book has been a journey in itself.

I hope my journey makes you feel something: joy, sadness, discomfort, hope, anticipation....perhaps even a little inspiration. Maybe all of the above. You may love it or hate it, but that's OK, as long as you feel something!
This is me!

Thanks

There have been so many people who have helped me on my journey, in so many ways. I'd like to thank you all by name, but I guess you know who you are. From my sponsors:
Motobreaks, Halvarssons, Kriega, Giant Loop, Honda UK, to the people who offered me a bed for the night or a shoulder to cry on or company over a beer. To all my friends who have always encouraged me to be myself!
Thank you.
This has been far from a solo journey.

Thank you also to all those who helped put this book together. Ian Marchant, Pete Bog, Paul Blezard, John Pearce, Dylan Chubb and Paul Thomlinson, not to mention all those who offered advice along the way.

****

Let everything happen to you. Beauty and terror. Just keep going. No feeling is final. – Rainer Maria Rilke

Contents

Before

The Journey Begins – Europe and Asia

Australia

Antarctica

South America

Central America

North America

Africa

Chapter 1
1995 – The View

I knew it wasn't real, but I refused to blink. Instead, I let my vision blur and my imagination take over. If I tried really hard, I could edit out the razor-wire and the unremarkable tree beyond would become a mighty baobab on the plains of the Serengeti. The really old kind; a thousand-year-old monster with a trunk in which you might imagine an elf had made a home. On good days, I could paint in a leopard dozing in its gnarly branches. It helped to imagine. Imagination, and the odd dose of heavily-cut smack that had probably made it through security in some haggard old con's dry crotch, was all I had left between me and insanity.

Tonight was noisier than usual, because the wing had just received a delivery of new recruits; 'fresh meat' to intimidate and help us pass the time. New inmates also brought with them the glimmer of a possibility that one had managed to get a stash in past the screws, and that always got the wing hyped. It reminded me of my time working at a zoo when the chimps would go crazy at the sight of their bedtime Horlicks; a treat we would feed them in milk bottles through the bars before settling them in for the night. They'd jump around, bang on the bars and let out high pitched grunts, then press against the bars with bottom lips protruding in a cup shape, ready to be filled with warm, milky goodness. They'd do anything for that Horlicks. Pulling back from the window, I realised my face had been pressed so hard against my own bars that there were two big red lines on my sickly white skin. So this is what it felt like to be on the other side. The chimpanzee in the cage. It sucked.

The calls and screams were getting harder to block out. These new girls were in for one hell of a night. Worse still, I was going to have to listen to their whimpering. If one of them did have smack, holding out from their new 'roomies' would have serious consequences. Those pack hunters would tear them apart if it meant just a few hours between the hard walls of their prison cell, enveloped in the embrace of 'The Big H'. They could sniff out heroin faster than you could say 'de-crotching'.

Smackheads, junkies, trainspotters, whatever you choose to call them, they were easy to pick out. Thin and sickly-looking; lined faces, and often toothless, if they were old-timers. Withered, as if someone had sucked all the juice out of them. Childhood memories long gone and all hopes of salvation lost, they reminded me of that bit of mouldy spud you find behind your fridge; the one that escaped two years ago when you were making chips. Once fresh and full of potential, but now wrinkled and long past its sell-by date, with only the silverfish and the dust mites for company.

As my mind threatened to drift back to memories of my own 'welcome party' several months earlier, I forced myself to concentrate on the tree again; away from the reality of my prison walls and back to my conjured-up image of the African plains; (a slightly fantastical one, as I would eventually discover... many years later).

That night, as I roamed through the golden grass and sandy trails of my mind, I came to a decision. I don't think I really believed it at the time. My reality was a difficult one to escape, but I wasn't meant to be here behind the fridge, festering and forgotten. I'd always wanted to travel. I *would* travel. I would do all the things I'd dreamed of as a kid. I would be free, and I would go and find myself that baobab if it killed me. I pulled out my diary and wrote in big childish letters:

TO DO:

1. GET OFF DRUGS.

2. TRAVEL THE WORLD

Returning the tattered prison-issue exercise book to its hiding place in the bottom of my prison-issue cupboard, under my prison-issue slippers, I pulled out my stash of non-prison-issue heroin and did another line before the previously administered warm glow and feeling of hope wore off completely. Salvation seemed so far away, so impossible, but hey, at least I had my old friend 'H' to keep me company until I got there...

Chapter 2
My First Hit

By the time I was fifteen, I'd spent many nights hanging out in rural Welsh squats, smoking weed, and listening to Lynyrd Skynyrd and Pink Floyd. I didn't live in them as such, but found myself drawn in most nights by the loose and liberal environment of the Great Welsh Unwashed. My main haunt was a squat called London House in the tiny little village of Deiniolen, nestled at the foot of the slate quarries of Snowdonia. The house is perfectly respectable-looking these days, but back then it was a damp and dirty place, with posters of Che Guevara and Oscar Wilde peeling from the already crumbling walls. To most, it was a shithole infested with dropouts and a place to avoid at all costs. To me, it was sanctuary: my hide-out from conformity and the dreary existence of rural living.

I hung out with people who were, to me, glamorous in the roughest sense of the word. They didn't care. They saw no point in slaving to be like the last generation, whose values they (we) despised. We often ditched our Doc Martens and walked barefoot in the street; wore tie-dyed clothes and rebelled against all that was uniform or expected. We worshiped Bob Marley and stank of patchouli oil. I was always the youngest of any group and ridiculously naïve but, at the time, I felt I belonged to something profound. We were the next generation and we were going to start a revolution from our stained communal mattresses.

Six months after getting sacked from my first job as a trainee pool attendant in a 5-star hotel, I moved up the North Wales coast to Colwyn Bay, where I started in the zoo's Youth Training Scheme. Working with animals was my dream as a kid, so landing a job as a trainee zoo keeper was an amazing opportunity for me. I was still only sixteen.

It was a wonderful two years, and I still remember it the way others remember their university days. By day, I loved my job; I worked and studied hard. By night, I partied just as hard to rock bands in the local smoke-filled pubs. Here, I felt alive and part of a culture I could relate to – all leather jackets and long hair. I loved the smell of leather, the taste of snakebite, and that strangely comforting feeling of sticking to the carpet if you stood still too long.

It wasn't long before I fell pregnant to a Scotsman I'd met in a pub over a game of pool. It was December 1992, and I was seventeen. We'd been dating for three months, and were on the brink of splitting up when I realised I'd messed up with my pill. We talked it through. We tried to make it work. It lasted about three days as I recall, before he walked out. I held the door wide open for him, resisting the temptation to slam it into the back of his musty dreadlocked head. *We* were not meant to be, but the baby? I didn't see what else I could do. I would keep it, of course. I would keep it and I would stick it in a rucksack and I would travel the world. Just me and my baby, barefoot and carefree. We'd follow the yellow brick road all the way to Kansas, skipping and clicking our heels to the sound of Pearl Jam and The Red Hot Chilli Peppers.

I went into labour at 3 a.m., alone in a pokey little flat on the third floor above a newsagent in Colwyn Bay. It was September, 1993. This was well before the omnipresence of mobile phones; when a landline was a luxury I could ill-afford. After taking a warm bath, I packed my things, slowly crept down the first set of stairs to the flat below, and used their phone to dial for an ambulance. Then I called my mum.

"Hi, Mum. Just thought I'd let you know I'm in labour. I'll let you know how it goes."

"Okay. Do you want me to come to the hospital?"

"No, I'll be fine."

It seemed important that I should stand on my own two feet, but by the time I got to the hospital, she was already there, and I was very glad to see her. I suddenly felt like the foolish and scared child that I was, as reality came banging on my uterus-shaped door in the form of severe and regular contractions. I needed my mum.

The whole process lasted sixteen hours and I was pathetic throughout. I screamed so loudly that the midwives told me to keep it down as I was scaring the other mums-to-be. Then my belly went flat and jellified as one final push delivered a large and healthy baby boy. Eight pounds, ten and a half ounces. They handed him to me and everyone smiled down at us as if it was the happiest time of my life. I looked at this bundle in my arms. It was covered in blood, it had a pointy head with jet black hair stuck down to its scalp as if had just arrived through some space vortex from another planet. What was I supposed to do now? What was I supposed to say? The midwives looked expectantly at me for a reaction. I gave them what I guessed they wanted. I smiled and said something like,

"Oh, isn't he cute?" That seemed to do the trick, as they quickly whisked him away from me and left me to sleep for eighteen solid hours. When I woke, they brought him straight back and placed him in my arms. I looked down at his now clean yet scrunched up face and waited for those motherly instincts people had told me so much about to kick in. "The most natural thing in the world," they'd said. They didn't come. Instead, I felt overwhelmed with fear and confusion, and, worst of all, I knew I couldn't show this to anyone. I couldn't show them I was unnatural, not a real person. I was a freak of nature and clearly, I was the only person in the world who had ever felt like this. Completely alone. Kansas and the yellow brick road drifted far, far away, to a place that would remain buried and forgotten for many years to come.

****

All too soon, I was left alone at home with my son. I kept smiling and nodding at those who cooed over the pram, but in truth I felt numb. What I didn't know then was that I was falling deeper and deeper into a severe postnatal depression. At eighteen, what did I know about mental health issues? I didn't even know it was 'a thing', let alone that one in ten mothers suffer from it. Instead of getting treatment, I continued to believe I was simply a terrible person who'd just had an alien dropped on her with no instruction manual. I couldn't imagine there was any treatment for that. All I knew was that nothing was happening as I was told it would. It didn't come naturally *at all*. I was silently screaming and didn't know how to reach out to anyone, so I just went through the motions of what I thought was expected of me. Clearly, I was a freak of nature. A failure at being human. Telling anyone would be the worst thing I could do.

I considered ending my life many times, but the alien needed attention. Who would look after him if I wasn't around? I thought he might be better off without me. Fortunately, a strong instinct of loyalty and protection, that I didn't understand, forced me to hang around.

Six months later I struck up a friendship with Dewi, the long-haired boy next door, with big brown eyes. Gentle and caring, he had an exceptionally well-organised flat, a little blue budgie named Gizmo, a Yamaha RD250 motorcycle, and could name the year of any reggae or rock song ever released. I found him interesting, and was amazed when he took an interest in me and my little alien. I followed his lead when it came to his patience and caring for my son, Nathan. There was a glimmer of light emerging from this thick blanket of depression. It was muffled, but it was there. In the evenings, once Nathan was in bed, we would chill together and listen to his vast record collection. My condition didn't disappear, but I felt a little less freakish and distracted from my inner thoughts. Maybe even a little hopeful that I wouldn't have to do this alone.

We soon moved in together and for a while, we felt like a proper family. Dewi had a lot of biker friends – black leathers, mullet haircuts, self-styled, matt-black choppers, and loud exhausts. I loved them all, and whenever I could, I would join them on rideouts and camping trips around Wales.

I mostly rode pillion on the back of the boys' bikes back then, although I had a acquired a Yamaha MZ 50 of my own. There were some women with their own motorcycles though and occasionally I'd get on the back with one of them and we girls would leave the boys behind and head out, all black leather and exhaust fumes, for a weekend of clubbing and camping. There weren't that many female bikers then and seeing ten of us rolling into a town had the effect we wanted. We would keep our leathers on as our identity, hitting the local clubs and really giving the towns something to talk about.

Dewi had another group of friends who would turn up occasionally. I had no idea they used heroin until I walked into the bathroom one day and found Lucas there, 'shooting up'. He was a laid-back kind of guy with scruffy hair who always wore a retro green mod jacket, and he was on the floor with a belt around his arm and a needle stuck in his bulging vein. He looked at me with glazed eyes and muttered something like "Oh, sorry Steph," as if I had just walked in on him having a pee. This was *way* out of my league. I screamed at him and threw him out.

Until that moment, I thought all heroin addicts were nasty vermin who'd crawled out from the gutters; wasters you could spot a mile off. This was not your stereo-typical junkie – Lucas was a really nice guy who didn't fit my bias. *Did I have it all wrong?*

Dewi and I fought until he admitted that he too, "occasionally dabbled" in the drug. He was surprised that anyone was 'shooting up' in our home. He agreed to stop and to ban anyone associated with it from the flat. That didn't last long.

Dewi missed the warm and cosy feeling that smack gave him, filling that hole that most of us have to some degree or another. In some, it's a vague feeling of something missing. In others, it's more. If we're lucky, we find a way to fill it. Some find God. Others find a person, or travel, or gardening. For Dewi, it was smack.

My first 'inhalation' was a spur-of-the-moment decision. By now, people were smoking in front of me, and I grew used to having it around. I also grew curious and perhaps my first time was the classic "Fuck it" moment; a quick fix; a holiday from myself and from the constant noise of a depressed mind. I didn't think about my responsibilities or what might happen later, I just did it. There seemed more reason to try it, than not.

As the smoke filled my lungs, the dark cloud in me instantly melted into the warm rush that came over me… a feeling of love and of being loved… a simple clarity that life was going to be OK. It neutralised my pain, and like a super-efficient fairy godmother, it swept me up and transformed me into a better person; a person capable of loving and of a greater empathy than I had ever felt. It brought out the best in me. I became the me I wanted to be in a world that was soft and warm and inviting. Then I threw up – a common side-effect, as it turned out.

That night, the four of us in the room talked openly and honestly about our lives. We listened and cared about each other. I felt a togetherness and affection for my fellow human beings that I had never before experienced.

Dewi and I lived our secret lives for over a year, with apparently no one suspecting we were heroin addicts. Many commented at first on how good I looked, and asked what diet I was on. Stick-thin was the way forward in the social circles of the day – especially among supermodels, to the extent that the term 'heroin chic' became common parlance, even though cocaine was more likely to be their drug of choice. I don't remember how long it took from that first glorious hit for me to need it every day, but it wasn't long. Heroin is exceptionally good at taking control of you.

As long as I had it, I could function normally. I became a master in the art of portrayal, as I held down a full-time job and played the role of mother. No one would have guessed, except for a few tell-tale signs, that this mother had a little helper. We eventually went out less with the biker friends, and spent more time with the people who were part of our secret society. The heroin side of our life slowly took over. What started as a seemingly harmless occasional pastime became our lives. Our passions and interests faded away along with any desire to chase our dreams. What dreams? Nothing was more important than that little £10 brown bag; it was what we lived and breathed for. Every week we crossed another boundary; took another step down the ladder to get that fix. Belongings sold; deals made; actions taken; £20 bags required instead of £10 ones as our tolerance for heroin increased; caught in a vicious circle of filling the void at all costs, which soon became a downward spiral – and some of us paid with our lives.

Overdoses were common in the secret society. So common, it became normal to us. Some came back from the brink to continue where they'd left off. Others did not. It wasn't just us of course. Many lost their lives in the late '80s and early '90s as the UK suffered a plague-like epidemic of heroin addiction fuelled by two aligning factors: sky-high unemployment and a flood of high-quality brown, smokable heroin from Iran and Pakistan. It spread out of the music scene and into every corner of the country, with the poorer areas of Scotland and Wales becoming a breeding ground for the disease.

By the late 1990s, over 400,000 people, later known as 'The Trainspotting Generation' were receiving treatment. During this post-

Thatcher, post-punk, high unemployment era, many people chose heroin over the grim realities of life. The majority of them never escaped. Today, about 300,000 receive treatment in the UK, most of whom are middle-aged. As the scars began to show on the ageing users, the drug became 'dirty', turning the next generation to ecstasy, cocaine, and the rave scene. I was a 'trainspotter'.

Dewi and I never injected. In this way, we protected ourselves from overdosing and from needle-related diseases like AIDS. Instead, we would smoke it in the ritual commonly known as 'chasing the dragon'. You heat the heroin powder on some tin foil until it turns to oil, then keep it moving in lines so it doesn't overheat and burn up too quickly. You then chase and inhale the smoke with a foil tube. The ritual itself can become addictive, and it was a long time after I kicked the habit before my senses stopped going crazy whenever I heard the crinkle of someone unwrapping a Kit Kat. Back in 1995, I was spending about £20 a day on heroin.

Whichever way you take it, the craving for that soft, dream-like world where everything is perfect, becomes insatiable and, at the same time, harder and harder to reach. The fairy godmother now wants more pay for less magic, and soon your body physically needs it to function. Heroin is so similar to the natural opioids in our brain that our body soon gets lazy and stops creating them for itself, instead sending out the craving signals to seek out what it needs.

You tell yourself every day that tomorrow you will find a way to stop. Tomorrow will be your last hit. Tomorrow you will find the strength to beat it. But by now, your body and mind are working against you, breaking down your self-control and rational thought. All they want you to do is get more. Only a small, powerless part of you remains, trapped inside a demanding vessel. Once you've had the sharp edges removed from life, adding them back in is cruel and unbearable – not to mention physically painful. Then I started dealing and getting my supplies wholesale, and my tolerance increased massively.

****

One evening, in the summer of 1995, I was driving home from one of my weekly trips to Liverpool for supplies. As I rounded the last corner to home, I spotted roadworks outside the house and one of those little workmen's tents right where I usually park. I slowed to see if there was any space for me, as they seemed to be clearing up. One of the workmen – in high-viz vest and dirty steel toecap boots – gestured to me as if offering me a space where he was removing the cones. I smiled at him and nodded. He moved the cones out of the way, and I remember feeling a little bit of pride as I reverse parked beautifully in front of all these men. I kept my heroin in my hand (this gave me a chance to throw it if I was pulled over) and as I got out of the car the same guy came toward me,

"That was good timing, wasn't it, love?" As I replied,

"Yes, it was. Thank you," he kept coming closer. I remember thinking, "All right, mate, no need to come that close," and being vaguely annoyed at his use of the word 'love'. It was then that he grabbed me by the neck and threw me to the ground. Suddenly, all six men were on top of me. Knees digging into my legs. My torso. My arms. Voices shouting,

"POLICE! KEEP STILL!"

Sharp pain as my tail bone digs into the tarmac. No meat left on the bones. The full weight of six policemen on top of me. Pumped. My head forced to one side. My breathing laboured. Panicked. A voice in the distance. Getting louder.

"Let it go, Steph. LET IT GO! IT'S OVER!"

In that instant, the realisation of my situation dawned on me and my world came crashing down. Thoughts raced through my mind so quickly, that it was as if time stood still. I squeezed the drugs so tightly in my hand that they could not pry my fingers open.

I needed more time to think; to work out how I was going to get out of this. It wasn't over. There had to be a way, surely. This could not really be happening to me. I found the strength to keep my fingers shut. The only thing between me and a prison cell. I knew it. I could already see it as I lay there in all the commotion, staring at the detail in the road that I was forced against. The little cracks; the edge of the pavement; and the gutter I was heading for.

But it *was* over. Soon everyone would know what I was. What I had become. There was no escape.

"What have I done? This is not me. How has it come to this?"

"THIS IS NOT ME. THIS – IS – NOT – ME."

But it was me. Then all the fight left me. I loosened my grip and let the drugs slip through my fingers. It was time to let go.

Chapter 3
Grisly Risley

"Seriously?" I asked incredulously, as the prison officer stated the number that would replace my name, "You're giving me an ID that sounds like an STD?" She grabbed my arm and checked for the tell-tale track marks before ordering me to get behind the curtain and strip. I was now known as Prisoner VD3713, and I placed what was left of my dignity in a marked jiffy bag with the rest of my belongings, hopefully to be reclaimed on the way out. "Can this day get any worse?" I wondered, as I hauled my skinny torso into the cubicle, and stripped. Yes, it could get A LOT worse.

That morning I had left home, bags packed, prepared for my inevitable sentencing (I had already pleaded guilty). I'd been warned to expect time behind bars when my case had been elevated from magistrates' to Crown Court, several months earlier. Today was merely a question of 'How long?' That morning was also the first time I met my barrister. A solicitor had done the 'leg work' and then passed the baton when it came to the trial and the sentencing in the Crown Court.

Now, I sat at a desk opposite this stranger, in a small interview room below the court. He pulled out my file and scanned it, seemingly for the first time. There it was. My life in black and white. The good, the bad and the indefensible, all neatly laid out in Times New Roman font. I wanted to scream. Instead, I reached for my cigarettes and observed the shake of my hand as I brought one to my cracked lips. The filter stuck as I pulled it away and drew a little blood. I registered the pain with a minute twitch above my left eye. "Can I have some water please?" I asked with my most polite voice. I wanted to make an impression. I wanted him to see more than those inkjet lines. But he was seeing nothing. He hadn't even made eye contact and, continuing in this vein, completely ignored my question and got straight to the point: "You're looking at five years, if you're lucky."

It was an agonising couple of hours before the judge called me into court. A pair of security guards led me up the stairs and into the dock. My dad was there – I didn't like to think what was going through his mind. Little was said in my defence, and the judge delivered his verdict very quickly. He said I was a disgrace who deserved five years, but he would take time off because I'd pled guilty. He added that I didn't deserve the reduction, and I couldn't really argue with that. So my sentence was three and a half years; if I behaved myself, I would spend half that time behind bars.

Until that moment, I had not fully understood the true power of words. The sheer gut-punching terror they can evoke. The way they can hit you with a physical force so strong as to buckle your knees and empty the wind from your lungs. Just three little words, "Take her down."

Handcuffed and losing altitude fast, I was grabbed at the elbows by the officers who flanked me and guided me gently out of the dock and down the narrow staircase towards the underground cells. My legs were working desperately to rebuild their molecular structure as my brain refused to process this new reality: my life is over.

I was now a convicted drug dealer. I was officially the dregs of society, and no amount of talking was going to get me out of the sentence I had just been dealt. It would be all over the local papers the next day:

SINGLE MOTHER CONVICTED OF SUPPLYING HEROIN

Cold, hard print for all to see. By then, I would be tucked up in my prison cell, joining in with the chorus of calls and wails from the forgotten, the desperate, and the incarcerated.

Upon arrival at Her Majesty's Prison (HMP) Risley, the three other newbies and I were processed. Then the stereotypically butch prison officers at 'Reception' led us, each carrying a bundle of blankets and prison issue clothing, to see the doctor for the standard checks before they took us onto the wing. My whole body was SCREAMING for heroin and I felt very weak; 'cold turkey' – withdrawal symptoms – had started. It had been several hours since my last 'hit', and I was starting to feel clammy and achy all over. Yup, this was gonna be a long night.

Risley is just how you might imagine prison to be: steel bars, netting between the sterile landings to break the fall of the unfortunate, and bunkbeds seemingly made of stone. There are many kinds of prisons and Her Majesty eventually had the pleasure of putting me in four of them, each with a different amount of oppression in its regime.

Back then, Risley was worse than your average prison. It was a remand centre, so everyone was in flux. It was a non-place; a crossroads for the embittered and nihilistic; a waiting room for those still to be sentenced, or, (as in my case) the newly-sentenced, who had yet to be allocated their final destinations. Nothing there is permanent and as a result, people were more unpredictable. People become prey to their environment very quickly – either as bullies, or as the bullied. Also, for various reasons, some of the nastiest and most disturbed inmates end up staying much longer at Risley than most, as I was about to find out...

I felt clumsy as we slunk in following the screws. My brain desperately tried to get a message to my feet to turn back; my legs feeling heavy in the confusion of mixed messages. All my senses were on high alert; my heart rate quickened and the hairs on the back of my neck stood to attention with a prickly sensation warning of my impending doom. A bead of sweat began to work a path down my furrowed brow, yet I felt deathly cold. Helpless, I walked on, fighting against my urge to scream while dragging my limbs forward to the sound of the jangling keys and the combined smells of damp, despair, and stale cigarette smoke – a scent that will stay with me forever.

The girls on the landings above began shouting the usual unrefined taunts, designed to ridicule the powerless newbies and assert their authority from the jump-off. This was not my first time on the wing, so I knew what was happening. I had been remanded here six months earlier for a couple of weeks before the court set bail for me until sentencing. Back then, I was bullied quite badly by some of the feral gangs who found strength in togetherness. Alone, I was powerless and quickly intimidated into becoming a submissive wreck who dared not fight back. Fear of the unknown can be powerful, but this time it was fear of something I desperately wanted to forget. There was no way out, and loneliness painfully and slowly sank deep into my very soul with such force that I felt it would ultimately be the death of me. It was a wound as undeniable as a blade to the heart.

As the screw led me into my cell, I pleaded with her to lock me in.

"You know I can't Steph, it's Association. You can stay in here, though, and close the door if you want, but I can't lock it." We were locked in for twenty-two hours of each day unless we had a job, and Association was the only time we were allowed to 'socialise' or watch TV on the wing. I didn't want to socialise; I wanted the safety of my locked metal door.

The door suddenly opened and some familiar faces walked in – four of the wing's top dogs – or rather, top bitches. I recognised Gail first. Gail was always the loudest, most aggressive, and least predictable of the gang – the one everyone tried to avoid. Big, black, and addicted to crack, she was someone with whom you avoided eye contact at all costs. Rumoured to be in for stabbing her own mother over a bag of crack, Gail was a master – or mistress – of intimidation. Each week, she bought talcum powder from the prison shop and ate it in a big, dramatic production, saying, "Any white powder is better than nothing."

Once, she got a rubber gorilla mask from one of the screws and put it on with her hood up. It was very spooky; she ran around the wing during Association making ape noises and loving the attention. To Gail, any attention was good attention. I've forgotten the names of the others, though I do remember that each had the "Don't fuck with me" look down pat.

"What did you bring us Steph? Got any boot?"

'Boot' is prison slang for heroin. I knew I was in trouble because no matter what I said, there was no way they were going to believe me.

"I swear. I didn't bring anything. I didn't want to risk it."

You could almost taste the adrenaline as it seeped out of every pore of every woman in the room. The atmosphere was pregnant with the promise of violence; the pause before the final pounce. I stood frozen to the spot, barely able to breathe, scared to move even a fraction, in case it triggered the attack. The single second it took them to reply felt like hours, "You gonna hold out on us?"

I recoiled in terror at the rubber gloves in Gail's hand. They were probably stolen from the hospital wing and are used for 'de-crotching'. A plastic spoon stolen from the kitchen is also sometimes used.

'Crotching' occurs when an inmate smuggles drugs vaginally or rectally inside condoms or in the cut-off finger of a rubber glove. This is

15

done either upon entry into the prison, or after being handed over by a prison visitor. This could succeed because the prison is not allowed to search women internally and sniffer dogs are unable to detect drugs through the skin. 'De-crotching' is what it sounds like and is often brutal, causing internal injuries. Women can be vicious – especially when it comes to feeding the insatiable appetite for drugs and the escape they bring from reality.

Like a cornered rat, I ran for the largest gap between the four women and tried to get out the door. There was no escape. They piled on me, slamming a hand over my mouth to stifle my scream as they wrestled to get me on the floor. Just as my knees were buckling for the second time that day, the door behind us flew open and the screw who'd earlier seen the fear in my eyes stood in the doorway, "What's going on?" Before she even finished her sentence, we were all upright and facing her. "Nothin' at all, Boss. Just messin' about." With that, they made a sharp exit. "You okay?" the screw asked softly. "Yes." I replied, staring at the floor. "Did they think you had drugs?"

"No, Boss, we were just messin' about." She knew exactly what was going on, and said, "I'll lock you in for the rest of Association if you want." I nodded thankfully.

"Grisly Risley" was living up to its nickname. I might have done anything to get out of there that night, but would've settled for a bag of heroin. 'H' can calm nerves, even as the realisation sets in that you just narrowly avoided painful violation *and* you have two years of this ahead. In its absence, I slid down the wall, curled into the foetal position, and sobbed with all my heart. Right there and then, any remaining hope I had of surviving this ordeal slipped away, making room for the onslaught of fear, helplessness, and worst of all, cold turkey.

Chapter 4
Prison Life

The heroin withdrawal was intense: vomiting, cramps, sweaty sleepless nights; pain that felt like my legs were being ripped apart from the inside; jangling nerves screaming for one more hit; and a deep sense of bereavement because I knew I wouldn't get it.

A typical day went like this…

As I lie on my bunk, allowing the sheets and the skinny mattress to soak up my sweat, a prison officer – a 'PO' – walks in,

"Come on Steph, time for your medication." Skinny and weak, I'm helped up and over to the medication line for my one-paracetamol-a-day.

"This is pointless" I sigh, as I collapse against the wall at the end of the queue, unable to hold up my own pathetic body mass. This is my daily routine now.

There is a cold, clammy feel to my skin and I wouldn't look out of place on a mortuary slab. Dark circles underline my lifeless eyes and my hair hangs limp and abandoned around my drawn white face. I am not the worst case, not by a long way. Yesterday, a woman collapsed, convulsing and foaming at the mouth as a male PO laughed and mimed filming it with his hands – the most vulgar example of desensitisation I've ever seen. We are all prone to it. I'd seen so many overdoses by now that it barely registered on my radar these days.

The line takes a step forward as the next girl is served her medication. As I concentrate on lifting a foot and moving a leg forward, the alarm bells ring and all hell breaks loose. Opposite me, a door to one of the lower cells is being flung open by a frantic looking PO screaming,

"She's hung herself. Get over here, quick!" I can see directly into the cell, and I watch from the queue as three more POs come running. The girl, with long brown hair and porcelain skin, jerks in her bedsheet noose, dancing mid-air to a silent drumbeat as the red-faced PO runs over and grabs her legs, trying desperately to hold her up.

"Another one for the hospital wing" says Miss Uninterested in front of me as we all take another step towards the counter. It's just another day in Grisly Risley.

It was easy to get cynical and angry in a place like that, but I had the energy for neither. I chose 'hopeless and lost'. I gave up on life and all that went with it. I ate the lukewarm slop that was put in front of me every day without complaint and kept my eyes to the floor – all in the hope of going unnoticed.

Two weeks later a newbie PO that I didn't recognise came into my cell,

"Prisoner VD3713! Pack your belongings. You're shipping out in ten minutes." I was ready in five. Anywhere else had to be better than Risley, although I'd heard some horror stories from other prisons such as HMP Styal in Cheshire and Holloway in London.

Further into my sentence, I would see women so desperate not to be shipped to one of those two that they would barricade themselves into their cells, only to be forcefully removed by screws in riot gear wielding batons. I was once sent into a cell with a mop to clean the blood off the floor and walls after a woman slashed her wrists. It may have been that she would rather have died than go back to Holloway, though more likely, it was a desperate attempt for her plea to be heard. It didn't work. After a few days on the hospital wing, she was shipped there anyway. Prison life was hard and unforgiving, with no room for sentiment or sympathy.

It took no persuading to get me out of Risley that day. I was still weak from the 'cold turkey', but I was putting on weight and getting stronger every day. Those two weeks remain the toughest of my life to date, both mentally and physically.

It was another half hour before someone came to get me. I was pleased to see it was Miss Burrows. She was one of the good ones, doubtless destined to remain at the bottom of the ladder, passed over for every possible promotion, due to her kind and gentle nature.

"Are you ready?" she asked with a smile.

"I'm ready!" I exclaimed, getting up off the bed and grabbing the clear plastic bag with 'HMP Risley' printed on it in royal blue. "Where am I going?"

"Brockhill." she replied. "Redditch."

I was put on B Wing at Brockhill, where I very quickly sensed a slightly less hostile environment. The wings were smaller and more enclosed. People had routine. It felt safer. I didn't feel I had to watch my back all the time. Keeping busy was key to staying sane, so I took whatever was offered. As well as my daily job as a cleaner in the officers' mess – a trusted 'Redband' position that came with an elevated £10 weekly wage and included cooking fry ups for the governor – I enrolled in a business administration course. I also went to the gym, where I tried to get some muscles back and lose some of the massive amounts of weight I'd gained after rediscovering my appetite, previously robbed by the H. Locked up and fed stodgy food, I very quickly ballooned to twice the size I'd been when I arrived. With my wages I would buy tobacco, chocolate (like gold in a women's prison), and stamps from the prison shop, which was open once a week.

Heroin was still available to me here, and although my body was now recovered, my mind would take much longer. I still took it when offered and felt that I would never be able to say "No." This frightened me more than anything. At least I wasn't physically addicted now, because I couldn't get it all the time. The strongest grip though, was on my mind.

My neighbour on B Wing was Rita and we became good friends of sorts. She spoke with a strong Lancashire accent, undeniably of Bolton. She had a borstal spot*, and short cropped hair; a bit longer at the front and parted over one eye, which made her carry her head at a slight angle to keep it from getting in the way. I can't remember what she was in for. These things were not important inside unless you had hurt a child, and that lot were often segregated for their own protection.

I liked Rita. She had a lopsided, boyish grin and was always happy to share whatever supplies she had. If I ran out of tobacco, she shared hers; if I needed weed, she found some. I learned that even with the roughest of prisons and the roughest of people, there was often a kindness to be found – if you bothered to look. Rita was fairly quiet and unassuming; not a bad person, just a person caught up in her own private set of circumstances. Despite our promises to stay in touch, we never spoke again after she left. Perhaps in another life, she would have been a nurse or someone's wife. I hope today she is.

As we sit together playing our daily game of backgammon in her cell, Mr. Wake comes wandering in. He is one of the nice POs who often potters around and chats to the girls while turning a blind eye to the odd joint being passed around. We think nothing of his arrival until he says,

"Right, Reet, get your stuff together. You're being released."

"WHAT?" She jumps up and knocks over the board.

She is being released four weeks early due to overcrowding. Rita packs her stuff in quick time, gives me her remaining supply of biscuit rations, and turns for the door. Her days of counting the minutes are over. I keep a smile plastered on my face as I walk with her all the way down the corridor. We hug at the gate, and I wish her luck before adding,

"Now go, and I don't want to see your face in here again!" With that she turns and leaves. I go back to my cell and cry. The elation I feel for her makes way for self-pity. I'm going to miss Rita a great deal. A few days later, Mr. Wake came to me and suggested that I become a Listener.

"What's one of them, then?" I ask, only slightly interested and looking up only briefly from my game of solitaire.

"It's a group of volunteers who help other prisoners who are struggling. It'll keep you busy. You'll get training from the Samaritans and be 'on call' 24/7 to go and listen to people when they need someone. I think you'd be good at it."

I didn't see myself as a helper of people, certainly not the Samaritan sort, but I did want to keep busy and this might help me take my mind off my own problems. Not least, I was missing my son dreadfully, though I was one of the lucky ones. At the age of two he was blissfully unaware of all the shit I had made of our lives so far, and was happily settled into a routine with my parents. Maybe helping others would help me fight my own demons and

FOOTNOTE: *A borstal spot is an indian ink tattoo. Just a dot placed below one eye. It is used to signify that you had been to borstal or Young Offenders Institute. Usually seen on men.

take my mind off the overwhelming guilt that permeated my mind since the drugs had worn off.

My first 'customer call' came at midnight, the day after I finished my training. One of the screws came to my door,

"Steph, wakey wakey. Your services are required."

"What the…?"

"We've got a newbie on C-wing screaming the place down saying she's going to kill herself. Let's see if you can talk her down. We've got no room on the hospital wing and if she doesn't shut up soon, we'll have to put her down in the seg before EVERYONE kicks off. She's keeping the whole wing from their beauty sleep."

"The seg" or "the block" is where trouble prisoners are sent. Here, they are segregated from the rest of the community, either for punishment, their own safety, or for the safety of others.

"How long did she get?" I asked as I stood up and searched, bleary-eyed for my prison-issue pants.

"Seven days."

"Oh, for fuck's sake. That's it?"

"It's all relative," he said with a smile. I raised a sarcastic "Oh, really?" eyebrow at him as he held the door for me and I walked, barefoot, into the empty corridor.

"Um, you might want to take your shoes. I think she's been spitting on the floor." I turned and feigned defeat as I walked back with hunched shoulders.

"Remind me why I agreed to this?"

This was the first of many 'house calls' I made. They locked me with the prisoner in her cell, and I just listened. I was not there to advise or counsel as such; it was more about being an understanding ear and a shoulder to cry on. Many were new arrivals; the newly sentenced, terrified and alone in a hostile environment. Some were looking at a few days; others at a life sentence, and everything in between. Some had just received bad news from home; husbands left them, kids put into care, or someone died. The frustration of being stuck, helpless, behind bars while their life outside fell apart was often unbearable.

21

No matter whether it was seven days or seven years, the emotions and subsequent behaviour were the same. I learned quickly that a 'mine's bigger than yours' approach was not helpful when it came to problems, and over time, my initial "Get over it. It ain't that bad" attitude softened, eventually turning to genuine empathy – even for those who would be out before my biscuit rations were through.

After almost a year at Brockhill, they moved me to Foston Hall, my penultimate prison. It was still under construction when I arrived; we were the first inmates to inhabit the new facility. The cells were more like Swedish holiday cabins than cells – long, red, wooden structures sitting on a manicured lawn. Inside, there were laminated floors, proper bathrooms with in-room showers and a bar-less window. After the first two prisons, it was almost divine.

I don't remember much about Foston Hall, though I do remember that it was there, while having a shower and listening to the radio, that I heard the news that Princess Diana had died. I vaguely remember crying at this news. No doubt some of it was self-pity – an excuse to release some of that held-in sadness at my own pathetic situation. I also remember a fight with a fellow inmate who attacked me with a chair while watching Gladiators on TV. The screws dived in and took several minutes to get her off me. She apologised later, saying she had been suffering from PMT.

I was there just a few weeks, when I was given the opportunity to go early to Askham Grange, near York. Askham was an open prison – no bars and no fencing. It was a beautiful old building in pretty grounds and slap-bang in the middle of a picturesque village, at the heart of which was a large duck pond right outside the entrance to the prison. It seemed an extremely unlikely place for a prison, yet, if it weren't for the sign outside saying 'HMP Askham Grange', you'd never guess it was home to several hundred female convicts. It was more St Trinian's than Alcatraz.

Shortly after arriving, I became friends with a lifer named Sue who'd been inside seventeen years for murdering her husband with a bread knife. Fifteen years after her sentencing, the courts accepted domestic abuse as a legitimate defence – sentences in cases like hers could be reduced to manslaughter. It was all too late for Sue, though; she took the full brunt of the law.

She and I often sat together on a garden bench in the grounds, discussing what we would do when we got out. More often than not, there were three of us on that bench. Like me, Sue was a great lover of animals. One day, she found an injured young raven in the grounds. The guards let her care for it in her room until it was well enough to be set free. The raven chose to hang around and visited her daily on the bench, nuzzling up to her as she spoke gently to him. She called him Billy and he clearly loved her; returning her gift of life and freedom by making her lack of it more bearable. True friendship like this was a joy to behold.

Sue had three years to her parole date when I left. I often wonder what happened to her – and to Billy. Being a 'freebird' is often harder than it sounds. Many, like Sue, suffer from institutionalisation; the thought of making simple decisions for yourself in a world that has changed so much since you left, is frightening. Nothing is the same and everyone's gone; either moved on or died. You have no home and belong nowhere. The freedom you wished for every minute of every day is now a terrifying prospect. Some reoffend just to get back in because it's all they know. Maybe that's the real reason Billy stayed so long, and the reason he and Sue got on so well – they were kindred spirits. Fear of the unknown and the unfamiliar can be its own prison.

One Sunday, I decided to go to listen to some gospel singers who'd been invited to the church from 'the out'. I am not religious and didn't normally attend, but I do love music and gospel can conjure up such a wonderful energy, I knew it would lift my spirits.

They were about twenty-strong, dancing and belting out the numbers with great passion. I love to see passion in any form. Like a smile, it's infectious, and this was no let down. A young man stood forward ready for his *a cappella* solo finale. 'I believe I can fly'. The lyrics were cheesy, but right there and then, the way he sang them, he touched a chord with everyone in the room. His voice so angelic; his face so full of passion.

From the weak to the strong; from the bullies to the bullied; everyone fell silent and focused solely on him as the rest of the choir joined in with background humming and hushed vocals. By the time the chorus came, there was not a dry eye in the big house. At the end there were audible sobs from the women, who were now hugging each other. For a moment, everyone was united in their sorrow. No arguing. No violence. Just understanding and togetherness.

After the concert, I went back to my room and found my roommate bent over the sink with her exceptionally long ginger hair tied back in a ponytail and her fingers rammed down her throat. I didn't need to ask what she was doing. She had just had a visit and clearly it had been fruitful. As I walked in, she gagged, and up came the jackpot – a nice little bag of brown.

"Oh, good timing," she said, looking up and wiping her mouth like a pro.

Tracy looked so innocent with her freckle-sprinkled button nose. Her hair, when loose, went all the way down her back and beyond her bum in waves. She often tied it in pigtails, which made her look like she could just click her heels together and get us all the way to Kansas if she wanted to. I often thought about asking her if she would try, but she probably would've punched me in the face.

It was cellmate etiquette to share your gear. You couldn't hide it, so there was no other way, really. In the more 'traditional' prisons you might even share it with your friend (or the bully) in the cell next door. People were always on the lookout for those with pupils like pins, so it was hard to hide it. During lock-in, the drugs, cigarettes or contraband would pass from one cell to another by hanging a line (often made from bits of sheet) out the window and swinging it, building momentum until the person in the next window could catch whatever was attached to the end.

Tracy was in a very good mood. This little bag signified a happy and content night ahead – an evening off from all that was harsh or worrisome, safe and secure in the warm embrace of the big H. It never failed to seduce and deliver. She rummaged around in her hiding places for the foil she needed, and quickly got the proceedings underway. She hadn't even noticed that I stood frozen to the spot.

"I don't want any." I said. It was barely audible. So much so that I wasn't even sure I'd said it out loud.

"What?" she asked, only vaguely listening now as there were more pressing matters at hand.

"I don't want any." Louder this time.

"Ha ha, yeah right. Stop fucking about. Have you got a pen?" She wanted to use the pen to roll some foil around it and then slip it off to make a tube. I was gaining momentum now, but still didn't quite believe I was saying it.

"I mean it. Do what you want, but don't give me any." I walked over to my bed and stuck my head under the pillow. "Let me know when you're finished." Tracy couldn't believe it, but was not going to argue. This meant more for her, so she carried on.

I couldn't help but hear the beautiful sound of the foil crinkling as she smoothed it out to make a runway for the heated oil. I began to sweat, and my gut began to twist in the usual excitement of the promise. I drew my knees up to my chest and put my fingers in my ears. Then the smell took its turn, attacking my senses as she lit the flame and began the chase. I concentrated on the sound of my own laboured breathing and tried to take myself away to another place. *The baobab. The Serengeti. Think! What do you see in the tree? What is all around you? Feel the warmth of the sun on your face, the stroke of the long grass on your legs. You're nearly there. Just four months until you get out. You have to stay strong!*

But I did not feel strong. How easy it would be now to just sit up and take one last hit. Who would know? What did it matter? Surely it was understandable in these circumstances? Who could blame me for that? Come o-o-on. Just *one* more. This is cruel to deny yourself. Just wait until you get out and *then* you can really get sorted. Everything will be better on the out. Easier. I loosened my grip on the pillow and began to sit up.

"I knew you'd change your mind," she said in a deep voice as she held the smoke deep in her lungs. I could almost feel the warmth of that smoke in my own chest and the feeling of enlightenment as she slowly let it out. But something in her voice; the sarcasm; the pleasure in my failure that proved her right, made me stop. Instead, I turned over, faced the wall, and began counting in my head. When I got to 100, I started again, and again, until she was finished.

"All done," she said and lay back to enjoy the warm rushes of pleasure

I knew would be enveloping her. Had I really done it? Was it over? All my muscles that had been tensed relaxed simultaneously and I let out a sigh of relief.

"Gotta hand it to you girl," said a now very distant Tracy (she may not be in Kansas but she was certainly miles from here), "You're a stronger person than me."

Winning the battle that day was all I needed to keep saying no. I never touched heroin again, and unlike most, I never wanted to. There was no counting the days of sobriety or calling myself a 'recovering addict'. From that moment on, I knew, without a shadow of a doubt, that for me, the war with heroin was over.

Chapter 5
The Journey Begins

23ʳᵈ March 2014

"A deal's a deal," Pete said, slipping £50 into my hand as I headed through the crowd towards the stairs and walked up to the roof of the Ace Café, the legendary bikers' gathering place in North West London on what used to be the North Circular Road. Two nights earlier, I'd been in the London Studios having a live satellite interview with Australia's Morning Sunrise TV, and it had gone quite well. My chance to win Pete's bet of getting his word of choice into the interview came with seconds to spare as the presenter asked,

"So, will you come and see us when you make it to Sydney?" I smiled and replied,

"If I make it to Sydney (pause for effect and to enjoy the smug moment), I'll be doing *handstands* into the studio."

Ker-ching!

I'd originally met Pete at an off-road training weekend back in 2008. We and several others were heading for a motorcycle adventure together in South Africa that would prove both terrifying and exhilarating in equal measure; one of those adventures where half the time you wonder why you're doing it and the other half wishing it will never end. That trip had been a pivotal point in my life and now, six years later, we were standing together again, in very similar adventure-anticipating circumstances, except that this time, I was going to be having a much bigger adventure, and I was going to be doing it on my own.

I looked down from the roof of the Ace Café at a sea of faces all looking up at me; bikers I'd never met, friends, and family. *Surely they don't think I can actually do this? I'm not sure I can even make it past Germany. Yet here I am, with all my worldly possessions reduced from a three-bedroom cottage in Wales, to whatever I can carry on Rhonda, my Honda CRF250L. Today I am setting off on a mission to become the first person in the world to circumnavigate the globe and ride a motorbike on all seven continents.*

My original plan hadn't been quite so grand. It was more of a general plan really. I wanted to see the sunrise from different angles; I wanted to see an orangutan in Borneo, and more than anything, I wanted to keep that promise I'd made all those years ago of seeing a baobab tree. But the seed had grown. With the right lighting, and enough tequila, my confidence grew sufficiently to tell anyone listening that I was going to ride around the world. Eventually I had one of my classic, "Hold my beer, watch this!" moments and announced: "I'm going to ride to all seven continents." SHIT. Why hadn't a responsible adult tried to stop me?

At the Ace Café I remember the strangers in the crowd cheering me on; the mixture of pride and fear etched on my parents' faces as I was swept along and onto my bike before I was ready; the official bacon butty ceremoniously handed to me in a silver foil wrapper.

What on earth have I done? Am I really the sort of person who can pull this off? There must be a reason some better person than me hasn't done it already. I don't have enough money and I have no idea how I am going to make it to Antarctica with my motorbike.

This was a journey into the unknown on so many levels; but then this was *not* the first time in my life I'd taken a leap of faith. Like the proverbial cat, I was incurably curious, though I couldn't help but wonder, at that moment, just how many of those nine lives I'd already used up.

The last eighteen months had been a whirlwind. Once I made the decision, I closed my off-road motorcycle business and honoured the last bookings with the support of the remaining team. It helped that my business partner at the time, a Dakar Rally rider called Mick, had decided to…well let's just say 'do things his way' – in a big way (I'll spare you the gory details)! Mick and I had met by chance and within three months had set up a fairly basic off-road school together. With no cash flow and no premises, we operated out of an old container tucked around the back of an understanding bike shop owner's showroom. Even the jet wash was borrowed. We were spending more than we were making, but pushed forward regardless, ignoring the obvious hurdles in our way. The blind leading the blind. The only thing we *could* see was the end goal, our shared vision: to create the best off-road school in the country. With this combination of brute force and ignorance came a passion and a strength that was to pay off.

Within six months we'd gained a contract from Honda for thirty bikes and a 4x4, and within twelve months, we were running what was probably the biggest off-road school in the UK as well as motorcycle tours in Morocco – which is a whole other story!

Working with Mick was, in itself, an adventure. He was a chancer, a risk-taker. He was both the toughest and the softest guy I'd ever met. He also had a focused mind that at times led him to extreme selfishness.

After five years of ups and downs together, we went our separate ways. The end was messy and I was left to clear up the shit he had left behind. It was the end of an amazing time in my life, brought about through an unfortunate turn of events over which I had little control, but the untimely death of that chapter made way for the birth of the next.

Like many chapters in my life, it may have come with an aggressive prematurity, but what I *could* control was how I reacted to this new situation I found myself in. Rather than trying to rebuild my business, I made peace with Mick, turned to a blank page, set a date and started planning my Great Escape. I was going in search of baobabs!

In the months before my departure I had worked hard to gain support and raise the profile of my trip. Getting sponsorship is often a thankless task. You have to be pretty thick-skinned to take the knockbacks, and downright persistent to get any kind of outcome. It helped that I worked in the motorcycle industry because I had already formed some good relationships in it. People knew me and knew I was good to my word and I was shown support by many companies – including Honda – who kindly supplied the CRF250L, along with a small contribution in good, hard cash.

Cash flow was still an issue though. A big, ugly, sarcastic smirk kind of an issue, staring at me in my sleep and whispering "You fool. Who do you think you are? You've got no chance!" With just a few months to go, I was getting a little stressed. The enormity of the challenge ahead shadowed my every thought. The ping-pong ball in my head was bouncing around as if it was trying to cut loose, tired of all this nonsense.

As I was clearing out my attic one afternoon, preparing for my departure, my phone rang.

"Hello?"

"Hi. I believe you are attempting to get your motorcycle to Antarctica?" I didn't recognise the voice.

"Um, yes. Yes, I am. Can I help you?"

"Actually, I'd like to help you."

"Really? In what way?"

"I'd like to give you the money to do it. How much will it cost?" At that, I stood up in the crawl space and banged my head on the beam.

"Ouch... *Damn it*... Sorry, what did you say your name was again?"

"It's Tim. Tim Carrier. I got an email from you asking if I wanted to advertise on your website." I had been emailing all sorts of motorbike companies for my new website business which offered advertising and reviews for biker cafés, tours, B&Bs etc.

"Well Tim, I am not entirely sure. I'm guessing around £6,000," plucking the figure from my overloaded brain.

"Then I shall transfer the money to your account in a few days. I love what you are trying to do and this is my way of helping you achieve it."

"And in return?" I asked hesitantly.

"Oh, I don't know. Perhaps you could put a sticker on your bike for my B&B in France, Motobreaks?"

"*Really*? I don't know what to say." My excitement quickly turned to worry. I had to admit to him that I hadn't yet figured out a way to make it there, but I was working on it. In fact, all my enquiries so far had led to a dead end. "What happens if I don't find a way?"

"You will," he said, "But if you don't, then spend it on Africa. You're brave enough to try and *that* is what I admire". On that he put the phone down. No one believed I could find a way to Antarctica, and yet here was this man, a total stranger, believing in me enough that he was willing to stake six grand on it. The money arrived in my account five days later.

Despite this 'win', self-doubt and insecurity crept in like a cancer. Many times I wished I could just be happy with a simpler life of working 9-5 and going home to watch the soaps. Why did I keep putting myself in these stressful situations; jumping into the unknown with no idea of where I would land? Why couldn't I just be happy watching Opra or something equally mind-numbing? Sometimes I envied the people who could.

After prison, I'd wanted nothing more than normality. I wanted to be a normal person with a steady job, a pension, and a sofa bought on sale from DFS with interest-free credit. I wanted to buy a house and spend my weekends doing DIY or staying up late watching the Australian Grand Prix live. I wanted a vegetable patch and a dog I trained *not* to shit on my cabbages.

I wanted a life without drama, and if I had all of that, I thought I'd be happy forever. I got it all, too. The good job, the mortgage, and the dog. I even got the veggie patch, though the dog *did* occasionally shit in it. Why couldn't I just be happy with that? 'Steady away' worked well for me until 2008. I'd owned three houses and been a human resources manager. I'd even been a mortgage advisor. I was a *proper* grown up for a while, with suits and high heels and everything. As my son grew up, I managed to maintain a fairly steady path in life with all these things that, at the time, represented security. Was it *that* bad that I kept running away from it? It wasn't, but there had always been a baobab-shaped itch that needed scratching, and I couldn't even stop there. I wanted to see the sun set on every continent. I wanted to take back the time I had lost and double it. I wanted to *live*, and be damned with the consequences.

Now, at the Ace Café, with Tim there to see me off, with my friends and family and total strangers all believing in me, I felt like a fraud. I had convinced everyone I was strong enough for this; that I could do it. Surely, this had gone far enough... but it was too late to turn back now.

As I started the bike and wobbled towards the edge of the car park, I felt somehow trapped inside myself. I looked up briefly at the crowd and my eyes met with those of my lodger and friend Tom, for just a second. In that second, they pleaded to him, "What the hell is happening? I'm shitting myself." His eyes replied, "I can't believe it either. *Shiiit!*"

Mark Wilsmore (the Ace Café owner) stopped the traffic by waving an enormous Union Jack, and a stream of bikers filed in behind me ready to escort me out of the metropolis, for the first few miles of the trip. Some I knew. Some I'd never set eyes on before that day. My pride and nerves were battling for pole position. As Mark lowered the flag, the crowd cheered and our wheels hit the road.

The pressure was intense. Perhaps *I had* bitten off more than I could chew. Suddenly the oversized map with my route so carefully drawn out to cover all seven continents now just looked like an egomaniac's wet dream.

Of course, telling everyone was always part of the plan. It was my own personal insurance policy to stop me backing out – and it had worked. I was past the point of no return and it scared the shit out of me.

I felt sick as I approached the Eurotunnel to France. Looking at my heavily laden bike, I wondered again, what I'd got myself into. The last of my bravado disappeared as I shakily followed Pete and his BMW R1200GS onto the train. Pete was my personal escort out of the country... probably making sure I actually left. As we rode in and through the train as instructed by the arrows, I made sure not to scrape my video camera off the top of my helmet on the low ceiling. It probably wasn't that low, but at that point everything felt like a disaster waiting to happen and I was ridiculously over-cautious. We parked up, removed our helmets, and prepared to be carried under the English Channel and into France in just twenty short minutes.

"How are you feeling?" Pete asked in his usual caring voice with a smile to match.

Pete really is one of those guys whom everyone loves, and I was so glad he was with me at that moment. If anyone could help keep me calm, it was he, with his caring nature a balm to my fragmented nerves.

"If I could just get rid of this sickly feeling, I'd be okay."

"I'm so proud of you," he said, with enough sincerity to stop a steroid-fuelled cage fighter in his tracks.

"Oh, pack it in, Pete Bog," I said and turned to reach in my tank bag for my roll-ups before the tears had time to form. I always called him Pete 'Bog'. Ever since an off-roading incident many years before when we found ourselves stuck in a bog, or 'the bog of eternal stench' as it became known. As one of the guys spun his back wheel in an effort to escape, big clumps of it roostered into the air. When we looked around, there was Pete, giggling, with a big lump of it planted firmly on the top of his helmet. The name stuck as firmly as the bog.

That March day in 2014 was the 70th anniversary of the real-life event that inspired the film 'The Great Escape'. It seemed strangely fitting, almost romantic, to me that seventy years ago to the day, under a moonless sky, 76 men had escaped from Stalag Luft III prisoner of war camp and, of course, that, in the fantasised film version, Steve McQueen attempted to make it across the Swiss frontier by motorbike, in true Hollywood style. It gave me a glimmer of strength, a feeling of hope, as my mind skipped around from one thought to the next until my overactive negativity gland reminded me that all but three of those who escaped were recaptured and fifty of them were executed. I snapped back to reality and realised that Pete Bog had been watching me with a smile, as a variety of expressions passed over my face while making my rollie and apparently whistling the theme tune to the film.

"You'll be fine," he said, knowingly.

The first few sunny days riding through France and Belgium were a breeze. The butterflies in my stomach had finally fluttered off to bother some other poor soul, and I felt freer than I had ever felt before. Pete headed for home after we visited the Nürburgring Motorsports Complex in Germany, where we wisely decided against taking our bikes out on the legendary *Nordschleife* circuit and opted instead for a night of steak and Belgian beer.

The next morning, I said to Pete,

"You leave first." knowing that would give me time to gather my thoughts and prepare myself for being alone. We said our goodbyes and he rode out of the carpark waving, only to find he had turned into a dead end. We waved again, laughing, as he rode past in the right direction this time. I heard him giggling in his helmet as he rode off, then out of sight and thought: "Who knows when I will see him again?"

I was now alone with my trusty champion, Rhonda the Honda CRF250L. It actually felt good; scary, but good. I felt hopeful. It didn't matter that she was 'only' a 250cc or that she was ever so slightly too tall for me. It didn't matter that people said I needed a bigger bike and she wouldn't make it. It was a romance novel and I was in love. We were a team, and together we would take on the world!

Ten minutes after setting off, I dropped her. In a lay-by somewhere in Germany, the enormity of the challenge ahead reintroduced itself with an "I told you so" smirk on its face. I misjudged my carefully executed 'getting on procedure' and we toppled over. Bike, luggage, and rider hit the tarmac with a thud and a yelp. Right on cue, the rain came. For as long as I dared, and before some poor misguided passer-by called an ambulance, I lay on the ground next to Rhonda, looking up at the sky and wondering, "How bad would it look if I just turned back now?"

"Are you okay?" asked a German-accented voice, in English, from a nearby parked car I hadn't noticed before.

"Yes, yes" I replied, jumping up, dusting off my bum, and making like it was all part of some plan he was just not privy to. "I'm riding around the world!" I said. Because *that* explained it! It felt good to try it for size though.

"Good luck with that," he said, with a "that's what you think" look on his face as he got out of his car to come and help me lift the bike.

Chapter 6
Across Europe

After riding across Germany into the Czech Republic I spent my birthday couchsurfing in Prague with an English teacher named Steve, then continued on to Krakow in Poland, then Budapest in Hungary, before heading to Romania for a couple of days in the Transylvanian hills. It all still felt like a holiday, but I didn't hang around. I was being pulled by the urge to chase those lines in the road; to see what came next. *What happens tomorrow? What will the next country be like?* At this rate, I'd be home by 7pm, never mind seven continents!

Romania felt nice! Once off the main, truck-infested route, it felt beautifully familiar. The backroads heading towards Apuseni National Park reminded me of Wales. After stopping to adjust my playlist to something more conducive to twisty roads, I quickly forgot my line-chasing agenda and the time of day. *Ah, yes. Good old Foo Fighters. You'll do nicely.* Carefree, I sailed past Albac, where I had planned to stay, and headed on into the park. I completely ignored the facts that I had nowhere to stay yet, no local currency, and my eyes were feeling quite dry and irritable for some reason. I put it down to tiredness – which I also ignored – and carried on. I could find a camping spot later. First, there were twisties to be ridden! Life felt too good to listen to good sense. The only important thing at that moment was to enjoy the road, the evening sun shimmering through the leaves above, and the winding empty tarmac that could have been built for this very playlist! This was what I had worked so hard for. This was what it was all about.

These brief moments in time have the ability to bring you right into the 'now', with neither past nor present being of any consequence. *This was what I needed.* With the right playlist, a sunny day, and a good set of twisties, you are truly free. If only Zen Buddhists had turned to two-wheels rather than temples, it would have saved them years. One hundred books by

Eckhart Tolle will not give you this feeling. Books cannot explain how to live in the now as well as a motorbike can show you. Riding twisties is the true path to happiness and enlightenment – if only for that moment; a taste of real freedom without either the hangover or the years of abstinence.

I climbed slowly up into the hills, turning off to little adventures on dirt tracks that weren't shown on the GPS, until I reached the snowline. Then I turned around and did it all in reverse, back towards Albac. Night was approaching, and both Rhonda and I were close to empty. It was time to come out of the 'now' and think about the very near future which apparently consisted of an empty fuel tank and an empty stomach! Dropping down from enlightenment mode, I also became aware of the fact that my eyes were worse. They now refused to be ignored. I stopped and looked in the bike mirror to find they were red and quite badly swollen.

"Hmmm! What the hell is that all about then?" I asked my reflection. Thankfully, there was no reply. A good sign! I decided that camping was probably not a priority tonight (and if I'm honest, I started the trip quite lazy, unsure of myself and my decision-making, often convincing myself to go for the easier option). In any case, I was still, amazingly, within my budget for food and accommodation. As this was a mere £10 a day for accommodation and £5 for food (average), I was quite pleased with myself. My budget wasn't based on any real knowledge; rather, it was a case of just dividing my total available sum by the 18 months of my original timeline for the trip; then estimating my daily mileage and fuel costs, taking what was left and dividing it up between food and shelter. And *voilà,* I had my daily budget.

Choosing what to pack on the bike had been a similar process. I had no real knowledge as I'd only ever travelled with a support truck behind me before. It was a case of listing what I thought I needed, in order of priority, then crossing half of it out when I realised how much was actually practical. In the end, I took one change of clothes, three pairs of knickers, a washbag, a pair of flip flops, a Kindle and the all-important Shewee (For those not familiar with it, a Shewee is a funnel-shaped device that allows a woman to urinate while standing, without removing any clothes – just like you men!).

The rest of the space was taken up with a few spare parts (brake and clutch levers, brake pads, air and fuel filters), a tool roll, camping gear, and lots of gadgets to record the journey on my blog, as much as I could – including MacBook Air laptop, camera, GoPro and tripod. This wasn't just for my regular readers, but for me too. I knew there would be far too much to remember organically.

I decided the priority was to get some food and petrol, then find a place to rest, where I could shower and properly clean my eyes. There were bound to be cheap rooms around, and probably empty at this time of year. It was still only springtime, so a bit too early in the year for tourists, really. Any other year, this area could have been covered in snow, but Mother Nature was kind to me and gave me the best spring she had to offer: unseasonably warm during the day, and crisp at night.

Still glowing in the aftermath of my 'twisty hit', I closed my visor and pottered down to the first petrol station. Upon arrival, I was rudely shaken back to reality when I discovered that no one took cards in that part of the world; I mean no one. In my haste to have fun, I hadn't bothered with such trivialities as money! To conserve fuel, I stuck to the edge of the road and slowly made my way out of the hills, stopping at every garage en route. Each of them turned me away. The sun was sinking quickly now and just as I considered a Plan B (camp on the side of the road and figure it all out in the morning when my brain was not so tired), I spotted a policeman. He directed me to a petrol station three kilometres away that would take cards. It was a long shot and a very slow crawl all the way, but I made it. *Phew!* By then, Rhonda must have been running on nothing but goodwill and fumes.

Now all I needed to do was find somewhere to rest up after eleven hours in the saddle with eyes that looked like they'd been punched by an angry Buddhist who'd just had his life's worked trampled on! It was a shame not to camp. I felt a moment of guilt, almost as if I was cheating by not 'roughing it', but there would be plenty of tent nights, looking at the stars, to come, so I let it go.

An hour later, after stopping at what originally looked like a deserted hotel, I found myself checked into a massive and slightly damp room that smelt vaguely of stale cigarette smoke and mould. It reminded me of the squats I frequented back in the day, minus the Oscar Wilde quotes and Che Guevara posters. Indeed, my Doc Martens and tie-dyed leggings were now replaced with bike boots and Gore-Tex pants with armour, but the smell was the same and the wallpaper undoubtedly pre-dated that era.

I bathed my eyes with cold water and teabags, then headed for the bath. *Damn it. No plug!* Thankfully, my deodorant bottle cap was exactly the right size. A little pat on the back and a note to self: Survival Tip Number One: – If no plug, use a deodorant cap. It wasn't quite Bear Grylls' "sleep inside a dead camel" kind of advice, but it was a start! By 8pm, I found myself in the hotel restaurant, sitting on a white plastic garden chair on a swirly red carpet, being served free wine with my not-quite-like-home spaghetti bolognese, courtesy of the hotel owner, who was dressed like, and may have been, the mayor! He seemed to be hosting a wine tasting evening for the locals, and was serving up lots and lots of it to his guests whilst telling us all (in Romanian) what each one was (I guess)! I sat in the corner nodding and smiling and drinking! They stopped including me after glass number four. Perhaps I didn't look the part in my well-used Ace Cafe T-shirt and helmet hair? Or maybe it was the hooded and bloodshot eyes? I took my cue to retire to my grotty-yet-guilty-pleasure hotel room.

The next morning, I headed for a border town ready to cross over into Serbia. 'Serbia'; for me, the name conjured up mental images of bombs and chaos from NATO's 1999 air attacks on what was then Yugoslavia, to drive Serb forces out of Kosovo; images that were on the news constantly for a few short months. Those images etched in my mind, but were long forgotten until now. Did I really have anything to worry about today? This was a country that many had warned me was "still dangerous", suggesting that the police could not be trusted and I should not stop for them on the road. This was obviously dangerous advice to follow, but how should I separate a negative bullshit comment from one that might be worth taking notice of? The problem was that it was coming at me on a daily basis:

"Don't take that road, it's dangerous."

"Don't trust those people, they'll kill you." Or just "YOU'RE GOING WHERE? ON YOUR OWN?" No wonder people stayed at home. We had been brainwashed into living in fear.

I picked a name on the map and headed for it. Sadly, the town was not as I envisioned. It was a dump, with nothing but boarded-up hotels, stray dogs everywhere, and quite a few truck drivers parked up on the side of the road. Yet it looked so promising on the map – situated on the river and surrounded by forest. Sometimes things only look good on paper. This was one of those times!

I didn't want to stop, but I had to get my bearings and revise my plan. I also needed a roll-up. I have always prided myself on being able to get on with most people; not judging too harshly at first glance. Yet here, I felt jumpy and vulnerable with all my belongings on display and no back-up. It was getting late and the shadows cast an eerie malevolence on what was already a dark and dingy place. Rubbish blew across the empty streets as the last of the sun tucked in behind the tallest buildings. The truck drivers eyed me with expressionless faces from their cabs. If this was a horror film, this would be the time you are shouting at the screen saying, "Don't stop you fool! GO!". *Was that what the silent truck drivers were thinking? Why was none of them out stretching their legs? What did they know that I didn't?*

The thing about feeling that vulnerable is that you don't really want to take your helmet and gloves off, because you know that if you do, you can't just jump back on and ride off in the event of trouble. Of course, by now your overactive mind is imagining every possible situation – all the ones that end badly, anyway. I felt the same later in the trip when taking forestry tracks alone in the Canadian Rockies where I knew there were grizzly bears! Oh, for a flip-front helmet right now. The irony was not lost on me that smoking might just kill me, either by proxy – or by grizzly bear.

Back in Romania, the desire for a smoke overcame the wish to be able to make a speedy exit, so I removed my full-face helmet. As I lit my cigarette and tried to maintain an air of "I do this all the time," coolness, I heard a shout from behind me,

"Hey!" It was a group of kids who were jumping over a wall and making their way towards me. *Okay, just kids. Hmm, kids with sticks. Several kids with sticks. Shit! What do I do? Run? Don't be stupid. No time. Stay cool. They're just kids (with sticks).*

"Cigarette," demanded the tallest boy holding out his hand as they approached.

"How old are you?" I asked raising an authoritative eyebrow, trying to sound unconcerned and, to my surprise, actually succeeding.

By now there were five of them around me, of varying sizes, ranging from about ten to sixteen.

"Cigarette!" he repeated. I decided they were harmless, but gave them a roll-up to share behind the bus shelter or wherever it was teenagers went to smoke in Romania.

They skipped off, happy with their loot. I stood there giggling at myself and saying out loud,

"Will you just *chill out*?"

I set my Garmin to find a hotel and ended up heading back up the dirty riverside road the way I had come and towards the last town, which now looked inviting despite my thoughts as I had ridden through it thinking, "You couldn't pay me to stay here!"

It took a couple more failed attempts before I found a hotel in working order. With my heart sinking and my brain doing its best to compensate with positive chants, I finally came across a fully-functional hotel with a bonus – secure parking for Rhonda.

Once again, I was the only dinner guest in a large dining hall, complete with artsy cornice and chintzy chandelier restaurant. As I searched the menu for my dinner-for-one on the big, round, once-white-clothed table for eight, a band walked in. They weren't really going to play just for *me* were they? I was tired; dog tired. So tired that I almost skipped my usual end-of-ride beer. Yes, it was that serious. Since I have that typically British 'polite' gene, I knew that once they started, I could not just get up and walk out without feeling responsible for all the no-shows this room had the potential for. My brain then flipped to "but what if they *want* me to leave so they don't have to bother playing?" It was a real predicament and my face might have looked distorted with this thought process as the waiter brought the tray. *Why do I worry so much? So much effort to avoid offence.* The band kept looking over and smiling as they set up. I smiled back. It was exhausting.

My food arrived just as I was making a roll-up. It was bland, and I got through it quickly. As I lit my cigarette and sipped on my beer, the pasta began to sink through my stomach lining and settle into the very crevices of my soul. Then the second song started and I planned my escape. The food coma was imminent and eye contact was now unbearable. I couldn't look interested for much longer. As they paused for the next song, I got up and walked out. My plan was not to look back, but that would be rude. I turned and waved my thanks with a polite smile. They all waved back with what might have been a slightly disappointed look on their little faces.

As I climbed the well-trodden stairs, their third song began. It was *Hotel California* by The Eagles. I stopped mid-step and giggled as I listened to the familiar lyrics. *Was that for me?* I certainly had been on a dark highway, although it wasn't a desert and I had no idea what 'colitas' smelt like, or indeed what they were. Still, the song felt like my anthem. How could I not answer the call? I ran back down the stairs and 'stood in the doorway' just as they finished singing that very line. There was no 'mission bell' but it was close enough for me. We smiled at each other – a genuine smile now and I found myself singing back at them, arms out towards them. We sang our way right through the chorus and all the way to the end for a perfect finish. Across the empty room we laughed and clapped for each other. It was a beautiful moment.

That song came on in so many random places as I travelled further around the world, as if it were following me; this night was one to remember. I stayed for another beer, and then another. When the band finished their set, I was still there, shouting "Encore!"

The band joined me at the table-for-eight and we had a makeshift conversation over beers; part English, part Romanian, and part sign language. Beer, it seems, crosses language barriers quicker than you can say "Google Translate".

Back in my room, tea bags with cold water was the order of the day. Not to drink, but packing them onto my eyes seemed to help keep the swelling down a little. They still weren't right, though, not by a long shot. I spent the last hour before bed awkwardly sitting on the grotty carpet near the door where I could get just enough Wi-Fi to research what could be wrong with my irritated eyes.

It's never a good idea to ask Google to diagnose your medical problems. This research, however, led me to an eye specialist on my route. Based in Istanbul, he did regular slots on breakfast TV about people's health problems, specialised in eyes, and, I would later discover, was also a biker who loved the Red Hot Chilli Peppers. That night, I emailed him, explaining my situation and symptoms, then went to bed and slept long before my Radio 4 'Drama of the Week' podcast drew to its conclusion.

By mid-morning, having navigated my way along the quiet river road and through the Serbian border with ease, I found myself in a small town just outside Belgrade. Here, I met Sean, Dejan, and Marko, who welcomed me with a coffee and, on their Vespas, escorted me into the city. Sean was Canadian and had travelled the world on his Vespa and was the first person to contact me before I left the UK, offering a refuge en route. I'd put a pin in the map and actually adjusted my route to include Belgrade for that very reason. Sean and his girlfriend, Tamara, had a great flat in central Belgrade with a basement car park for Rhonda, a spare bed, and a very warm welcome for stray bikers like myself. I was looking forward to finding out all about Sean's own world trip, completed two years earlier; eighteen months on the road with only a fully-loaded Vespa and a heavy dose of optimism.

Checking out his photos, I found one wall full of his trip memories. They filled me with a warm glow as I imagined my own photo wall at the end of my trip. It was a long way off for me, but I was already building the memories. The wall helped me imagine that it would actually happen for me. I then wondered where those walls would be, as I no longer owned such things, and didn't care to worry about it. I was free and the end of my trip felt like a very long way off. Besides, not knowing was all part of the fun. Sean had probably never imagined that his journey from Canada would lead to a life in Belgrade with a beautiful Serbian fiancée. Two years after I left, they were married and had a new baby.

That night we headed out to meet some friends and eat in a rather trendy restaurant serving Serbian food with a modern touch. The conversation flowed easily as we consumed copious amounts of meat, wine, and cigarettes. By then, I was smoking only infrequently –the odd smoke with a beer, on occasion. Eastern Europe, however, had other ideas. People smoke there with monotonous regularity. In fact, for some time after, people would see me pulling out my fags and say "Oh, you smoke?" in a slightly disapproving way, and I'd reply "Yes, I blame Eastern Europe," and change the subject.

It felt deliciously naughty, although with everyone doing it, it was actually overwhelming at times, even for me; someone who can usually embrace a dirty habit with the best of them.

The next morning, I woke up feeling pretty rough. A mixture of rich food, booze, and nicotine indulgence. Riding three thousand miles in just the past two weeks was also catching up with me. I decided that if I was to keep up this pace, I would have to take the opportunities to rest as they arose. This was one of those opportunities; I turned over and went back to sleep until midday.

That afternoon, Sean suggested we get some fresh air. Still tired, I asked if I could go on the back of his Vespa. My first ever Vespa experience was on his red mean machine called Visnja (meaning 'Cherry'). I quickly realised that Vespas are the best way to travel in Belgrade. We nipped in and out of traffic, mounted pavements, and parked her up next to cafés as we checked out some of the city sights and drank lots of milkshakes and coffee.

Belgrade is a city that remembers. 'NATO' is a swear word there, and was still a topic of conversation. This much I'd discovered at the restaurant the night before. As we rode past the bombed-out military headquarters, Sean explained that it was left just as it was, as a reminder of the events that took place over a three-month period, just fifteen years earlier. We drove over the cobblestones and arrived at a trendy café on the walls of the city's fortress overlooking the River Danube. We ordered a coffee and Sean continued,

"During the attacks, cruise missiles could be seen at this level, following the line of the river en route to their targets, and teenagers came to this point, where we are now sitting, to drink beer, get high, and watch the display. There was not much else they could do as there were no studies in the city during that time so the only thing left was to party".

When in doubt, get stoned. I imagined myself as one of those teenagers and pictured the missiles in front of us. It's amazing what can so quickly become the norm during times of utter chaos and how we react to situations to see us through. We never really know until we're there. So often I hear people say, "Oh, I couldn't do that," or "I don't know how you coped." But we do. We have no choice but to find a way through. We survive. Belgrade has seen so many changes and troubles over the centuries, having been destroyed and rebuilt forty-four times in its two and half millennia of history. Today, it is the very definition of a cosmopolitan global city, shaped by people, food, and cultural adaptations from all over the world, in a very chilled atmosphere. It has a good feel to it – not nearly as hectic as I expected; it's one of those places you want to keep to yourself and not tell anyone about to prevent too many people coming in and ruining it.

That night, we had dinner and headed out to the Bigz building and Cekaonica ('The Waiting Room'), one of several clubs in the former printing press complex. The Bigz is now a 'derelict' high rise building that looms over one corner of the city, covered in graffiti and full of dark corners.

We clambered around the rubble that was strewn around its dark exterior, and into a dimly-lit hallway. Next to an incredible graffiti image of Kurt Cobain's face, was a rusty old door.

Behind the door was the lift. I was surprised that we were actually going to take it. Looking inside, I felt we had a better chance of making it if we did a 'Harry Potter' and just threw ourselves against the wall until it relented, taking us where we wanted to go. It looked decidedly unpredictable and downright unsafe. Still, up we went, while the lift complained every inch of the way. If you listened carefully you could hear it saying,

"Life? Don't talk to me about life!"

"Wow!" I said as the door opened on the top floor, revealing a stunning little jazz bar sprawled in front of me. The live music featured a handsome sax player and seduced my ears. There was a large terrace overlooking the twinkling city with what had to be the best view of this place that repeatedly refused to die. We sat drinking beer (later switching to tequila), looking out at the lights, sharing travel stories, and discussing life in general. It was the perfect way to say goodbye (for now) to this great city. Around 2am, I reluctantly gave in to the tiredness and headed for bed.

The next morning, I felt good. After a shower and a great breakfast of eggs Benedict, I hit the road towards Bulgaria with a full belly, new friends, and my Buff neck-warmer smelling of fabric conditioner. Life could not be better.

Bulgaria won the award for the biggest potholes and the most bugs of any European country. My visor became a morbid mess of foreign bodies of varying shapes and sizes as I travelled along dodging those missing bits in the road. It was bitter and wet by now and I stopped only briefly to visit a few more bikers en route before crossing the border into Turkey. I guessed things were going to get *really* interesting from here on in.

****

My Garmin sat nav died. Most times, this would be a minor glitch in the plan. Unfortunately, I was about to arrive in the centre of one of the largest cities in the world – Istanbul – and the road was no longer a gentle trickle, but full-on, unforgiving rapids. The incomprehensible road signs might as well have read, 'No rules apply', as I was sucked into the biggest mass of unrelenting traffic I'd ever experienced. One minute I was happily riding along minding my own business, the next I was in the thick of it and being shunted around like a puck on an ice rink. The trip was clearly dishing out its next lesson: 'Turkish Roads for Dummies'.

The panic-stricken voice inside my helmet said,

"OK, stay cool. Find somewhere to pull over and figure this out." I had to strip off some layers, lose the earplugs, get my bearings, and unpack my 'bring-it-on' head. I prayed that a quick reset of the Garmin GPS would get it working again, but that was the least of my worries. There was now only one way to go, and that was with the flow. Once I got into the city, I had to find Kemal, a Turkish biker who had recently contacted me via my blog and kindly offered me a bed for the night.

Cars and lorries from all directions suddenly had me pinned in the centre, forcing me from one lane to the next. In an attempt to assess the situation, I took a precious moment to stand on the pegs and look ahead. We seemed to be heading for a bottleneck of some description; maybe a toll? Everyone was determined to get there first and as a 250cc motorbike, I felt pretty insignificant. Like it or not, I was being taken for a ride. Inch by inch, I fought my way to the hard shoulder. At least there, I couldn't be attacked from my right…

So far, the smell of this historic city – the former Constantinople – was not of spices, mint tea, fish markets, and tulips as I'd imagined, but of thousands of overworked and dying diesel engines, coughing and spluttering their way into the metropolis, mixed with burning rubber and warm tar. As the heat became almost unbearable, the fumes formed a thick black coating on my face and seemed to force their way into every orifice. Later, I excavated enough black dust from my nostrils and ears to start a coalmine.

I pulled in just before the toll, threw off some layers, took a deep breath and, against my better judgment, dived back into the chaos that was Istanbul's Friday afternoon rush-hour. Istanbul is one of the most congested cities in the world and riding into it felt gruesomely unremitting. I made it to within five kilometres of Kemal's place before giving up and pulling into a petrol station to call him. After a few failed attempts at sourcing and describing some points of reference, I was unable to describe my location sufficiently for him to figure out where I was. Finally, I handed the phone to one of the greasy petrol attendants,

"Please tell my friend where I am," I said, in a very patronisingly British way, despite clearly being the only idiot in the conversation. I didn't intend to be patronising, but my request came out louder and more 'pidgin' than I had expected – as if shouting slowly would help make it clearer. I would later laugh at a Kiwi friend of mine whose 'Pidgin English' always came out in a French accent in such situations, despite being from New Zealand. The plan worked though, and in less than fifteen minutes, Kemal showed up on a Honda CRF250, a cousin to Rhonda. While waiting, I had managed to navigate my way through a bag of cheesy Doritos and a Red Bull while the bemused attendants looked on. The alien Welsh girl had landed.

That night I was treated to fresh pyjamas, slippers, and knickers courtesy of Kemal's sister, who had gone out and bought them especially for me. I was then given Turkish coffee, which I drank to be polite. Before this, I had retched at the taste of *any* coffee, let alone the strong Turkish stuff. Somehow, perhaps because of a combination of the environment, the right light, and the 'exotic goggles' I was wearing, I liked it. In fact, from that day on I've become a bit of a coffee demon – proud to be a member of the international group of caffeine addicts. I'd always liked the idea of ordering a latte or a mochaccino and now, thanks to Kemal, I could.

That night I pulled out my Macbook Air and updated my blog before falling into a restful slept on the sofa with Kemal's black labrador.

****

As I sat looking out at the busy Bosporus Straits that connect the Black Sea with the Sea of Marmara and tucked into the best kofta kebab I have ever tasted, I watched the dolphins playing between the boats under the shadow of the Bosporus Bridge. It dawned on me that the other side of that bridge was in Asia. I'd made it through Europe alone, and in doing so had notched up a little more confidence in my ability to actually see this adventure through.

Although it was only thirty days since the big send-off from the Ace Café, it seemed like a lifetime ago that I'd followed Pete Bog into Dinant, a little town in Belgium with saxophones on the city walls, (in honour of its most famous son, Adolphe Sax, inventor of the instrument). It was a quirky little place, where we'd spent the night scoffing pizza and drinking Duvel, a ridiculously strong Belgian beer and an education in itself. I smiled as I remembered Pete randomly waving to the cows on the way in, his whole body jerking in a giggly spasm each time he amused himself along the way. I missed his infectious laughter and was already reminiscing about the 'easy days' when it felt like one big summer holiday!

Sinking my teeth once more into the delicious kebab, I found myself in contemplation mode as it dawned on me just how lucky I was. My first 3,500 miles through ten countries had taught me a lesson or two and so, despite time flying by, it also had a depth that simultaneously made it feel like years. My travels through Europe had been an education and one that would stand me in good stead as I ventured on through more challenging lands.

"Just keep going; that's the trick," I thought, as I caught a dribble of garlic mayonnaise attempting to escape down my chin. "Easy as that."

Europe was my training ground for solo camping, 'wrong-side' driving on the right, and making myself understood with nothing more than a few gestures and facial expressions. I enjoyed the morning ritual of packing up and checking Rhonda for 'road sores' – stray nails in her tyres and suchlike. I took on the bad drivers and challenging potholes with a new-found patience. My daily helmet hair was a triviality and my lack of clothing choice was liberating. No choices meant no stress. As a lone biker travelling around the world, I had permission to look scruffy and wear T-shirts for every occasion. It was all part of living on the road, and I took to it like a beggar to a bag of chips!

Despite the nerves, I was learning to relax and trust people more. I learned to say "Yes" to that cup of tea on the side of the road and to have a chat with the locals who questioned me on everything from my marital status to my inside leg measurement. I took up invitations to stay with local bikers who spoilt me with hot showers, local beer, local knowledge, and best of all – friendship. Looking back now at some of my earlier fears and uneasy moments, I have to laugh at that person I left behind.

The hospitality afforded to me was humbling, and both time and agenda became less relevant by the day. Slowly but surely, all my grown-up worries, constraints, and paranoia were stripped back to reveal my inner child. I realised that I felt happier than I had ever felt in my adult life. I had never felt so truly alive and aware.

They say that when you crash or witness a shocking incident, time stands still, and you experience it in slow motion as your brain takes in a lot of information all at once. I wanted to know everything; to soak it all in and remember every detail of my journey as if it were a great big and prolonged car crash.

Chapter 7
From Dr Smellgood to Mr Stinkypits

I sat nervously in the splendid waiting room of the eye specialist who had agreed to see me when I emailed him from Serbia. As mentioned earlier, in Romania, my eyes had started hurting. And when I'd looked in the bike mirror they were red and quite badly swollen. I was quite surprised that he had replied. I'd included a description of the problem; a brief of what I was doing to try to alleviate it and photos of my irritated eyes. Fortunately for me, he turned out to be one of the motorcycling clan and the biker bond moved him to offer to see me for a consultation – with no charge.

"You have a rare condition that can be brought about by stress or dust," he said as he stood over me putting on his latex gloves.

"Yes to both" I said. "I was quite stressed leading up to my big 'jump' as I call it, and it has been pretty dusty on the roads." The doctor – whose name I cannot recall – was dishy in a suave upper-class Turkish kind of way. He had a confidence about him that can only come from private education and smelt of just the right amount of aftershave – the expensive kind that seduces your nostrils in a subtly flirtatious way. I breathed it in as he leaned over and prodded my eyelids.

"There is a blockage here. That is why you have this fluid build-up. It has now stretched the eyelids also. I will have to operate to unblock them and cut some of the eyelid away."

"Lucky me."

"This condition normally shows itself first in childhood," he said. "Then, as I said, it goes away until it is irritated."

"Well, I did have a bout of it in childhood, but we thought it was mumps" I said, happy that he was on the right lines and not just touting for business. After all, an operation was surely going to cost me, or at least my insurance company. A doctor he may be, but a well-paid one by the look of his office, and that could mean he's just a well-educated salesman.

My cynicism was quickly checked as he said,

"Now sit here," pointing at the comfy chair in the corner. "Let me get you some tea, and you can tell me all about your trip. We can operate the day after tomorrow. There will be no charge, biker to biker, but there is one snag."

"And what is that?" I asked dubiously as I settled into the plush leather chair as instructed.

"If we go to the hospital to do the operation, it will take longer and I can put you under a general aesthetic, but it will cost you money for the use of the hospital and the drugs. OR, I can do it here for free and you will be awake."

"AWAKE?!" I squirmed in my chair as the image of a scalpel in the hands of a cocky, albeit dishy, Turk flashed through my thoughts. He took the opportunity to call for tea to be brought in. There was a smirk on his face as if he were enjoying my discomfort.

Putting the phone down, he explained that he had done this many times before and that I would not feel a thing. He would give me many injections to numb the area and it would be over before I knew it.

"I'll take it?" I said in a tone that implied a question rather than a statement.

"Great. Now tell me about your plans to sail to Antarctica with a motorbike. This is amazing." The tea came in as if on cue, and we spent the next hour or so chatting about my plans and his own dreams of taking off on such a journey one day. It felt like a little bonding time for which I would be very grateful soon enough.

The following morning, I decided to move from Kemal's to a hotel. I later regretted this as Kemal and his family seemed genuinely upset at my departure. I explained that I didn't want to be a burden on them after the operation, and that I would rather be in my own space. After all, although we had only just met, I knew they would wait on me hand and foot while I recovered; the perfect hosts. They seemed confused at my explanation, as if they could not comprehend what I was saying. It was not the language barrier. To them, my trail of thought was just plain lunacy. This was an opportunity to learn from each other, and about each other; an exchange of cultures that they had been looking forward to. Being British took a bit of cracking. I was too polite, and truth be told, I was less than comfortable in other people's homes. I found it quite exhausting, even back in the UK. Learning to relax in others' space and understanding that it was a two-way exchange, was a life skill I needed to get to grips with fast.

Kemal could not convince me to stay, and I felt his disappointment so strongly that I found myself in tears that night in my hotel room. Actually, I think it was partly that, and partly due to tiredness. It had been a fast-flowing few weeks and all of a sudden, I was blubbing and muttering things that would have given Joy Division enough material for a new album. Instead of being free I had *"given up my wonderful life and can no longer return"* and I had *"burned all my bridges"*. Oh no *really*...I mean, instead of being the tough and independent road warrior I hoped I was becoming, I now needed my dog, a big hug and quite a substantial snot-rag. Kemal's kindness had struck me down with a bout of classic homesickness, although I don't recall ever suffering from it again. Perhaps the past month was just catching up with me. The constant processing of new things. New places, new faces, always having to think for myself, and constantly make decisions was tiring. Socialising with new people was both pleasant and exhausting. Perhaps these were just the symptoms of a freshly severed umbilical cord.

The following morning, I was on Dr. Smellgood's table, prepped for my op, and giggling. The giggles were partly due to nerves, and also because the doc was quite a character. He and his nurse buzzed around me with gowns and trays of menacing sterile implements. The scalpel seemed to wink at me as it was moved, and the bright lighting above caught its shine for a pregnant second that coldly said, "I'm coming for you, ready or not." The lack of empathy in that metal implement hung in the air as I lay there gowned up and ready to go.

"Do you like music?" the doc asked cheerfully, moving towards the stereo.

"Of course," I replied, "In fact, that's a great idea. What ya got?"

"Chili Peppers OK?"

"Perfect."

He turned and walked over to the neatly organised tray of needles as The Red Hot Chili Peppers started up with the unmistakable intro to *The Zephyr Song*. To my surprise, the doc began to giggle as he picked up a needle from the end of the row in his latex-covered hands.

"What's up, Doc?" I asked nervously; partly giggling and partly fighting off the flight mode that had just kicked in. Several images from the awful B movie horror film 'The Human Centipede' suddenly flashed through my mind. A crazy doctor kidnaps some people and sews them together in a room not unlike this one. If you haven't seen it, DON'T. In some small way, you will never be the same again.

"I've just realised," the doc continued, "This must be awful for you. You're alone in a strange country, lying on a table, and some butcher is about to cut into your eyelids." He laughs properly this time; thankfully, not in an evil 'mad professor' kind of way. Just a man happy in his work. A rather sociable chap who, in another life, might have been a jolly waiter. Instead, he was the guy who now loomed over me, blocking the light. His silhouette, and that of the needle, growing larger and glowing at the edges.

"Don't worry," he finally added, switching to a more gentle and professional tone now, "You won't feel anything. I'm just going to start by giving you an injection in your bum that will help you relax. This is probably the most pain you will feel." I didn't like the 'probably', though I was keen to feel the relief of a warm opioid bath wash over me. Like a cup of good strong tea, but straight to the veins. I was sure things would all seem a lot less dramatic after that.

He was right. The operation was not as bad as the anticipation. It was certainly not pleasant, with several numbing injections (the first of many throughout the journey as it turns out) around my eyes, a lot of poking around, and then some stitches to finish. My guess is, it took no more than an hour – all done to the sound of the Chilis and the doc's slightly muffled dulcet tones as he sang along under his surgery mask. I left with instructions not to ride for a week and to make the most of Istanbul while I waited. I would return in a week to remove the stiches, and, assuming all was well, would then hit the road again.

This was *not* the best start to Asia. It certainly wasn't what I expected, but then I really wasn't sure *what* I had expected. For the first of many times over the next few years, I found myself placating my thoughts with the phrase, "It's all part of the adventure."

My week of recuperating largely involved wandering around Istanbul while sporting a rock star-sized pair of sunglasses to hide my black eyes and Frankenstein stiches. A few days into the week, Dan arrived. A lovely bloke from Essex, whom I had met a month or so before I left home. He contacted me asking if he could pick my brain/compare notes as he, too, was planning a big road trip on his BMW Dakar 650. We met up in a McDonald's on the A55 in Wales and spent a good two hours discussing visas, nerves, and motorbikes. Dan left a couple of weeks after me and had now caught up, ultimately heading for Australia.

We moved to the cheapest hostel we could find in the city and shared a room together for a few days. Me recuperating, and Dan paying many visits to the Iranian embassy trying to arrange his visa; something I did before I left home. In between we would wander around the city together and check out the sites. Putting the bikes aside and just being a 'tourist' was a nice change.

It's a strange thing, this labelling. No one wants to be a 'tourist'. Already, a month in, I saw myself as something different from your average traveller. I was an 'overlander' – "a totally different thing", I told myself. Instead of flying from destination to destination, we make our own way through the good, the bad, and sometimes, the exceptionally ugly. Or dull. We carve a path dictated not by public transport or beauty spots, but by our imagination and our ability to keep going. This was never truer than on a trail bike because there is little limit to where we can go. Sounds romantic, right? And it was, although the early challenges of a long journey are often forgotten when you look back with a heavy dose of nostalgia. As someone once said, "Nothing is more responsible for the good old days than a bad memory."

I remember those moments where my hands were so cold I could barely hold onto the bars any longer, or the times I'd got myself lost as dark was falling. The dust, the dirt, the traffic. But this truly *was* 'all part of the adventure', and it's what made me appreciate the sweet moments and the little things all the more.

For a couple of days, Dan and I indulged in some plain old pottering about, people-watching, soaking in the atmosphere of the city that is literally torn between two continents, and drinking coffee next to the gaping, yet beautiful, 'wound' that is the Bosporus Straits. We sat enjoying an Efes beer or two on our last night in the city and I found myself opening up to him. Brought on in part by the beer oiling the hinges on that dusty old skeleton-filled closet; in part because I felt far enough removed from real life that it didn't really matter; and in part because Dan was so easy to talk to. Of course, I also had the added bonus of being able to ride away in the morning if required.

It seemed the right moment to remove another block from the wall I had built up to protect myself and in doing so relieve a little pressure from the guilt I carried. It was something I never talked about, but suddenly I found myself telling Dan all about my past as a heroin addict. The shame I had brought on my parents, how I had failed at being a mother and I told him of the guilt that ate away inside me each day.

Dan listened with an honest and kind face. Never interrupting. When I was done, the first thing he said was,

"Wow! I have so much respect for you. You are an amazingly strong person." That was *not* the reaction I had expected! Dan later surprised me by sharing his own demons. My willingness to speak first had opened the doors and almost given permission for him to be vulnerable too. Two near-strangers, baring our souls over a beer; rewarded with a lighter load to carry, and a meaningful friendship that in another time and place might have taken years. I decided that night that it took courage to be imperfect; to bare all; to be vulnerable, and that might actually be the very thing that finally set me free. OK, maybe that was courage, a couple of beers and the tequila talking, but it meant something at the time, and I know it helped us both that night.

We hit the road and left the city behind the next morning, on a mission to check out some of the west coast of Turkey. We'd had our fill of 'tourist' and now wanted to get our 'Mad Max' groove on again – dusty, dirty, and ready for anything. First, though, we had to make a quick stop at a Honda Dealer on the outskirts of the city who had contacted me and offered to give Rhonda a good once-over on my way through.

The once-over became a full service, a thorough washing, some additional lights and reflectors "because," we were told, "the Turkish drivers are crazy." I was then given a business card with a number to call if I got into *any* bother at all, before then being given the honourably title of 'daughter' by the owner. I was now 'part of the family'. We left the dealership very late in the day, yet with still enough light left that we could at least make it to the ferry between Eskihisar and Topçular before finding a place to camp. Dan had been happy to just 'tag along', but after the ferry, the day was drawing to a close and we still had no place to camp. We were tired and dusty, but the sun has no empathy for tired limbs. When it's clocking-off time, no amount of bribery can convince it to work overtime. That's the thing about nature, it's predictably unemotional. Deal with it.

I could sense that Dan was not happy about 'dealing with it' right then, so I pulled over.

"It's nearly dark," he said. "I don't ride in the dark. I promised myself." I saw his heart sinking along with the unrelenting sun, but these were the conditions we found ourselves in and there was little we could do but roll with the punches. I was feeling quite chirpy, considering, and found it all quite exciting, but Dan was right. We did need to find somewhere, sharpish. Turkish roads were no place to be in the dark on a motorbike.

"Okay, let's take this dirt road here and see if we can find a place to put our tents," I said in a voice that suggested "no problem here, just a minor hiccup, all part of the adventure". Dan's face didn't look particularly happy under the thick layer of dust, but he nodded his agreement due to lack of any immediately obvious alternative.

The sandy track was surrounded by what looked like an orchard of some description. I guessed they were olives, but it was really dark by now. We stopped on a corner and looked around.

"What do you think? In those trees?" There wasn't much space and we couldn't see beyond the first line of trees. Everything else was just black. It could have been right in front of a farmer's house for all we knew. Dan seemed less stressed now. In the zone and ready to deal with whatever came our way.

"Yup. To hell with it. Looks good to me"

That's when we heard the sound of dogs. At that moment, in the dark, on a narrow sandy track with no quick way out, it sounded like someone had released the hounds of hell.

"Shit. Sounds like they're coming right for us" I said with a marginally less 'no problem' tone now. For all we knew, we were trespassing on their land. Or it could be a pack of hungry strays. We imagined the worst and knowing it would take some manoeuvring to get out, we quickly decided to heed the warning and move on before they reached us.

Jumping off our bikes we tried to manhandle them around in the small, uneven and soft space we had. In the rush, Dan dropped his BMW. *Shit!* How far were the dogs now? I put my bike on the stand and helped pick his up. There was a real sense of urgency, probably both imagining the same thing: savage salivating dogs that would rip us limb-from-limb if they found us on their patch. Rocks and shouting might well have scared them away of course, but we were in the dark and on unfamiliar ground – two rookies of the road still full of images of the bogey monster (I later learned that a revving bike is also a great dog-scarer). Shadows from the gnarly branches reaching all around us, and the sound of a pack of angry dogs heading our way, was *more* than enough to have us believe that they were on a hunt; and probably for us. This was no time to rationalise with our overactive imaginations. They could be right, after all. This was a time for flight, *not* fight! These images gave us the strength we needed to lift the heavily-laden 650 Dakar off the ground and pointing in the right direction.

"Right. Let's get the hell out of here," I said breathlessly, with a hint of a nervous giggle. I didn't need to say it twice. We were off and out of there faster than you could say "Mad Dogs and Essex boys".

We travelled for what felt like an age down those dark country roads, pot-holed and unpredictable; too nervous to stop and too fuelled from our close encounter to be tired now. Eventually, we came across a small hotel near a lake. It seemed an unlikely place to find one, though in reality, we really had no idea where we were. It felt like we were far out from civilisation, but the dark and lack of any other traffic will do that to you. We were both on a tiny budget. Still, we agreed that we would go in and negotiate the best deal possible for a twin room. We succeeded, and sleep came quickly that night.

The following morning we found our way to the Island of Cunda, where we spent the next few days camping on its beaches and riding miles of windy dirt tracks through olive groves and along its coastline. We climbed its hills and savoured its delicious views, stopping only for drinks and a dip in the sea to cool off at lunchtime. We finished covered in mud and smiling from ear to ear, before pottering back to our tents and cooking dinner overlooking the still and tide-less waters of the Aegean Sea.

The next morning Dan headed back to Istanbul where he would spend the next few weeks battling to get his Iranian visa. It seemed I would be one of the last Brits to be granted entry without a guide. Since receiving mine, the rules had changed, and as Dan had arrived so soon afterwards, there were no procedures in place to deal with the new policy. Our hopes of seeing each other further down the road quickly faded as the time and distance between us grew. We would not see each other again until I got back to the UK.

I continued south to Selçuk, then east to Didim, and eventually to Bodrum. All this time I had been working every chance I got on finding a lift to Antarctica. Whenever I had Wi-Fi, I was researching, emailing random people, and chasing all leads. I contacted shipping companies who delivered supplies to the scientists or to the navy there. I spoke to cruise ship operators who are allowed to go out so many times a year and take no more than five hundred passengers at a time. All were under strict rules and a set agenda; there was no way they were going to allow me on board with a motorcycle. Eventually I decided to try private expedition boat owners. Most didn't reply. Finally, a self-confessed world class sailor with many race wins under his belt agreed to take me, only to change his mind a week later after telling me how amazing he was and how dangerous it was to cross the Drake Passage. I was mortified. Still, I persevered and eventually found an Australian-owned boat called *The Ice Bird*. The couple had been across the Drake many times and saw my request as an added bonus to their adventures. The email came back reading:

"Why not? Sounds like fun! We're in! We won't be able to bring the bike back though but we can figure it out when you get here. Just make sure you are here by the 2nd Feb at the latest. Oh, and we have nowhere to put the bike except on deck so it will be open to the salt water."

I fought back my doubts and replied,

"Great. I'll see you in Ushuaia."

It was April. There were only another eleven countries and two continents to get through. How hard could that be?

Bodrum is a port city in southwestern Turkey on the Mediterranean coast, and I was there to meet Fuat for a bit of Antarctica education. Fuat managed a sailing club and invited me to stay on one of the yachts for a night so I could "see what it felt like." I doubted very much that a moored boat on calm waters did anything to prepare me for what was to come, but it was a start.

"The Drake is a big deal for sailors," Fuat said, "It's like Everest is for climbers." He then showed me videos with waves the size of ten-storey buildings and *The Ice Bird* bobbing around amongst them like a dinky toy. This was not helping. Eventually, he got to the good stuff: photos and videos of the crew kayaking amongst the icebergs; whales swimming just metres away from them as they surface and dive, flicking their enormous tails as they go. Daring to dream, I imagined myself there with Rhonda. *Could it really happen? Could someone like me really do this?*

The thought of crossing the Drake frightened me. I may be the granddaughter of a Merchant Navy captain, but it clearly doesn't run in the blood. I am pretty nervous of *any* sea to say the least, *but the Drake? With those waves?* Still, it was so far away that it was easy to convince myself it would all be fine before filing it away in the bottom tray of my mind – along with that idea I once had to ride a camel across the Nubian Desert – for future reference. I took comfort in the fact that I probably wouldn't get that far, so the chances of it seeing the light of day again, were slim.

From Bodrum, I headed for Lake Köyceğiz. I was determined to avoid the dual carriageway, so, despite my sat nav insisting I go that way as there were no other roads marked, I eventually switched it off and navigated by sight and 'instinct' – as I liked to call it – imagining myself as some kind of female bike-riding Dr Livingstone or something equally sexy and adventurous.

I found myself on the smallest of roads, meandering through fishing villages and along the coastline. The road was broken up and turned to dust as the bypass had clearly seen it long forgotten, but it was beautiful; taking me alongside secluded coves, around switchbacks, and up into the hills with breathtaking views of the coastline I had just ridden.

Early in the afternoon, I set up camp next to the lake. It was wonderfully peaceful as I sat writing in my diary and glancing up to watch the goats on the cliff face across the road. "Aren't they amazing?" I thought to myself as I watched them hop around on the rock face with ease, "You never see one fall." No sooner did I have that thought than one of the goats fell to the ground from about twenty feet up. Thankfully, it got up, shook its head, and wobbled off. No permanent damage, it seemed, but the timing was comedy perfection.

As dusk fell and the colour changed, the light cast eerie shadows through the dried-up shrubs around me, changing the entire look of the land. It was beautiful to watch it evolve; to be part of it and amongst it. With dusk, though, came teenagers. I was pretty close to the road and, I was now learning, the layby was a place for teenagers to meet and drink out of sight of the grown-ups. So much for the perfect spot. I decided it was time to get into bed and switch on my podcast, hoping they would go away before too long. I made a mental note to always camp further from the road in future, and out of sight where possible.

Camping was easier for me now as I developed my routine, although I still camped with pepper spray I'd bought in Istanbul and my Leatherman multi-purpose knife under my pillow. Each night got easier, and learning to make friends with the local stray dogs helped. They would sit around and guard my tent if I shared my sausages with them in the morning. It was a basic survival technique developed by our prehistoric ancestors. That, and the sound of the shipping forecast on podcast drowning out the noise of the monsters in my head. In a tent, a tiny movement outside can grow in your mind to become a bogeyman and it's hard to shake once the seed is planted. If you can't hear it in the first place, it doesn't exist. The chances of it really being a bogeyman or an axe murderer are pretty slim, I reckoned, and grew happy in the knowledge that it was best to just block it out before the idea could germinate. If it ever really *was* an axe murderer, well what was I going to do about it anyway? And how did they know the axe murderer wasn't the woman in the tent? No, it was best all round to stick my head in the proverbial sand and relax to the wonderfully relaxing sound of the BBC's Radio 4.

This didn't always work of course; sometimes there were no dogs to bribe. Occasionally I would hear voices outside my tent. Nothing ever happened, but those were sleepless nights when I really had to face my fears. Mostly, though, I didn't. Mostly, I just cowered in my tent and hoped whatever it was would go away. I truly felt like a coward on those nights. I wished I could be stronger. I wished I had listened to my friends who told me to take self-defence classes. I wished I could be a proper adventurer, fearless in the face of fear. The truth was, I was a quivering wreck those nights. Was I a fraud? The impostor complex was still strong in me, just like the day I left, just over a month earlier when I wobbled out of the Ace Café car park thinking. "Any minute now these people are going to realise I have absolutely no idea what I'm doing."

I continued through Turkey at a steady pace, taking in the culture change and often comparing my surroundings to Morocco; a place I had visited often when taking customers on tours through the Sahara. The calls to prayer from the mosques were distinctively different in Turkey though; they sounded much more exotic and were a wonderful reminder as you woke, that you were living your dream. It always put a smile on my face.

Being on my own was also very different. It felt so right when the sun was shining and my brain was taking it all in, not yet desensitised to my environment. I was aware of every little detail, and I wanted to know everything. What is the name of this tree? How old is that building? Where are all the women? Why do men seem to have all day to sit around and drink tea? So many questions, I soon realised, I did not ask in my own country. Familiarity seemed to blind me whereas change opened my eyes, my senses, my thoughts. Travelling alone let me hear those thoughts in what I decided to call my 'helmet time'.

I didn't expect much of Turkey. It hadn't excited me like some other countries when I'd spent hours back home pouring over the map. I'd imagined it as a touristy place where Brits went for cheap holidays; a place I had to ride through to get to the really exciting stuff. I underestimated it; Turkey was turning out to be a delight. My little trail bike made it easy. I could get well away from the touristy areas with their McDonald's and Starbucks.

About three weeks into Turkey I found myself camped in an empty campsite on the outskirts of a small village. It was still out of season and clearly a sleepy little place at the best of times. The owner saw me and quickly made his way over to greet me. A middle-aged man with bad teeth framed by the best comedy moustache I'd ever seen and pinstriped shirt over a white-ish vest. He spoke a little English and once we agreed on a price, he invited me in for dinner.

"Thank you, I am fine" I said. Tired and keen to chill in my tent. "I have some pasta. I will cook that and then sleep" I continued, making hand gestures and speaking in my best pidgin English.

"No, no, you must come. I have fish. Best fish. Then you sleep". I could see there was no getting out of this one, and so I agreed and thanked him for his kind offer.

"Okay. You put your tent up and then in one hour you come," he pointed to his little restaurant at the front, which was clearly closed.

"Okay," I resigned, "See you later".

There was something about him I didn't like. Something that made me feel slightly uneasy about the situation. He was greasy. And I'm not just talking about his shirt. I could sense a sliminess to him already. Nothing obvious, maybe subtle eye movements? The way he held my hand a split second too long when we shook. The tone of voice perhaps. It didn't help that I could smell his armpits long after he walked away, lingering in the air and hanging around like a guest who's outstayed their welcome and won't take the hint. He scurried away and back through the orchard, kicking a chicken in his path. I watched until he disappeared around the corner and then pretended to gag at the smell. *Jeez. This is going to be an interesting night. I should offer him my pepper spray for a deodorant. Maybe he'll get the hint.* Despite my attempt at amusing myself, I was also making a mental note to put my Leatherman in my back pocket – just in case – as I finished erecting the tent.

The restaurant *was* closed. However, when I arrived, I found a table for two and a waiter who looked slightly awkward, and not really like a waiter at all. He looked like he had been dragged in at the last minute to fill the role. There was an intimate candle on the table and it looked all too much like my host's intentions were to impress me for reasons I was sure I would not be impressed with. He'd even cleaned himself up, and as he came over to greet me, I noticed a new white vest under his partially unbuttoned shirt. It was the same shirt, but clearly efforts had been made. "Clean underwear, eh? Hoping to get lucky, are we?" I thought.

At that moment he glanced down at his chest in what seemed like acknowledgment. He had to puff his chest out a little just so he could see clearly. Almost like he was proudly thinking, "Ah! So, you noticed. Yes, this is my special occasion vest!"

I tried to think of a quick excuse to leave just as he placed a glass of raki in my hand. It was late and my tent was just outside. I wasn't going far tonight, and had clearly arrived for dinner. What was I to do? Walk out with no explanation?

Ah, raki. True, traditional Turkish 'rocket fuel'. This was not my first encounter with this anise and grape mix. It is the national drink of Turkey and a revolting combination if, like me, you hate anise. It's also known as 'lion's milk', and twice distilled, it packs a powerful punch. This was not my first encounter with a bloke who thought me naïve, or a soft touch, either. I was in this game now, like it or not, and I never could resist a challenge. When an animal sees you as prey, sometimes the worst thing you can do is run. I raised my glass to his.

"Serefe" we said in unison as our eyes met. The game was on!

The fish was delicious, and Mr Stinkypits was a delightfully attentive host. Wine was poured and more raki was being placed in my glass at every opportunity. As he reached over to get us more food, wine, or cigarettes, I would pour some of the raki into the potted plant next to me. He was getting sozzled before my eyes, but I was still pretty sober.

Lighting another cigarette he hands it to me flirtatiously, except by now he's struggling to deliver. I take it quickly before he takes my eye out. I decide now is probably a good time to leave. He can't see straight and is pretty malleable, I reckon, but just as I am about to make a move, he grabs my knee. I pull his hand away firmly but politely.

"Dance," he says grabbing my hands and trying to pull me up. More firmly now, I say,

"No. I am going now. Thank you for dinner". I glance around. No sign of the waiter.

As I make my way to the door he grabs me and pushes me against the wall.

"NO!" I say pushing him firmly away. He steps forward again and grabs my breast. This time I push harder and shout again "NO!" wagging my finger at him and using that tone I normally reserve for naughty dogs. He stumbles against a chair and falls into it. I feel a jolt of righteous indignation, and then maybe a tiny glimmer of pity as he struggles to stand. I ignore it. "Good night," I say, and walk out.

I put on my head-torch, grab my Leatherman out of my back pocket and walk purposefully through the pretty little orchard, back to my tent. I sit waiting – cross-legged, head-torch on, cocked Leatherman in one hand, pepper spray in the other. There will be no shipping forecast tonight. There will be no victims here tonight. Except, perhaps, for that poor house plant. I doubt it survived the night.

Chapter 8
Who Needs Drugs?

I clock the open window as the dishevelled man leads me from the dark corridor into the small room. It will be my escape route, should I need one. My eyes take a few seconds to adjust. The room is straight out of the Dark Ages and smells vaguely of damp straw. It has three rusty iron beds with ancient mattresses and grey army surplus-style blankets that have that just-slept-in look, like the itchy ones used in prisons. The chubby guy smiles, revealing tobacco-stained teeth, and turns to leave. As he shuts the door behind me, I find myself listening for the click of a lock. It doesn't come. Frozen to the spot, I listen to his footsteps, as he walks back up the corridor and closes the second door leading to the garage front where Rhonda is parked. I wait. And listen. Still no locking sound. *OK Good!*

My attention turns to the job at hand. I quickly take off my boots and trousers and slip on my waterproof liners. Then the jacket. I try not to rush because that would only reinforce the feeling of dread. I don't hang around to zip the liners in place though, and just put my boots back on before making a hasty exit.

I had stopped at this little garage in the hills intending to put my inner waterproof liners on in the loo, but the man led me to this back room instead. I guessed it was the staff quarters and had more room than the malodorous little toilet hut outside. As I walked out of the room and back into the light of day, I found a cup of tea waiting for me on a silver tray and encouraging smiles from the guys I had just walked past. They all looked the same as the 'gate-keeper': portly, round and leathery. Imagine an old leather football with a head, legs and a large moustache and you'll be close to the mark. The football nearest me immediately stood and gestured for me to sit. They had looked so serious when I walked in. The lack of any hint of even a smile had unnerved me, as I had been intensely looking for any signs of warmth. Now, it was a completely different picture. Instead of villains, I saw someone's dad or grandad – warm and welcoming.

"Where are you from?"

"Wales." A blank look.

"Next to England," I explain. "Ah, England"

"Yes."

I drank the sweet tea from the little shot glass, thanked them and rode on, laughing at myself in my helmet. *Not human traffickers then*!

"Don't flatter yourself," I say out loud in my helmet, "Not much call for middle-aged scruffy biker women. Who'd buy you?"

A few days later I had a balloon ride over the 'fairy chimney' rock formations of Goreme and a visit to the underground city in Derinkuyu where I was trapped by a load of senior citizen holidaymakers in what seemed to be a queue for the Gates of Hades. Then I found myself in a cave hotel in Urgup singing Dusty Springfield's *Son of a Preacher Man* into a microphone. Accompanied by my new friend, the hotel owner, and the barman playing a Baglama, a Turkish guitar, we were apparently 'entertaining' the French cyclists who had been forced to stay and listen, despite trying to make a bolt for the door on more than one occasion.

I woke up the next morning feeling a little disconcerted. I blamed Radio 4 rather than the hangover. It's not often you can accuse the BBC's most serious radio station of depriving you of your fortitude, but it was definitely the catalyst in the string of events that led me to feel so unnerved that day. Despite the spirits I had consumed (my payment for being part of the entertainment), I had done my usual bedtime routine and put on a Radio 4 podcast before drifting off slowly to *Drama of the Week*. This particular drama ended quite violently and left me having a horrible nightmare in which I was held hostage at gunpoint. As I woke in the middle of the night, the man and the gun disappeared, leaving behind the very real emotions of fear and confusion coursing through me. Even after my brain clicked back to reality and I could deduce that there was no man and certainly no gun, I still felt very unnerved and uncomfortable, alone in a cave room in the middle of Turkey heading for the Kurdish mountains and the Iranian border. It was silly of course, but not even the hearty breakfast of cheese, egg, and fresh bread could shake that unsettled feeling.

My plan was to have a bit of a 'transit day', riding quickly and without deviation to cover the three hundred or so miles to the base of Nemrut Dagi, or Mount Nemrut, one of the highest peaks in the Taurus mountains. The following morning, I would climb the mountain and race the sun to the top before watching the dawning of the day over the ancient stone heads that lay there. Common wisdom says the statues were erected to protect a royal tomb and, at some point, the heads were removed from their bodies. It would be a great place to watch the sunrise. Tomorrow's sunrise was my focus for now. I just hoped the weather improved. My last few days had been soggy to say the least, and it wasn't looking any better out there this morning.

After a slow 450 kilometres of cold and waterlogged riding – and already picturing my wrinkled bath-hands wrapped around a warm brew – I decided to check my map. I must confess to a few swear words when I realised I had messed up. In the crease of the map was a line of mountains – with no through road! *Shit!* A quick calculation deduced that it was 200 kilometres around them. *Double shit!* I could handle long days in the saddle; even my narrow saddle. What I couldn't stand was thinking I only had thirty kilometres to go and then finding that I had to add another two hundred. *Okay fine. It is what it is! Reset the clock and knuckle down.* There was still plenty of daylight and I should make it in about three hours.

Thirty kilometres later, the wind and the rain took a sharp turn for the worse. It was that powerful sideways type that all bikers hate. Rain-carrying, gusting wind that battled for control over the bike. I just wanted to go forward and straight! *Was that too much to ask?* But I was losing the battle, relentlessly forced to the other side of the road and having to fight my way back, only to start the process all over again. I was cold, tired and downright grumpy now. What was I doing out here in the elements, battling this bullshit anyway? I could be at home with my dog in front of a warm fire. Sometimes I just didn't understand myself at all. This was one of those times. Just as I was considering stopping and just pitching on the side of the road in a huff, I spotted a truck stop. A petrol station with lots of well-used parking spots for the truckers, a 'greasy spoon' café, and what looked like a few rooms for rent. Everything was covered in oil or dirt – or both – but it looked like heaven to me!

It didn't matter how bad the rooms were, I was taking one. I pulled in, parked up outside the front door, and approached the guy behind the counter who was clearly surprised by my presence. He looked confused as if I might have walked into the wrong place, and no matter how many charades we played to get through to each other, he was just not getting it.

"Room. Sleep. How much?" I repeated over and over. Truck drivers walked past and stared in at this spectacle. The pasty white girl, her sodden clothes dripping a pool of water over the well-trodden tiles, possibly speaking in tongues? I might as well have been. It was way beyond his grasp that I might actually want a room, despite not being a hairy-arsed Turkish bloke with a big truck and a belly to match.

In desperation, he called for the guy from the café next door who, as it turned out, spoke about ten words of English. It was all we needed. We established, much to the owner's surprise, that I was to sleep there and that the cost would be thirty lira (around £4). Relieved, I reached into my jacket and accidentally pulled out the wad of cash I kept stashed in my liner along with my passport, for all to see. Then, I nearly handed over 300 Lira instead of the thirty. *Great! Go Steph!* I couldn't wait for this day to be over.

As I was led through the narrow corridor to my room, I was horribly aware of all the eyes staring at me. Many truckers in their vests watching me from their doorways. Three of them were sitting in the corridor playing cards and drinking tea like a picture postcard cliché. I carefully stepped over them, smiled as sweetly as possible, and repeated "merhaba" ("hello") to each one while trying to look less awkward than I felt and aiming for a confident 'I do this shit all the time' look.

Once in the room, my mind pressed 'Play' on all the worst-case scenarios again. That feeling from this morning's dream came back and I imagined them all out there, conspiring. A lone woman surrounded by vest-wearing Turkish truckers. I'd heard many a story about Turkish truckers and now they all seemed quite believable and distinctly worrying. *What would I do if someone knocked on the door? Why would someone knock on the door? Just don't answer it. But they know you're in here, you can't not answer. Don't be silly. No one is going to knock on the door.*

I had a shower and put the little black-and-white telly on to drown out the argument in my head. George Orwell's 1984 was on. It was the first time I had watched it, having only previously read the book.

"The book's better," I said as the credits rolled.

Then it came. A knock. *Who the hell is that? Someone coming to mug me? They know I have money; I'd made sure of that. A gang coming to rape me?* I was trapped. Nowhere to go. What could I do? *Ignore the door. Just don't answer the door.*

But then, my British 'politeness gene' kicked in again and I opened the door. Of course I did. How rude to leave your rapist waiting! I grabbed my Leatherman and slipped it into my back pocket. Bracing myself for the worst, I opened the door.

There stood the waiter from the greasy spoon next door brandishing no more than a cup of hot chocolate and a big smile.

"For you," he said, "Help sleep. Tomorrow you come eat". Now *that* I had not imagined.

I was going to wear myself out if I carried on letting my imagination run riot. I had to find a happy place between vigilance and trust if I was going to make it alone. I had to desensitise my gut! "Trust your gut,". It's such a throwaway comment. The gut requires training and regular stock-taking. Contrary to popular belief, 'gut instinct' does not come naturally, but being alone was definitely a fast track to fine-tuning.

I went to bed with a warm glow and a feeling of security. My perspective on the entire situation changed from that single kind gesture. It was another ring of the bell and a hearty meal for Pavlov's dogs!

The morning saw an improvement in the weather: no wind, a little drizzle, and the sun breaking through. I could cope with that. First though, I had to pay a visit to the waiter next door, buy a breakfast, and thank him for his kindness last night. I doubt he had any idea what it had meant to me. A tilt in my perspective that tilted even further with the morning sun and his refusal to take payment for my breakfast.

Now all I had to do was make it the last few miles to Mount Nemrut, then over the Kurdish mountains to the border of Iran. I was warned to be careful in this area, as it was the stamping ground of the Kurdish rebel group, the PKK. I was advised not to stop for anything and definitely not to camp alone. I should only stop around other people or next to one of the many military guard-posts. Well, it certainly didn't sound dull. In fact, I was almost buzzing with anticipation – or perhaps it was the coffee. I was still getting used to the caffeine hit. Either way, I was ready to take it on. Who needs drugs when you've got the PKK and the Iranian border to look forward to? *This was a proper buzz!*

Chapter 9
Iran

Despite the locals' warnings, I found myself surprisingly at ease as I travelled through the Kurdish mountains. Even when I ran out of fuel and got rescued off the side of the road by the people who would apparently 'kill' me if they found me.

"Kurdish, NOT Turkish," said the guy proudly and pointing at the three of them. "Welsh, NOT English," I replied poking myself in the chest. They understood and we laughed out loud, proud to be the minorities. The underdogs.

My new Kurdish-not-Turkish friends took me to their nearby home-in-the-hills, fed me, and sourced some fuel fifty kilometres away before escorting me back to the fuel station and then taking great delight in washing Rhonda down for me. They seemed fascinated with her.

"How much? How fast?" they asked as they lathered her up and hosed her down with love. Then, with confidence growing, they asked, "Are you married?"

"I'm married to my motorbike" I replied, to a round of girly shrieks and high-fives.

I arrived at the Iranian border hot, tired, and just a little bit agitated; not a good state of mind for borders, where the art of patience – and a packet of biscuits – really comes in handy.

The last few kilometres through the misty mountains of Kurdistan were a little rushed as the day had threatened to close in on me, along with the border. Overwhelmed by the beauty of the mountains, I had meandered along with rubberneck syndrome, failed to check my map, missed my turn and went twenty miles out of my way before realising anything was wrong. The route was heavily patrolled by the Turkish armed forces with tanks, trucks, checkpoints, and even helicopters. Otherwise, all I saw was the odd raven or two tugging away at roadside carrion, and an open road that cut an easy trail through the limestone palisade; it was easy to get lost in my 'helmet time'.

The border crossing was much smaller than I expected. At the front of a very long queue of ageing trucks was an old rusty gate. It looked closed and, possibly locked. Had I missed it for the day? I rode to the front and parked my bike awkwardly on the uneven surface, instantly drawing the attention of the officer on the other side. It didn't take much, as he was already eyeing me up in a way that said, "Is that a *woman?* On a *motorcycle?*"

In fact, as I looked around, I realised *all* eyes were on me and I now wished my entrance and dismount had been more elegant. *Damn!* I'll never get *that* moment back. The officer opened the gates and gestured for me to come in. *Did I want to come in?* I'd come all this way, but it felt uncomfortably like the point of no return. What was Iran really going to be like? There was only one way to find out. I took a deep breath, smiled, and walked in. The gate closed behind me with a cold metal clang. "The last time I heard *that* sound", I thought, "it didn't end well for me". I smiled again. This time an audacious grin that said, "Check you out. You go, girlfriend!". It was a timely reminder of how far I'd come – and I owned it!

Before arriving, I had ingeniously thought to put a silk balaclava under my helmet so as not to show any hair. It covered all of my head apart from a window for my whole face. Taking my helmet off now, I felt quite ridiculous. I hadn't felt this silly since 1990 when my friend dyed my hair bleach blond and – on my insistence – cut it diagonally across the back so it was long on one side and short on the other. Neither were looks that I could pull off with any confidence. Still, I was stuck with it and so I began working my way through the processing lines.

After some initial confusion, I came across Hossein, my host from the guesthouse where I would be staying for a couple of days to get my bearings. Hossein had thankfully decided to come and meet me, making the process very easy. After thirty minutes of running back-and-forth between desks, having my fingerprints taken, and using the first page out of my *carnet de passage* (a temporary import document for Rhonda), I was in The Islamic Republic of Iran – the ancient land of Persia. I'd actually made it.

The next morning, I found myself in Urmia with Hossein, adorned with hajib and black, loose, long-sleeved top bought especially for the occasion. We were milling around, checking out market stalls, and listening to the voice booming out of the crackling public address system all around us. Hossein explained over a lemonade that this was the supreme leader, Ali Khamenei,

"He's mostly saying bad stuff about the English".

"Oh" I said struggling to find some other appropriate words to follow on from that. I was already feeling a little paranoid as I was sure people were tutting at me.

"Well, I'm Welsh" I continued with a smile, trying to make light of the situation and to ignore the disapproving looks that seemed to be following me.

"Oh, by the way. Your top is too short. That is why they are staring." he clarified. All matter of fact. *Oh gawd! Let the ground swallow me whole.* So, I wasn't paranoid. My top was in no way revealing, but now I felt naked,

"Well, best we get to a clothes store fast then eh?". Hussein shrugged and led the way.

Money was my next issue. I found rials quite a confusing currency. So much so that I still can't tell you whether I paid a thousand or a hundred thousand of them for a loaf of bread; and was it Tomans or Rials? (Hmm. Google tells me ten rials equals one toman). The most unusual part of the paying process though, came when I tried to pay for my 'appropriate' top. The lady behind the counter refused to take my money. Now, I had vaguely heard of a custom called 'Taarof' before. I had been told that in this situation I must keep offering up to three times. If the money was not taken on the third time, the item was mine for free! But this was a clothes shop, in a mall, and the lady behind the counter had just refused three times! Surely, this wasn't a gift. I looked at Hossein for help.

"Contrary to popular belief, Taarof does *not* have a set amount of refusal times" he explained. "You have to keep insisting until they take the money. This will happen everywhere you go." It would seem then, that many travellers had misunderstood this process and were getting it *very* wrong. The poor shopkeepers must think Ali Khamenei had a point. Although, it took another three offers before she finally accepted my money, and without Hossein being there, I might have just given up and walked out with the top for free.

I had my new top, and my hijab (actually a purple sarong I'd been carrying since home) was staying in place thanks to my good friend – and girly advisor – Jenny. She had sent me off with an emergency pack including stockings, paperclips, elastic bands, and hairgrips. The hairgrips were now hijab-holders.

That afternoon, suitably dressed and bags packed, I said goodbye to Hossein and his family before setting off to meet up with some rather familiar faces. Between closing my business and setting off from the Ace Cafe, I had worked with a guy called Nick Sanders, an extraordinary motorcycle adventure rider and author. (Nick had several laps of the globe under his belt, by motorcycle, and two before that by bicycle!) I'd originally contacted him asking if I could interview him for a magazine piece. I wasn't working for any magazine at the time, but I was practising my writing, and I definitely was *going* to be a journalist of sorts in the near future, so I guess it was more of a 'stretching of the truth' than a lie. What I actually sought was information and inspiration from a man who'd been there, done that, and written many books about it.

"I can give you an hour," he'd said. "Come tomorrow at 1pm."

Two hours later, I left with a job offer and an invitation to dinner to meet his girlfriend the following week. "I like you," he'd said. "You're different." Different from what? I wondered. Other journalists? Other women? Either way, I agreed that I probably was.

I ended up living with my dog, Chui, in a cabin on Nick's land near Machynlleth in one of the coldest Welsh winters on record. In addition to helping him set up his adventure centre, I also became his support truck driver on his tours, taking bikers across the US from coast-to-coast and back again. The previous holder of this job gave it up after he was kidnapped on the most recent trip to South America.

Nick and I remained friends after I left and when he needed someone to take on the job of support driver for a motorcycle trip to Mongolia, I suggested my dad, Peter. Now Nick, his fifteen customers and *my dad* were in Iran and carrying a spare set of tyres for Rhonda. I was pretty excited. It had now been two months since I'd left the UK and I was looking forward to catching up with them all; mostly my dad, whom I was hoping would fix my GPS and pay for his daughter to have a nice hotel room for a change! He did both.

We rode together the following day, and after stopping in a village for lunch, we found ourselves mobbed by the local youth, who clearly loved the bikes. The *woman* on the bike though, was the real draw and I quickly became the centre of attention with people queuing up to have their picture taken with me. *A girl could get used to this kind of attention.* Trying to get out was difficult though, and in the end, a couple of our guys came to my rescue, flanking me either side on their bigger bikes to escort me out of the crowd. We were then escorted out of the village by the lads who had jumped on their own bikes and raced in and out of us cheering and filming as we left.

After several hours on the road we followed a likely-looking track up on to a hill with 360 degree views and started setting up our camp. Within fifteen minutes we had been spotted by the local chicken farmer who raced up with his friend to tell us that we were actually camping on a minefield and that this spot had been a disputed border with Iraq. We mulled it over and decided to stay as we had already ridden over most of it so it couldn't be that bad. Could it? They seemed satisfied and raced off again only to come back half an hour later with two dozen fresh eggs, some bread, pans and a big pot of tea!

Nick had an Iranian guide with his tour (the new visa rules) and so I made the most of him as interpreter, sitting with our new friends sharing facts about our different lives and learning as much as we could from each other until the tea ran out and they left. I headed to my tent wishing I hadn't drunk so much tea! Fortunately, no one was blown up that night.

Despite enjoying my time with the guys, it was important to me to ride Iran alone. So the next morning, I said my goodbyes and split from the group about sixty kilometres down the road and headed for Kermanshah. Just me, Rhonda (with her newly fitted tyres) and a couple of boiled eggs left over from breakfast!

Iran was not always easy riding. My first day alone saw the first of many stops by the police. As I pulled my dark visor up to reveal my face, I got the by now usual words of disbelief,

"Woman?".

"Yes I'm afraid so" I replied, waiting patiently for it to sink in as he scratched his head. It is illegal for Iranian women to ride motorcycles, so I was a bit of an irregularity for them. After a quick gawp and a quick conflab with his fellow officers, I was sent on my way again with a host of smiles.

As the day wore on, the heat became unbearable, reaching the high 30s centigrade (over 100° Fahrenheit) with little shade in the treeless landscape. I felt stifled by my hijab. The loose material wrapped around my neck and the extra layer over my head was not helping. I wet my hijab as often as I could, keeping me cool for a while, but always drying off way before the next opportunity to repeat the process. The traffic was chaotic; there seemed to be no rules. The people I'd met so far seemed fairly gentle and caring, but put them behind the wheel of a car, and we were playing a very different game. With that said, much of the danger I encountered was born from kindness. Keeping a steady 60mph down a long stretch of desert road that afternoon, a car came up alongside me. I looked over and saw the toothless grin of a middle-aged man who was neither keeping his eyes on the road, nor worrying too much about my personal space. Before I could wave him on, he unwound his window and produced a bottle of water before offering it to me with his outstretched arm.

"Take, take!" he shouted over the noise of both vehicles still purring along at 60mph.

It was worth the slightly precarious moves just to see the beaming smile as I grabbed the bottle and peeled off with a shout of gratitude.

Arriving in Kashan after a few hot and dusty days on the road, I planned to meet up with Mohammad, a couch-surfing host, who had offered some respite from camping, at his place in town. Kashan is one of the more strictly religious cities. In some places I found the hijab was worn loosely, barely covering anything at all, but complying with the law of having to wear one. This town had a more serious feel to it; all black, and all-consuming. Once again, I felt underdressed and self-consciously pulled my own hijab forward to cover more hair as I stood on the side of the road waiting for my host to arrive. Mo arrived minutes later on the back of a motorbike.

"Welcome, welcome!" he said as he dismounted, before reaching out to shake my hand. Not a done thing to women here. Still, couch-surfing was forbidden in Iran too, particularly for women alone, but that didn't seem to bother Mohammad, who proudly introduced himself as, "the king of couch-surfing".

Somewhere in his mid-fifties, Mo described himself as a 'free thinker' and had received more than two hundred guests since he started hosting three years previously. On arrival at his apartment, I was introduced to his friend who was waiting for us, then quickly asked if I would like a shower. Of course, I would! I was sticky, dusty, and smelly again. He opened the door to his bathroom and said,

"Surprise!" I walked into an all-pink bathroom, in very stark contrast to the dusty streets outside and indeed the beige room before it. I mean ALL PINK. It had pink walls, pink shower curtain, pink floor, pink bottles, pink slippers, and so on. So much so in fact, that I was a little freaked out,

"Do you have a daughter?" I asked hopefully. Mo looked confused and shook his head. "A wife then. You are married?" Again, the answer was no. But there were so many girly things! I don't mean to be sexist here. Of course, men can like pink too, but it was pink and fluffy and a room made for a little princess. Cuddly toys, the works!

I was suddenly aware of the fact that I was in a strange apartment with two strange men and no one knew where I was. It also crossed my mind that perhaps *I* was to become his princess. Had I trusted too much? Had my highly trained gut let me down? This was just a little weird, especially after coming out of a hot and beige desert surrounded by people in black.

It turned out that Mohammad just liked pink and designed his bathroom to make his guests feel 'comfortable' in his own special way. An oasis of girliness in a sweaty, dusty world. I'm personally not a fan of pink, though I appreciated the effort – once I got over the shock.

Iran is full of interesting characters. I wish I could introduce and describe them all in great detail to you in this book. So many of them have to hide their light under a bushel for fear of non-conformity in a country where individualism is not just frowned upon by the government, but downright illegal. Dancing in public? Forbidden. Dating? Don't be ridiculous. Mohammed was already technically breaking the law by having me in his house. He wasn't the only one who risked the odd law-breaking either.

The following day I made my way to Abyeneh, an ancient village in the hills of the Barzrud Rural District. Made of red mudbrick, and housing around three hundred residents, the village instantly takes you back in time as it has kept its originality over hundreds of years. After setting up camp, I took a wander around the narrow-cobbled streets, occasionally photographing the wonderful old ladies who – instead of black hijabs – wore white scarfs with printed or embroidered red flowers which covered their shoulders down to their colourful dresses beneath.

As I wandered, I met a lady and her husband who had come for a weekend away in the countryside. They were from Tehran, the capital, and a slightly more liberal part of the country. They invited me to join them on the terrace of their hotel that evening. We ordered some tea and sat on the colourful scattered cushions around the ankle-height table sharing a hookah with molasses-flavoured tobacco called shisha in the cool evening air.

"Would you like something a little stronger?" asked the lady, whose looks were those of a Persian princess. Her hair was a soft red and flowed freely out of her set-back hijab. She rocked the look, and the husband was her male equivalent. Maybe in their mid-twenties, the pair looked as if they'd just walked off an exotic film set. She pulled a brown paper bag from her handbag, glanced around and then quickly revealed the top of a wine bottle.

"Oh, yes, please," I said, without hesitation. I had been in this 'dry' country for two weeks now and was missing my daily end-of-ride drink.

We sat through the evening playing backgammon and sipping wine out of our china teacups. As I took my move on the board, she leaned in and asked again,

"Would you like something stronger? Maybe to smoke?" I nodded, and we headed up to their room where we shared a joint. The problem was, we were all now *really* stoned and the small amount of English we were getting by with became...well, less fluid. I stared blurry-eyed at a Persian rug on the wall and briefly wondered about the person who'd made it, before getting lost in the intricate pattern. Was that a flower or a jellyfish recurring?

"Say something" she demanded. My eyes darted to hers and found them looking back expectantly.

"Um…sorry…yeah...nice weed eh?" I failed to entertain, and we reverted to staring at our own personal spots in the room. We had broken our lines of communication. I decided to leave – awkwardly – and spent the rest of the night in my tent listening nervously to the noises outside until sleep finally came like a welcomed friend.

I left Abyaneh village the next morning and headed for Yazd. Yazd is the Las Vegas of Iran, in that it is surrounded by a large expanse of desert and is pretty much where all the tourists go, though not to gamble, of course. It was a 500 kilometre ride with little shade, and I was feeling pretty groggy from the wine and 'the sleep of the paranoid' the night before. About fifty kilometres out, I stopped to take a breather in an area where there were a few houses and a little shade. I felt dizzy as I got off the bike. I hadn't drunk enough water and recognised the early stages of heat exhaustion: confusion, feeling emotional, and just wanting to lie down and go to sleep. It was a rookie mistake, but there was no shade in which to stop and I kept telling myself I would stop as soon as I found some shade. The satnav was still playing up, so I got my map out and realised I had gone past my turn-off. *Damn it!* I was too tired for this now. I just needed to get out of this heat and off this stupid bike. I looked up and saw a car passing very slowly with the occupants, a husband, wife, and small child all staring at me. I forced a half-smile and continued scanning the map. A couple of minutes later they were there again. This time they stopped.

The husband wound the window down, said

"Nescafé?", and gestured for me to follow them. I quickly packed my map away, put my helmet back on, and followed.

We drove into what looked like a derelict estate full of rubble and broken-down houses. They pulled up outside a set of big metal doors that opened to reveal a small garage into which Rhonda and their car just fitted. They then led me downstairs to the basement. Most of the house was underground and, while simply decorated, had an order to it that was in total contrast to the outside of the building. Before I knew it, there was a fan on me and a glass of water in my hand. The husband then disappeared, leaving me with his wife and daughter. We couldn't communicate through language, but I remembered the newspaper article I brought with me for these very moments. It was from the *Sunday Times* and had my route plan with lots of pictures of places I was planning to visit. It did the trick and gave us something to focus on.

Twenty minutes later, the husband returned carrying two big burgers complete with gherkins and a can of Coke! He then made a fire and brewed some lavender tea for us all. There were no issues of trust, or questions of my needs, either. Clearly, this western woman needed to cool down, drink Coke and eat burgers. I must admit they were right, and they knew it before I did. Was I a stereotypical westerner?

After my burgers, the five-year-old took my hand and led me around the house. The basement was their living room and kitchen, with a small squat toilet in a cupboard size room under the stairs. Upstairs had no roof, and a tarp down one side where there was once a wall. There was another section a few more steps up, housing an old broken-down Honda, several pet doves, two live chickens, and several stuffed ones. After a couple of hours of respite and a declined offer of staying the night, the family jumped back in their car and led me to The Silk Road hotel in town. It was 9pm. Twelve hours after leaving the mountains, and it was still hot.

I settled into my bunk in the dorm (which cost a mere £4 a night), and the following day I decided to rest my steed and set out on foot to the Tower of Silence, one of the many historical sites in Yazd. A relic of the Zoroastrian days before Iran was conquered by the Bedouin Arabs in the seventh century and gradually converted to Islam over the next three hundred years. Today, a few pockets of Zoroastrianism still exist, mainly in India, where followers of the religion are known as Parsis, as in Persians. (Freddie Mercury's parents were Zoroastrians). The Tower was where they held sky burials in which the priest would sit and wait with the bodies as they were devoured by vultures. Nothing was buried as it was believed to contaminate the soil, while burning would contaminate the air. I was told it was a beautiful place to see at sunset; and it was.

As darkness quickly fell, I found my way back to the road and hailed a scruffy little cab with equally scruffy driver to take me back to base. As we took off, we chatted fairly easily. His English was not bad at all. Maybe mid to late twenties, he looked as if he'd been overfed by an adoring mother. Despite my instructions he took a different turn at the roundabout. I was sure of the correct route and could therefore see he was taking me in totally the wrong direction.

"Wrong way. We should have gone that way" I said pointing backwards. He ignored me. "That way!" I said more insistently this time, but all he did was shrug his shoulders. "Hey. Stop the car!" I yelled.

"No English" he replied, despite seconds ago, having a pretty good grasp of it. I banged on the dashboard,

"Stop the fucking car!". No response.

I took off my seatbelt, grabbed the door handle and opened the door, threatening to jump out. He finally looked at me and said,

"OK, OK" before pulling over.

As I went to jump out, he grabbed my hands and frantically tried to kiss them. It was a bizarre spectacle and perhaps highlighted the sheer frustration of a young man living in a severely repressed society, where touching, or even walking with a woman who is not your sister, wife, mother, or aunt is forbidden. In a way, I felt for him, but still, I scolded him (in the voice I normally reserve for naughty dogs and drunk Turks), slammed the door and walked the last mile back to the hostel.

The following day the desert wind was relentless, but I kept going, dodging the tumbleweed and trying to avoid the bigger dust devils until I could find some shelter. When I stopped in Firuzabad, just sixty kilometres short of Shiraz, the judge appeared. He spoke perfect English and looked forty-ish. Clearly well-educated and looking very smart in his well-tailored suit, I accepted his invitation to join him and his family for some respite from the road. He lived in a small house, with rugs for furniture, with his mum, two sisters and a cousin. I was so tired, that I rudely went to sleep on the floor for two hours almost as soon as I arrived. The oppressive heat, in all my riding gear was clearly still getting to me.

The judge later suggested a short walk in the cool evening air. I thought this might be a good opportunity to find out what an Iranian judge thought of the laws of his country and it wasn't long before the topic came up.

"Please remove your hijab if it bothers you". It did bother me, as I was constantly fighting with the wind to keep it on.

"Thank you" I replied pulling down the sarong and wrapping it around my neck. I took the opportunity to steer the conversation, "What do you think of your laws?" I asked with genuine intrigue. He paused,

"I am in a daily battle with myself" he replied. This took me aback. I had not expected such honesty. "By day, I agree with the government, but by night I have my own opinions". I was pleased to hear he did. He was a good man.

As we neared the road I replaced my hijab because I could see a police car parked up near ours.

"It is fine" he said, "Please leave it off". I pointed out the police car below us at the bottom of the hill. "Please do not worry. I am a judge. If anyone should bother us, I will have them in court tomorrow," and he laughed.

I liked the judge. We exchanged WhatsApp details before I left. No sooner was I out of sight, he proceeded to bombard me with images of his most intimate parts. Despite being a highly educated and polite man, he too had frustrations that he just didn't know what to do with. I didn't hold it against him. He *did* have the decency to wait until I'd left, and after all, who else can say they've had dick pics sent to them by *an Iranian judge?* It was priceless!

****

I left the small town of Lar, in the far south of Iran, at 5.30am and headed to the ferry port of Bandar Lengeh, my final stop in Iran. Half-an-hour later, after ensuring I was downwind of the rotting, stinking donkey carcass, I stopped to watch the sunrise. I took my time, wanting to savour my final sunrise in a country that had taught me so much. A culture that is easy to see as standoffish from a distance had shown me that reaching out was actually a big part of their culture. People give unconditionally and never pass by a stranger in need of a burger.

As the sun rose and warmed my face, I felt a tinge of sadness. In a few hours, I would be crossing the Persian Gulf to Dubai and saying a fond farewell to Iran. Who knew if I would ever get the chance to return? Despite the 'handsy' taxi driver and the judge's dick pics, I never once felt threatened or afraid there. This was in part, perhaps, because I was finding my mojo; partly because any advances, yes, even the taxi driver's, felt more like a teenage blunder than a serious threat. I felt pity if anything – living under a religious dictatorship, whatever your beliefs, must suck. It's hard to live under rules that don't allow any freedom of expression. Boy, did I know about that. I empathised most of all with the women who had once tasted a different life in their youth. I recalled Hussein's mum back in Urmia; when I'd asked her what it was like to live through the revolution, she replied simply, "Heaven to Hell." But like many before me, I had fallen in love with Iran and its people.

I entered the processing machine at Bandar Lengeh, preparing for the worst and hoping for the best. I got something in the middle. The Nokia ringtone will now forever transport me back to Iran. It was everyone's ringtone there and it was never more apparent than in the ferry port. Lots of people with lots of phones and no exceptions to this tune – it was even on the Iranian Nokia that I'd bought for myself.

Only eight hot and sweaty hours of walking from office to office and I was in the departure lounge, bike loaded, with only the small hurdle of customs to get through before I could get on the ferry. No sooner had I joined the queue for customs than I was pulled out of line, led to another door, and pushed into a small room. The room housed a short dumpy woman wearing a full black burka, a batoola, and matching black latex gloves. She looked so well suited to her dimly-lit environment that I wondered if she ever left, and if she was thrown the odd westerner for sustenance now and again. The batoola is a face mask traditionally worn by women of the Persian Gulf region, though these days usually only by the older generation. Hers, like many I saw that day, had a metallic look to it and could easily have been mistaken for a prop from *Silence of the Lambs*. If she was going for the intimidating look, she had certainly nailed it – full points from me. A bead of sweat ran down my neck and joined the party in my now sodden T-shirt. Why exactly was she wearing gloves? She grabbed my Kriega tank bag, (which matched her outfit beautifully, I noted) and rummaged through it. She then gave me a pat down, and, satisfied, pushed me through another door on the other side. One of the officials still had my passport, so I went back to ask where it was. I was told to keep moving. A little odd, but okay; *I guess I'll play the game for now and see what happens.* I had little choice at this stage.

I sat in the departure lounge and waited obediently before being called over to another office and another customs official. Holding my passport, he started questioning me with the appropriately stern look that men in uniform often adopt to highlight their own importance.

"Where did you go in Iran?" I listed the cities I had been to; at least the ones I could remember off-hand and under pressure. "And what do you think of Iran?" he continued, maintaining the serious face.

"I love Iran" I said with a genuine smile, "and I love the people." I might have overdone the enthusiasm.

Then came the next question. A question I had been asked many times in Iran,

"And what do you think of the government?" I thought for a second before replying. I'd had enough conversations with Iranian citizens to know that most did not like their government or its overbearing rules. Even as far up as the judge. I didn't want to say that to an official though.

"I think the same as most Iranians." It was now his turn to smile – a big, genuine, Persian smile that said, "I know exactly what you mean." He handed back my passport and wished me a safe trip.

I joined the long queue that had formed at the gate to board the ferry. After a few minutes of waiting I suddenly heard,

"Lady!" and saw the official at the front was looking at me.

"Me?" I asked, slightly bewildered. He gestured for me to come to the front and past the barrier. I obeyed, and as I did, he pulled up a chair and offered it to me. *What?* He smiled and gestured for me to sit. I sat but I felt extremely uncomfortable getting special treatment. Now I was the centre of attention with the best part of two hundred Iranians looking at me from the other side of the barrier. A woman from the middle of the queue decided this was not on, grabbed another chair, pulled it up next to mine and sat down. The room went eerily silent for a split second. All eyes on the security guard now. It felt like a Caesar-style thumbs up or thumbs down moment, and I swear I could hear my own heartbeat, adding a rhythm to the drama. Thankfully, he admired her tenacity, laughed and allowed her to stay. The rest of the crowd breathed out in unison and laughed along with the security guard. I was impressed with the tenacity of the lady underneath that all-consuming black burka. I smiled at her. Her eyes smiled back.

Once on the boat, I was directed to a long wooden bench seat away from the crowd. As others tried to sit in my row, they were moved on. I turned to the crew member who just gave me a thumbs up as if to say "You're welcome," and walked off. Once we had set sail and been going a couple of hours, I decided I could now take my hijab off. Could I? I guess I was out of Iran now, right? Of course, but it still felt strange. I removed it, only to find myself replacing it after a few minutes. In just one month, it had already had an effect on me; I felt naked without it.

After a few hours' sleep I went outside for some fresh air and let out a little squeal of delight. Literally. I could see Dubai; I could actually *see* Dubai. I don't mean as a pin on my map this time, but from a ferry on the Persian Gulf heading into the port of Dubai. *Was that the Burj Khalifa I could see in the distance?* It was the biggest moment of realisation so far.

Don't ask me why it hit me then. It was just like waking up after a long sleep (which actually I had) and finding myself still in the dream. I suppose I'd had to be at least mildly confident to start this trip, putting my fears to one side and saying, "I *am* going to do this,". I'm guessing a part of me never really believed it though, not even when I was doing it; then it would just hit me full-on in the face every now and again, without warning.

It wasn't a constant overwhelming high, it was a quick hit. A rush, that faded as the business of getting over my next hurdle took over once more.

On to Dubai…

Chapter 10
India

They told me you need three good things to survive on Indian roads: good brakes, a good horn, and good luck! It's all true, especially on a motorbike. This was the ultimate game platform – *Grand Theft Auto* on acid. And I enjoyed it.

Thankfully, I had made the most of my luxurious two-week stay in Dubai with my friend Martin. I got my hair cut, watched telly, lounged on Martin's luxurious sofa, and slid down his ridiculously large banister while he was at work – a sure sign I was recharged and feeling playful again.

After working, and occasionally lying, my way through the crazy Indian customs procedure in a time they told me was the fastest ever, I made my way into Mumbai and started working my way north. I had landed at the Mumbai/Bombay airport feeling a little disappointment at having flown into India and not come overland. My original plan to ride across Pakistan from Iran had been hastily changed because a cyclist, whose blog I was following and who was a couple of weeks ahead of me, was shot and several of his convoy killed, shortly after crossing the Pakistan border. The international newspapers went wild with the story, blaming the cyclist for being there in the first place. I had no desire to add to either those headlines or the troubles in Pakistan, and so took up Martin's offer. Now, in India, I was grateful to have had the respite.

Riding in India can feel at times like riding on a knife-edge. The near misses, the smothering heat and humidity and the cows that know they're sacred, walking smugly around the cities as if they own the place. I'd often find them sleeping in shop doorways or stepping out from behind parked cars causing me to swerve or slam on my brakes. Is it possible for cows to look self-righteous? There were also roaming packs of bike-chasing dogs, diesel fumes, wrong-side drivers, kamikaze truck drivers, rickshaws, motorbikes, and the sheer volume of randomly moving obstacles. Then there were the stationary ones; speed bumps that appear without warning and open manholes were the most common of many.

Sometimes, it felt like a great adventure. Other times, it felt like a battle to simply survive. I often struggled to keep my concentration and energy levels up in the mid-summer heat, having to stop every half-hour for water and rest before pushing on. You know the heat is getting to you when your mind tells you to just let go of the bars and let yourself fall into that lovely soft verge.

My initial fears of the country grew smaller every day, though. I even got used to being awakened by the odd mouse or cockroach running over my bed. One cheeky mouse woke me by nibbling on my toe (I'm not sure who was more surprised when I woke up!) – a cheap Indian hotel alarm clock. People were right; the roads *were* crazy. Within a few days, I slowly found the rhythm in the chaos. The method in the madness. Here, I could overtake any side I saw fit, ride on one of the few-and-far-between pavements, and, as long as I didn't hit a cow, I could pretty much do as I damn well pleased! In fact, it was imperative that I adopt a hooligan-style approach. Riding as I would in the UK would likely see me killed.

Gandhi once said, "At the age of eighteen I went to England. Everything was strange, the people, their ways, and even their dwellings. I was a complete novice in the matter of English etiquette and continually had to be on my guard. Even the dishes that I could eat were tasteless and insipid. England, I could not bear ..."

Unlike Gandhi, I enjoyed the culture shock immensely. The sights, the smells, the elephant rush hour, and most of all, the right to ride like a hooligan. India was punching through my visor and ensnaring all my senses without apology and enjoying it almost felt masochistic. When riding under bridges, the smell of ammonia from stale urine was sometimes so strong it would make my eyes water. The near misses on the road were so frequent that I'm sure my adrenal gland shut up shop, leaving a big sign on the door saying "Out of stock". At the same time, I also felt extraordinarily alive.

I'd been nervous about India from the day I started planning my trip. It didn't help that two days before I left Martin's I found an email in my inbox from an Indian resident:

Dear Steph

Please be careful in India. There are MURDERERS and RAPISTS walking around everywhere amongst the normal people.

Have a nice trip.

"Have a nice trip." I had to laugh, even though I was growing tired of all the people who seemed determined to keep me in a state of fear. This was far from the only 'warning' message I received and, although I knew they probably meant well, these warnings often felt more like threats.

I clearly had no right to be taking on such journeys as a woman alone. I was either completely naïve or downright crazy. I often wondered what they would say if I *had* been hurt or killed, "I told her so"? Would they find some satisfaction in my demise? Would they feed on that story like the flies on that donkey carcass back in Iran? I was going to make damn sure that didn't happen – if only to prove the naysayers wrong.

India is very diverse and continually surprised me throughout my eight-week journey from Mumbai to Ladakh. Rhonda proved to be a great choice, particularly in cities like Surat, where the traffic was literally touching, and I had to fight my way through the rickshaws. She is a small bike with the power to pull away quickly and get in front of the pack when the opportunities arise. I relied on her so much that I was actually starting to worry about how much of an emotional bond I had developed with her. Was it wrong to love a machine this much? And I wasn't the only one who loved her.

From Surat I headed for Vadadora, the third most populated city in the west coast state of Gujarat. It is on the banks of the Vishwamitri River. Like Surat, it is rarely frequented by visitors and once again, I found myself the centre of attention, or so I thought. As I unloaded my bike in front of my hotel, a crowd gathered. This was already the norm, but this time an argument broke out between the hotel security guard and one of the men in the crowd. The guard told them to move back and give me some room, concluding with,

"What is the matter with you? Have you never seen a human being before?" The man retorted,

"I'm not looking at her, I'm looking at the bike!" We all giggled. It became increasingly apparent that I was *not* the main attraction. Rhonda the Honda was the real star of the show. I was merely the sidekick.

I continued steadily north through Ahmedabad, Bundi, Udiapor, and Ranthambore. The heat was exhausting, and the last 2,500 kilometres was beginning to take its toll on me. I vowed that next time I would come in December when it's cooler than June and July. Rhonda and I had now travelled 14,500 kilometres together and I decided to celebrate with an email to my mum.

"Look Mum! We only went and made it half way across India. BTW, any news from Nathan?"

Not long before I left, my son had announced that he and his girlfriend were due to have a baby. MY FIRST GRANDCHILD. I was going to be a grandmother, and boy did the media love that. The 'warrior princess' look I had hoped to convey was swept aside in favour of headlines like 'Grandma rides solo around the world'. I was 39 for Christ's sake; hardly grandma material. Still, I was very excited and had been hoping for news any day now, so after the email to my mum I posted on my blog,

"SOMEONE PLEASE PHONE ME IF MY GRANDCHILD IS BORN."

In Ranthambore I did indeed get the news I had been waiting for. The day I went into the jungle and saw my first wild tiger was the same day I found out I was a proud grandmother to a little girl. A warm glow washed over me as I received the news from my son over Skype – a total rush of love and pride for both son and granddaughter. I was proud. I was happy. Really happy. So much so that I had to go out into town and tell everyone I came into contact with – the rickshaw driver, the man who smiled at me in the street and the waiter who served me a nice cold glass of wine (Champagne wasn't available!).

It was a magical day.

****

I smiled for another photograph in the street, put my helmet on, then politely but firmly asked the crowd to stand back a little as I swung my leg over the bike. I was getting used to the interest, and had quickly learned that I had to take control of the situation if I ever wanted to get anywhere. The matter of personal space often had to be addressed, and once I learned a little about the differences in cultural behaviour, this became a lot easier to manage. Personal space is not an automatic right in India, it's a luxury, so it's not considered rude to invade it.

Instructing the crowd is not considered rude either, and people are only too happy to get out of the way just as soon as they are told to do so. If you don't speak up, then they will assume there's no problem with them being just inches away from you as you try to smoke your roll-up, drink your Red Bull or get on your way. This was also true during times where I tried to make discreet use of my *Shewee*. I mean a white woman on a motorcycle drew enough crowds, but when they saw her peeing against a wall as well? Now that *really* got their attention! It can be a little overwhelming until you learn the rules, though.

Much as I still secretly hold onto my childhood dream of being a rock star some day, I now realised, as I put on my gloves and started the bike, that this amount of interest all the time just doesn't suit me. It soon gets tiring having every single move you make observed by so many. I kicked up the sidestand and decided that it's best to put the rock star dreams on hold and stick to karaoke.

I squeezed through the reluctantly-parting crowd, shouted my goodbyes, and headed down the dusty road, carefully avoiding the chanting Hare Krishnas who were meandering towards me in their orange robes and banging their drums. It was 6:30 a.m. and I didn't bat an eyelid as I weaved through them. Nothing was out of the ordinary for me in India, because it was *all* out of the ordinary. Therefore, by definition, it was ordinary.

I came to the 'Pink City' of Jaipur from Ranthambore. Now, having had my fill of opulent royal courtyards, gardens and museums, I was moving on to Delhi to meet with the Malik family.

Neeta and Ashu Malik are a brother and sister team who run a company called Destinations Unlimited. They had offered me lots of advice on routes in India, and now I was going to stay with them for a night or two at their family home before pushing on further North and into the Himalayas. A big Indian family, they welcomed me into their beautiful home like a long-lost cousin and spoilt me rotten for two days. Neeta ensured her lovely Nepalese servant, who had been with her since she was a child, served me breakfast in the luxuriously large bed (eggs as I liked them and coffee) and provided me with medicine for my nose. Riding in the constantly billowing diesel fumes burned my nostrils and they were now red raw with ulcers. After a good rest in such luxury, I was soon reenergised, and so they whisked me off to check out the Taj Mahal, in Agra, with their driver Anil, at the helm. It's about 200 kilometres from Delhi and I hadn't planned on going, but the sight of it simply took my breath away. Anil took several photographs of me and the Taj. In one of them, I sat on the same bench as the famous portrait of Princess Diana. I found it hard to leave the comfort and friendship I was afforded by the Maliks, but time was moving on. The mountains were calling me, and when the Himalayas call, you answer – no matter how comfortable you are.

I sang along to the tune in my head and used the beat to give rhythm to my ride as I weaved and squeezed my way out of Delhi. A thousand car horns blasted out around me in their own incessant song, and I settled into my daily rants of, "What are you doing you idiot?" and "You'd better move because I'm coming through."

As I approached the outskirts of the city, I spotted a young man standing next to a motorbike on the side of the road. He appeared to be waving me down, so I presumed he might be a biker in distress. As a fellow biker, this is something you do not ignore, and it's one of the reasons I love being part of this worldwide community. Besides, he looked pretty dishy in his black leathers with his sexy sports bike; a rare sight here in India. The bike, not the dishy guy. Most people here ride Royal Enfields.

"Are you OK?" I asked as I pulled up alongside him. To my utter astonishment he replied,

"Hi Steph."

"What the....um...hello. Who are you then?" I jumped off the bike and removed my helmet to reveal my already sweating face and totally unsexy helmet hair, *damn it*. Once my helmet was off, he reached into his jacket and pulled out a small gift-wrapped parcel.

"I was reading the latest update on your blog last night, and giggling at something you said. My mum asked what I was laughing at, and so I told her all about you".

"Okay," I said, wondering where this was going and how he knew I would be here at this time.

"I told her you had just become a grandmother." The words reminded me that this guy was probably only twenty-two, so I was probably old enough to be his mother. *Double damn it!*

"...and she insisted I come and meet you to bring you this gift for your grandchild. Alexis isn't it?". I was stunned.

"Um! Yes. Alexis. That's right". He handed me the gift with a show-stopping smile. "Wow. Thank you. Please thank your mother for me, and thank you for coming. That is so kind. But how did you know I would be coming through at this time?"

"I know you like to start your ride early and...."

"But I am late today. I had to meet some other bikers for breakfast. They wanted to meet me and Rhonda before I left and..."

"Yes. I have been here since 6 a.m.," he replied in his gentle voice. Matter of fact. Just like that. No annoyance. No frustration. It was 8:30.

The ever-growing biker community in India has an extremely tight network and as word spread, they continued to come out in support of my ride. I was joined on the road by many different bike clubs, and one gentleman turned up at my hotel with a single rose. When I asked if he wanted to come in for a drink (I thought it was the least I could do), he refused and said,

"No, Madam. I do not want to disturb you, but please can I just have a photograph with you and Rhonda?"

One day I rode with an all-female group of bikers called the Bikerinis. They looked stunning on their big bikes with their skin-tight jeans, un-smudged mascara, and long flowing hair. How did they manage to ride in this chaos and heat and still look good? It was clearly beyond me. In fact, all this attention was beyond me. Although it was all very bewildering, it was great to meet so many of my Indian brothers and sisters who clearly chose not to hide behind cool nonchalance, as we might occasionally be guilty of in other parts of the world. Instead, they came out in their droves to enthusiastically support and encourage me just as my energy for India was waning.

There was one problem with all this, though. Every time I stopped to greet more riders, I lost time. We would say our hellos, take a thousand photos, and then they would ride with me for a while before peeling off and going home – to some nice cool shade, no doubt. I was regularly left riding in the 40°C midday sun, fully clothed in my black armoured jacket and biker pants. A dangerous combination, especially when the traffic dictates that you cannot even speed up to increase the airflow.

The heat increased even more during the two days I spent riding towards the city of Amritsar. I just couldn't seem to get enough water in me, no matter how much I drank. I sipped constantly from my Kriega bladder pack, and yet I never needed to pee. Already run down, and suffering from nose sores and, I think now, head lice, my body was crying out for me to show it some respect and give it a break. I foolishly ignored it and found myself once again, in a black-faced and emotional mess in some dingy hotel in a shitty little town somewhere in India.

I don't even recall where it was but it felt like a shithole to me. As ever, there was no working shower, the fan was temperamental, and no one understood me when all I wanted was a milkshake. I would have killed for a milkshake. I sat on the bed, put my head in my hands, and cried. What was I doing here? Was there no respite from this heat? I knew even the night-time would be unbearable in this stinking, dirty room, and I couldn't even clean off the grime of the day. I pulled out my phone and recorded a video diary to try to explain the feelings of the moment. This was a rollercoaster journey and I didn't want the shitty stuff that holds all the exciting and dramatic stories together to get lost and forgotten in the selective nostalgia that would no doubt ensue when I eventually got home. Home! It felt so far away right now. As it turned out, this was just the beginning of a two-week period that would test my stamina and patience to their limits.

****

I was feeling a little stronger the next morning, so I decided to make a break and 'get out of Dodge' as early as possible. I had plans to meet a Sikh friend, Sukjeet, in Amritsar, Punjab – home of the famous Golden Temple, also known as the Harmander Sahib, the holiest shrine in Sikhism. Sikhs make up just 2.5% of India's population and are easy to spot because of the turbans they must wear. As I entered Punjab, it became clear that this was where most of that 2.5% lived. As I got closer to Amritsar, an overwhelming majority of the population wore turbans.

Sadly, the city is notorious for two massacres, six decades apart. The first was in 1919, in the days of the Raj, when Acting Brigadier-General Reginald Dyer ordered troops of the British Indian Army to fire their rifles into a crowd of unarmed Indian civilians in an enclosed garden, Jallianwala Bagh, killing at least 400 people. The second Amritsar massacre was in June 1984 when the Indian army attacked the Golden Temple complex with tanks and artillery to dislodge a group of heavily armed Sikh fanatics holed up there, after negotiations failed. Nearly five hundred militants and trapped pilgrims were killed and in the aftermath, many Sikh soldiers deserted their units and five months later, Indira Gandhi, the Prime Minister, was assassinated by her own Sikh bodyguards. That in turn led to the killings of more than 3,000 Sikhs in anti-Sikh riots.

I'd met Sukjeet many years earlier when he came for an off-road riding lesson in Wales. He had been a worrying customer, refusing to wear any protective gear – including a helmet. This had concerned me, so I looked up the law before allowing him to ride. Originally, when the helmet law first came in, back in 1973, Sikhs were obliged to wear them like everyone else, but a determined campaign, led by Gyani Sundar Singh Sagar, who had already succeeded in changing the regulations for Manchester bus conductors, managed to get Sikhs an exemption in 1976, arguing along the lines of, 'We fought for the British in two World Wars wearing our turbans, without tin helmets…'. There was a late debate on the subject in the House of Lords, in which the bravery of Sikh soldiers was referred to several times. I remember asking Sukjeet to sign a disclaimer anyway just to be on the safe side, and I was glad I did as he was not the most natural of off-road riders. He must have fallen off the bike fifteen times that day; sometimes spectacularly. He always got up smiling, though. He had moved to Amritsar eight years earlier, from Birmingham, to be close to the Golden Temple.

As I stood at the busy tollbooth and waited for Sukjeet to arrive, the heat again felt dangerous as I wilted in my big boots and armour. I wondered if I should have stayed at the hotel for an extra day's rest. I never gave myself quite enough time to recover, and it was too late now.

An hour past our agreed meeting time, I was just beginning to *really* curse him when he showed up, looking just as I remembered him. In his white robes, turban, and extra-long beard that had grown considerably, he looked less out of place here than he had in the Welsh hills. He still managed to stand out though, with his Ray-Ban shades and his Kirpan, which was now sword-size, rather than the little dagger I remembered. "A bit decadent," I thought, as I spotted it trailing from his belt and down behind his leg. But what did I know? He swung his Enfield around and parked up next to Rhonda. As he did, one of his badly loaded rucksacks escaped its binds and hit the deck, landing by my feet. Still as clumsy as ever then, I smiled. It was good to see him again.

Now I was on his turf and the pupil. With our roles reversed, I was glad I'd been a patient teacher back in Wales. Sukjeet was going to teach me what it was like living in a Sikh community. I had also agreed that after my visit, some of the Sikhs would ride with me up into the Himalayas. It all sounded very romantic – five Sikh warriors and me riding up to heights of 18,000 feet, ready for anything. We would stay in temples along the way and eat for free. How could a girl refuse warriors, temples, and free food?

An unlikely looking pair, we rocked up to the village and stopped outside a big house with large gates and a grand entrance. So it wasn't going to be quite as basic as I'd imagined... Sukjeet didn't actually have a home, as such, and for the last eight years had lived completely free in this community. He was given food, shelter, and water, and used his time to meditate and study to become a Sikh of a higher order. I still had a lot to learn. Later, I found out that the Kirpan length is determined by the religious convictions of the wearer. Sukjeet had clearly stepped up his game since I saw him last. I was going to be staying with the community leader, at whose house we had just pulled up.

Following Sukjeet in, I saw that the outside had given no clues as to the interior. The house was almost completely bare, with concrete floors and no ornaments, pictures, keepsakes, or even paint on the walls. The only furnishings were a couple of plastic garden chairs in the corner, a fan, and a large day bed in the middle of the living room facing the large flat-screen TV on the wall. On the day bed lay a large-bellied barefoot man with a white beard, white robes, and an enormous smile. Neither he nor his wife, who came bustling in a few seconds later, spoke any English.

I needed to clean up after my ride and my wait, so after greetings were complete, the wife led me to the bathroom. It consisted of a large bucket with two large taps above. The water in the bucket was used to wash, flush the squat toilet, and of course, wipe your arse. No loo roll, of course, but I was used to that by now. One tap was from the tanks on the roof and one was pumped from the spring below. The spring water was very cold and I opted for that; stripping down and pouring it over my head, relishing the short respite from the heat.

As we sat on the floor that night eating dinner, Sukjeet explained some simple rules I must follow if I was to stay in the community.

"Hit me" I said enthusiastically, inwardly hoping that it didn't involve praying or the sacrificing of small children. The latter I might have considered, but I was definitely not going to do any praying!

"While you are here you cannot smoke. No alcohol, no meat and no caffeine."

"Fair enough" I said. "I can live without those things for a while, and if I *do* want a smoke I guess I can just go for a ride."

"No," he replied, "This is not an option. You must stick to the rules at all times."

I was a little taken aback by this. I was happy to respect the rules of the house, but what difference did it make if I was nowhere near the house? It wasn't as if I was trying to become a Sikh, and I had already seen one person smoking in the village. I quickly stuffed some more bread in my mouth so I didn't have to answer.

That night I was taken to the rice fields to meet one of their horses. Sikhs have a close connection with their horses and, as a rule, don't normally allow others to ride them. For some reason, that night I was given this honour, and I jumped at the chance. I love horses, particularly spirited ones.

95

This young stallion was clearly highly-strung and ready to go. I jumped on and tried to steady him; every muscle in his body felt tense and ready to rock given half a chance. Conscious of the fact that I only had flip-flops on my feet and no helmet, I fed him only a little rein at a time, easing on the 'throttle' and finding a comfortable pace.

"He's really strong," I said to the owner as I came back around towards them, "it feels like he wants to bolt for home."

"Yes. he does," came the reply. "Let him go. He will show you the way."

I knew that as soon as I gave him an inch, he would take a furlong. The problem was I had no idea which direction he'd take. I've ridden enough spirited horses to know that they are very good at fooling you if they don't want you on their backs anymore; they will make as if they are going one way and then turn unexpectedly the other way.

"If I give him an inch…." I began.

"Then don't give him an inch," came the reply.

We started off at a prancing trot, fighting for power all the way. He was clearly heading for the village on the other side of the rice fields. *Sod it. You only live once.* With that, I loosened the reins and relinquished control to the beautiful steed. In a nanosecond, we were galloping at full speed across the rice fields towards the village. It was beautiful.

When we got to the road, he turned sharp right through a narrow alley, then a left, before he eventually pulled into a small yard. I ducked as he came to a stop under the washing line. In the yard, an old lady was weaving a basket. She looked up at me and smiled, without a bit of surprise. The rest of the family came out to greet me, then Sukjeet and the owner arrived by car, followed quickly by the local kids, women carrying babies, and anyone else who fancied coming to meet the strange lady who had just galloped through the village. Within a few minutes, the yard was a mass of smiling faces and colourful turbans. I got the impression they didn't get many visitors – the whole community wanted to come out and welcome me.

Over the next few days, I had many discussions with Sukjeet about Sikhism, and I began to wonder if he wasn't something of an extremist. He certainly took his scriptures very seriously and seemed far more disciplined and less tolerant than the rest of the community in many ways. I wanted to understand his beliefs and to respect them, but I found a lot of what he said to be quite contradictory, hard to follow, or just downright unbelievable. When I questioned further or asked him to explain he would reply with the age-old cop-out of, "You could not possibly understand," which I found really annoying and very patronising. How could I ever understand if he just kept cutting me off? I wondered if he really understood half of what he read, and I wished I could communicate with the wise old community leader. All he could say in English, though, was "Cup of tea" (which I never got). He repeated this several times a day with a cheeky grin. I'd smile back and repeat his words,

"Ah yes. Cup of tea." I knew my duty and he never grew tired of this game. His eyes were kind and all-knowing, though, and I both liked and respected him.

Sukjeet, though, wandered around with a swagger, demanding respect from anyone he came across. He got road rage in the car when anyone got in his way, and whenever he didn't get the respect he felt he deserved, the cool, calm, meditative persona he worked on so hard would slip quite dramatically. He seemed very young at times, although I guessed he was my age or older. I wondered if he had misinterpreted some of his scripts. He preached that all people were equal, though it seemed Sikhs were 'more equal than others' in his eyes. He claimed Sikhs were the first to promote women's rights, and yet the only people working and doing the housework were the women. They ate only after the men had eaten, and before this they would wash the men's hands at the table in little bowls. There may have been good reason for this and far be it from me to judge others' way of life. I've chosen many paths myself and most of them with far greater sins than this one. I just wanted someone to explain it to me without the air of superiority I felt Sukjeet often displayed.

I enjoyed a lot of what this community stood for and respected their discipline and values, but after a few days there, Sukjeet was annoying the hell out of me. This wasn't for me. I grew weary of our discussions and felt myself getting less tolerant by the day.

I swear it got even hotter. The heat was almost unbearable, and that was, of course, adding to the downturn in my mood. I was constantly soaking wet with sweat, day and night. I dived under that bucket, and doused myself in spring water every chance I got, but no sooner did I dry off than I was wet with sweat again. The humidity was stifling, and I had no energy.

Everyone else was the same, it seemed. They moved as little as possible during the day, and I could see now why the day bed was the most prominent feature in the house. Sukjeet would leave the house at 9pm and head for the Golden Temple to meditate and pray, returning around 4am. During the day, he would sleep or lounge around like everyone else. He only ate once a day and was quite thin. I wondered how he would cope in the mountains without his daily routines.

Despite the hospitality I was shown, I found it quite isolating as everyone spoke in Punjabi. Only Sukjeet spoke English and he was often upstairs sleeping. I may as well have been deaf for my lack of ability to communicate, and it felt very lonely. I needed a fag and some coffee, damn it! I also needed some chocolate and a good old chinwag. Most of all, I yearned for a cool Welsh breeze.

One night I was in bed, reading. I was naked with just a single sheet partly covering me. The committee leaders wife walked in unannounced and gestured for me to get up. I stood up and scrambled to find a t-shirt to put on. She stood by, watching and laughing. Apparently, my pale body was the source of some amusement to her. She changed the sheets, then made a cot at the bottom of the bed. Just as she was finishing up, the husband walked in. I wasn't quite sure I understood what was happening here, so I just stood there, half-dressed and smiling like the true idiot abroad that I was. They got into the bed and gestured for me to get into the cot! I eventually guessed that their fan must have broken upstairs, so they had come to share the one remaining fan in the house. I was soaked through now, with less fan and more clothes. As I lay there wide awake with eyes wide open, the husband turned in the bed, propped his head on his hand and looked at me with his usual smile. *Oh, God, no! Please tell me we're not going to play that game now!*

"Cup of Tea" he said with a grin.

Yup, we were playing that game. It was going to be a long night!

In the morning, I decided to take a ride around to find some chocolate and some ciggies.

"Everything in moderation," I whispered as I put my helmet on, "…
including moderation itself." I just needed some comfort, and what harm
could it do away from the house? I hadn't promised anyway. Still, it wasn't my
proudest moment. It was only 8:30 in the morning, and already unbearably
hot. I couldn't stand this anymore. The sooner I got into the mountains, the
better.

When Sukjeet got up later in the afternoon, he suggested we go to a
McDonald's in the city. I couldn't quite believe it. Was he actually feeling
some empathy and stepping down a little? I felt bad now for buying the fags,
but not for long. I was already excited at the idea of a nice juicy Big Mac, but
it turned out to be the only vegetarian McDonald's in the world.

On the way back, we had another debate about his religion. I really
wished he'd drop it now. I was growing tired of him telling me how little I
understood of the world as it really was; how Sikhs could perform miracles;
and how one Sikh can wipe out 100,000 men with his warrior skills.

"So why are you spending all your time meditating and sleeping," I
asked, "when you could be out there stopping war and famine? Bit selfish,
isn't it?" I instantly regretted it. I was losing control of what I said, instead
of just keeping my thoughts unspoken. This was not how a guest should
behave.

After this debate in the car, Sukjeet went to pick up his motorbike, so
I went for a shady cigarette around the back of the house. I kept the ciggies
away from the house out of respect, stashing them on the bike. Now was
my chance to go and have one. As I lit my cigarette and drew in a deep,
contented breath, I felt a thud on the back of my head.

"What the…?" I looked around and saw a large raven flying away,
then circling to come back for another shot. *I'm being dived bombed by a
raven? Really?* Of all birds and all times, you just couldn't make this shit up.
I took cover under the branches of a tree and laughed out loud. *Has Sukjeet
sent it to tell me off for smoking?* Apparently, he *was* capable of such things!
Somehow, I doubted it. More likely, it had a nest nearby, but the timing, not
to mention the breed, was almost enough to have me believe; if only for a
second!

We were due to leave at 5am the next morning. I couldn't wait, but I doubted very much the boys – sorry, the warriors – would be ready if their past timekeeping was anything to go by. Still, I would be ready. I now wished I was going alone. The romance of the idea had long worn off.

I had failed miserably at being a Sikh in Amritsar.

Chapter 11
The Hindu Pilgrimage and Nepal

The ride out with my five Sikh warriors did not go well. The riding felt a bit chaotic, and on Day Two, one of them got a puncture. While the boys stood around staring at the back wheel – perhaps hoping for a miracle – I grabbed my tool roll off the bike and went over to help, but they wouldn't let me get a look-in. I offered the tools instead; they were refused. Eventually I gave in and buggered off to find a shady tree to smoke under. My rebellious streak was coming back. Sukjeet rode off with one of the others to get help while the rest came and sat under the tree with me, not seeming to mind the smoke at all. It took three hours, two cigarettes and a local mechanic before we were ready to move again.

As I put my armour back on, Sukjeet suggested I keep it off as it was so hot.

"You put your faith in God," I said. "I put my faith in body armour and big boots." This was *not* up for discussion. Later that day he crashed. Bright red blood spilling from his elbow and over his white robes making it look like a scene from *The Passion of Christ,* and far more dramatic than it really was. Unlike the inner tube, he allowed me to patch up his grazes before we moved on.

I think it was Day Three when I lost my Sikh Warrior chaperones. It really wasn't deliberate, though I admit that I didn't bother looking for them, as it was, quite honestly, a relief to be riding alone again. I had planned on hitting the Himalayas with them, and now that I was alone, I wasn't completely sure what was ahead. That night, in Srinagar, I did some homework, checked my route, and by the morning, I was back on the road and once more singing in my helmet with all my optimism and excitement well and truly restored. With Janis Joplin keeping me company in my helmet,

I left Srinagar and began my climb into the mountains.

The traffic stopped, the tarmac ran out, and the faces changed as I headed further north through Kashmir and higher into the mountains. Watching the merging and transformation of terrain, weather, cultures and especially faces was fast becoming my favourite part of this journey. The changes were sometimes rapid, like a border – the magical artificial line that separates two different worlds. More often, it was a gradual process that flowed with the curves of the earth. Faces carved through generations exposed to their conditions and surroundings; telling the story of their lives and those of their ancestors.

I found myself surrounded by snow-capped peaks and glaciers, riding higher and higher on narrower and bumpier tracks into the cool air of the Himalayas. It felt very remote. Despite having seen no options to turn, I stopped to check my map, wondering if this was really the right road. Reassured, I pushed on, passing a few guys with AK47s, some horses, and surprisingly, a little guy sitting on a rock, selling nuts and water. This told me that, despite its appearance, this road was well-travelled; certainly, enough to make it worth his while to sit there.

Eventually I passed a sign saying: WELCOME TO THE REGION OF LADAKH.

Rhonda and I had made it all the way across India from Mumbai – over 3,200 kilometres with temperatures ranging up to 45°C (113°F!), and through enough chaos to wipe out a whole butterfly colony in New Mexico. We were now in bikers' paradise, and it felt all the sweeter for the journey.

No one would argue that the Dalai Lama is a nice guy, but he had arrived in the town of Leh at the same time as me, and he had understandably brought adoring crowds with him. I was done with crowds, so I stayed only a couple of days. Just enough time to grab some nit cream, down a couple of beers, and ignore the advice to take oxygen with me for the next stage. I was heading for the famous Kardung La Pass – the highest motorable road in the world at 18,350ft. I had also been told that I could not go up alone and would have to join a group for safety reasons. Instead, I blagged my way past the military checkpoint at the bottom, rode to the top, and celebrated with what is probably the record for the highest cigarette ever smoked. It took an hour to burn down and even longer to light in the first place, but it was well worth it for the sense of achievement. I camped up above the snowline for a night before coming back down the same way.

I was just getting into my stride now. This cool mountain air was a tonic to my soul, so I just kept going. First Tanglang La Pass, then the Rohtang Pass, across the Moore Plains, through some splendid rock formations, and eventually stopping to camp at around 15,000ft. It was freezing. I'd left optimum 'feel good' temperature much further down the mountain. Tonight was going to be a 'long johns night' for sure – but at least my head had stopped itching! That nit cream had worked a treat.

Around 4am the next morning, I woke to the ominous sound of heavy rainfall. I knew I had two more high passes to make, and with rain comes fog, snow, mud, and the threat of landslides. I did not relish my ride and briefly pondered the idea of a pyjama day, before talking myself out of my cosy – if slightly malodorous – nest. The rain finally tapered off around 7am, and that is when I decided to 'go for it'.

As I headed for the first pass of the day, I cranked up my heated handlebar grips and prepared to hit the snow. Before the snow, though, came the slippery, rutted mud. Amazingly, trucks used this track as a highway, and the hardest part of riding was getting past them; there was no room for error as the drops were deadly. Generally, I beeped my horn to let them know I was coming, and they'd give me a little space to squeeze through without shutting off my throttle, to maintain my momentum as we climbed uphill. However, on this occasion, I got it wrong. As I passed the point of no return in the gap that opened between the truck and the drop, I realised he was not moving for me at all, but swinging out to take a bend. His front end closed in on me, and my only chance was to make it through the gap before it closed entirely. I squeezed the throttle and screamed in my helmet, convinced I was going into the abyss. Just when I thought India had used up all my adrenaline, it squeezed that little bit more out of me. I made the gap with inches to spare. It wasn't even 9am.

I made my descent out of the thick fog and was granted only a little time to admire the glacier lake to my left before the mountains presented me with my next challenge. As I came around a tight hairpin, I discovered the track ahead was covered in rocks from what was clearly a very recent landslide. There were trucks parked on the other side and all the drivers stood by looking as if they were waiting for something. *Strange. Normally they'd be on it, shifting those blighters out of the way to make a path.* There weren't even that many rocks really. "In fact," I thought, "I reckon I could ride over those bad boys". A local biker had caught up with me by now. I glanced at him, we shrugged, and went for it.

As soon as we did, the truckers started jumping up and down and screaming something I couldn't understand. I lost concentration and went down, as did the biker behind me. He pulled himself up and came over to help me lift Rhonda,

"What are they saying?" I asked as we manhandled bike and luggage to an upright position.

"The rocks are coming," he replied simply. *Whaaat?* I looked up to see a shower of stones bearing down on us. A couple hit me, but my helmet took the brunt. They were small, but what was still to come? We should have run, but we didn't. Alone I would have, but Mr Biker here was not giving in, so I stayed and kept my eyes upwards as we lifted. As the last of the shale fell, the truckers came running over to help and eventually we were free. It wasn't even lunchtime!

After a quick stop for a Laughing Cow triangle of cheese and a packet of M&Ms I'd been saving for just this kind of occasion, I continued through several river crossings – my favourite being the one with a load of half-naked men playing in the water who cheered when I made it through – towards Manali and eventually to Shimla.

****

It wasn't as if it was an aggressive crowd; just a force and a single mind with a single mission – to get through and collect water from this spot on the River Ganges and carry it back to temples for hundreds of miles around. It was like a wave, and impossible to reason with. All I could do was brace myself and try not to fall over.

With only a litre of drinking water and a heavily-laden bike, I found myself in the centre of the biggest Hindu pilgrimage in India on one of the hottest days of the year. I had unwittingly found the Gateway to God, and God was in *very* high demand that day.

I'd left Shimla the day before and made my way to Rishikesh – best known for being the yoga capital of the world (and the place The Beatles came to in February 1968 to visit the Maharishi Mahesh yogi). Nestled in the Himalayan foothills beside the Ganges River it is regarded as one of the holiest places to Hindus. Sages and saints have visited Rishikesh since ancient times to meditate in search of higher knowledge. My needs were less ambitious. It sounded peaceful and I just wanted to check the place out en route to Nepal.

All the way, people dressed in orange walked along the roads. Some were barefoot, and several carried what looked like water carriers (or the old-style milkmaid buckets) over their shoulders. I also saw several motorcycles with orange flags and orange-clad riders. Something was going on but I had no idea what! It could mean literally anything in India, and at this point I was far more interested in the painted elephants and riders I kept passing on the road. The closer I got to Rishikesh, the busier it got and so it was no surprise to find all the hotels were taken. Eventually I found an adequate room to rent with firm bed and a cold shower. I was stripped out of my sweat-saturated riding gear and under that water in thirty seconds flat.

By 7:30 the next morning, I was under way with a full belly and a good mood. The roads in town were busy, really busy, but I pottered on, assuming it would ease as I got further out. I had plenty of time, so I stopped to take photos of the walkers and runners who crowded around Rhonda and me as soon as we pulled up. I soon came to a familiar-looking bottleneck; a scrum of bikes and cars and trucks all scrambling for space. *No problem.* I'd dealt with these before *– just inch forward bit-by-bit, find the gaps where you can, and take them. If they give you an inch, TAKE IT!* I thought it would last maybe half an hour or so and I would be free – back on the open road with only the odd street-smart macaque monkey to avoid again. Not on this day though; unbeknownst to me, it was the annual Kanwar Yatra pilgrimage in the name of the Hindu god Shiva, and this was no mere bottleneck.

During the festival, the streets around the pilgrimage are jam-packed.

Stampedes are a regular occurrence, and combined with the sheer volume of people in a fairly confined space, and the heat, tens of deaths occur every year as a result. The bridges across the Ganges are the most dangerous spots. I wish I'd known this at the time. I didn't, and so I just kept pushing on. Working my way through the crowd, I came to a small clearing. It looked hopeful. I followed the flow of motorcycles only to find myself in a muddy field, and once again heading for a bottleneck. Was this really the road? I wasn't sure, and since anything is possible in India, I kept going, wondering what was causing all the fuss up ahead.

Inching forward with the crowd, we crossed a stream, then headed up a steep rocky incline where, with short legs and a heavily laden bike, I couldn't keep my balance. I dropped Rhonda. Some of the crowd helped me pick her up and then steadied her as I got back on to try again. Next came a muddy bank that people were manhandling their bikes over. I went for it, ignoring the bare feet around my tyres. If I waited for a respectable clearing, I'd be stuck here all day and I really didn't want to drop the bike again. I made it and cheered, as did the onlookers. Then came a train track and a small gap in the fence which hundreds of bikers were trying to get through. *Okay, so it's not the road.* Turned out I was caught up in the flow of people taking a short cut and trying to cut off the corner of the road.

Until now, it was a kind of pleasure/pain situation. It was insanely hot and I was the only fool fully clothed, but I was covered in mud, the crowd was buzzing and chanting, and the sweet smell of two-stroke engine oil lingered in the air. It felt as if I was part of some wacky endurance race and I actually quite liked it. The buzz was infectious and, anyway, I still believed that I would soon be out of it and on my way to Rudrapur, about 210 kilometres away.

Back on the road again, it seemed worse now and the buzz soon faded. I was overheating; I was surrounded, and there was no way of even walking out through this chaos, let alone riding out. People on foot just stood amongst the thousands of motorbikes waiting to inch forward with the rest of us. Both ways were going nowhere fast. As we headed for the bridge over the Ganges at Haridwar, about fifteen kilometres south of Rishikesh, things went from bad to worse; the heat was baking and my litre of water was gone. I took my jacket and helmet off and switched Rhonda off so she didn't overheat or lose battery life from all the stopping and starting. It was time to get off and push; every inch was a win. It took all day – eight hours of chaos – to get as far as that bridge, and over it.

Packed in with thousands of baking bodies around me, I wanted to fall. Perhaps the smell – now less engine oil, more stale sweat and piss – was the only thing keeping me from fainting – like a great big smelling salt! Everyone was staring at me, probably wondering "What the hell is a white/red-faced woman doing in the middle of this chaos and with so much gear and clothing?" I couldn't take it anymore and – as they say in the military – I took a knee – in fact, I took two. I had long ago run out of water, and the heat was unbearable. If I fainted, I would be in serious trouble; there was no way anyone was going to get me out of there. A hand came through the crowd and put a bottle of water under my nose. Without even looking at who had offered it, I took it, gulped down a few swigs, and resisting the temptation to finish it, I located the hand that had shared his rations.

"Thank you," I said weakly.

"You should not be here," he replied. "This is dangerous."

"No shit, Sherlock. Any suggestions?"

Then came the first stampede. I could hear the whistles and shouting getting closer. Then I saw them – hundreds of guys jumping over bikes and cars, smashing windscreens as they went, and forcing their way through. I felt panic rising. I had to get out. I *needed* to get out. As they came through, they pushed me to the ground, along with several others, using whatever obstacles were in their way as stepping stones over the crowd. The noise was deafening and the feeling of being trampled was overwhelmingly claustrophobic.

When I got back on my feet, the young man who had given me the water asked if I was okay.

"I think so," I said. "But I need to get out of here." He pulled up Rhonda for me and suggested we try to make it back to the last turn. Five hundred yards down that road was a hotel where we could take refuge. Then he changed his mind.

"I don't think we will make it," he said. "Our only option is to go with the flow, but it will take all day and night to get out." I wanted to scream, but what was the point?

Just as I was ready to lie down and resign to my fate, my new friend spotted a ditch. He suggested we get Rhonda in it and then follow it back up to the turn by the hotel. He said he would help me, and then continue on his mission to collect his water. There was no way he could go back to his temple empty-handed. I could have kissed him. There was no way I was getting out of there alone. In fact, it seemed impossible that we would get the bike through, no matter how many people helped. We had no choice. We had to try.

We forced our way slowly, inch by inch, over to the ditch, where he got on Rhonda and started riding her over the rocks and through the mud. Until now, no one had ever ridden Rhonda but me. He had longer legs, though, and I was in no position to argue. He could steady her better while I pushed from the back and kept him from falling over with whatever strength I had left.

Just as I thought we were getting somewhere, the ditch was suddenly full of other bikes and more people coming in the opposite direction. We forced and pushed and rammed our way through, only to be greeted by another wave of stampeding runners. There was no stopping them, and Rhonda was forced to the ground with me pinned to the wall as they trampled her into the mud. As soon as they had passed, the oncoming bikes tried to ride over her. I shouted for the riders to stop, but to no avail. Eventually people helped pick her up while the bikes kept on forcing their way through, threatening to topple us over again.

Then, I spotted a policeman in the crowd. He was beating people with a big stick to try to get some kind of order. We caught each other's eye and I looked pleadingly at him, but there was nothing he could do. He was in his own hell, and it was every man and woman for themselves. Eventually we made it to the car park of the hotel, which was blocked off by ropes and bollards; the police had it barricaded and were using it as a base. I said goodbye to my saviour and he dived back into the crowd to continue. Then I pushed a few police officers out of the way and forced my way in. No one argued with me. The relief was overwhelming.

I checked Rhonda over quickly before checking in. Amazingly, she got away with a broken clutch lever and nothing more; (fortunately I'd packed a spare). After a shower, lots of water, and some food, all that was left to do was put AC/DC on *really loud* and dance around the room in my underwear rejoicing in the fact that I was alive, and to top it off, I'd captured some of it on video! They say if you are going to get yourself into a scrape, at least make sure the camera is rolling, right? What doesn't kill you will make a great blog post afterwards. Being on my own often made filming difficult, and it was something I really only enjoyed much later, but my helmet cam had been rolling that day, and so with Rhonda, me *and* video intact, life had never felt better!

By 10am the next morning, the traffic and crowds were mostly gone, and I could see how small an area we had covered fighting our way through to that hotel; yet it had taken all our strength. Most of the noise was gone, too, apart from one remaining converted dump truck full of speakers which sat in the surrounding aftermath, blasting what I can only describe as Indian techno music.

"Not my cup of tea," I said to Rhonda, reaching down and patting her on the tank, "Fancy some nice relaxing Paul Weller this morning?" With that, I selected my playlist, cranked up the volume, and set off for Nepal. The danger had been real though; later I discovered that in 2003 no fewer than twenty-three people had died at Haridwar during the pilgrimage, and five in 2010....

****

The western border crossing into Nepal was so relaxed, it could have been a hippie commune. Everyone looked mildly stoned and in no rush at all. There were no queues and no barriers; I wandered from one small wooden shack to the next getting my visa sorted and my carnet stamped. Before I could say "*Groooovy*," I was in Nepal. My 'groove' was quickly interrupted by a policeman waving me down. I'd been so relaxed; I'd not put my helmet back on.

"You must wear helmet," he said with the mock harshness of a kindergarten teacher. "Sorry, I thought I had more office stops," I said.

"Traffic is not good in Nepal," he smiled, "You must be safe."

"You should try India," I replied.

The Mahendra Highway cuts across the entire width of the country. It is a pretty road surrounded by lush green foliage and initially very easy, which was something I particularly appreciated at the time. The dramatic change at the border, from instant chaos to calm was like a balm to my nerves, which were now ready for a shot of something less aggressive. If India were a drink it would be absinthe; attacking your senses so quickly that you don't even feel it after the first hit. Too much of it, and you will find yourself on your arse with a spinning head and a distinct lack of clarity, wondering what hit you. Nepal, on the other hand, felt like a glass of Cobra with a slice of lime on a summer afternoon.

It was just a nice pleasant ride; the road stretched out before me in a smooth, inviting path through the lush green hills, dotted with Nepalese men and women smiling at me as they sheltered under their parasols, sedately walking their water buffalo on long reins. I managed a steady three hundred-mile day pottering along with the sound of Amy MacDonald singing "This is the life" in my helmet. This *was* the life, and it suddenly slowed down considerably. And there it was again. That dramatic change. What a difference a border makes.

I took an overnight stay in Butwal, where I was woken in the night by a sharp pinch to my arse. Throwing back the well-used sheet, I found my body covered in tiny little ants; hundreds of them were all over me. I had to shower to get them off, and then shake the bed down and sweep the floor to get them away. It was an unpleasant way to wake up. Butwal was not a pleasant place at all, really. The ants and cockroaches seemed to like it though, so the next morning I left it to them.

I was looking forward to getting to Kathmandu and meeting up with Dani, my friend from the UK, whom I'd met the year before while working on the Nick Sanders tour. She rode her Yamaha R1 across the US and back with a few other riders, while I drove the support truck. We had roomed together and got on well, and my 'roomie' was now flying over for a week to check out Nepal with me.

Dani is an excellent rider, and I was ready for some company from home, so she hired a Honda XR250 Tornado for our ride together. After a couple of days of taking in the clubs, shops, and bakeries of the popular Thamel area of Kathmandu, we hit the road. I ditched my heavy top box, and off we went with just the bare essentials and the promise of some great riding ahead.

Julian from Kathmandu Vintage Motorcycle Club had been following my blog for some time, and he joined us on his Royal Enfield. We decided to make our way from Kathmandu to Nagarkot, about 35 kilometres to the east. From there, we pottered over to Pokhara, and spent a few days doing the general touristy things – checking out bat caves, riding on the highest zip wire in the world, and seeing some great live bands in the many local pubs. Nepal certainly likes its live music and plays all the old favourite rock tunes. We partied with the locals, and boy, did the locals know how to party!

It was great to ride with my old roomie. We travelled well together, both happy to go with the flow, pottering around, and taking it all in with frequent 'pit stops' for tea. The views of the snow-capped mountains, the rice fields, and the rivers were intoxicating. It was hot, so we would stop and soak our Buffs in the cold running mountain water whenever we could, to keep cool. It felt like a lazy Sunday afternoon as we meandered back towards Kathmandu, taking it in turns to lead and wishing we had more time before Dani flew home in the morning. We agreed that we would see Nepal again one day, and hopefully, together.

About thirty kilometres from Kathmandu, as Dani led us around a long sweeping downhill bend, our dreamy ride was brought to an abrupt end by a local rider and pillion who decided to turn onto the road with no warning. Dani had no time to react and hit the local's front wheel. Her bike went careering across the oncoming lane and into a ditch. Dani left the bike as it went down and continued into the wall on the other side with a thud so loud it left me in no doubt that she would have injuries.

I pulled up as quickly as I could and shouted to Dani, who was now in the ditch, "Dani, are you OK?"

"I don't think so," came the reply.

"Stay there. Don't move!"

I jumped into the ditch next to her; she was looking quite pale. She needed a minute to get her head together, and so rather than diving in, Julian and I gave her a little breathing space and let her take stock. While I stayed with her, Julian got the bike out of the ditch and moved things into a safe position.

After a minute or two, we began assessing the damage: she could only feel pain in her shoulder, but I could see blood dripping out of the bottom of her trousers, as well as a crimson stain now starting to appear on the shin. The locals had gathered, and were helping Julian move the bike. Slowly, I lifted her trouser leg to find lumps of fatty tissue falling out and onto my hand. "Okay; this is not good," I thought, managing not to say it out loud.

"My leg is fine," Dani said as I lifted the trouser leg further. "It's my shoulder. Shall I get out of the ditch now?"

"Shut up and keep still. You make a terrible patient. Just do as you're told for once," I said trying to keep the mood 'normal' with our usual style of banter.

The wound on her leg was gaping. The skin and tissue had dropped down to reveal her muscles and ligaments underneath. It looked as if her stockings had come down, except she wasn't wearing any – it was her skin. I quickly dropped the trouser leg back down and used Julian's scarf to gather it up, apply some pressure and slow the bleeding. "Julian, call an ambulance. I think we might need a lift home," I said calmly.

Having run an off-road school, I am quite used to dealing with injuries – lost fingers, spiral fractures, and even someone who had a piece of branch wedged firmly in his groin. This injury was not the most shocking, though it was one of the most fascinating. It made my teeth water a little, but I kept a straight face as if there was "nothing to see here".

Dani piped up again,

"Hang on. Not yet. I think I'll be okay."

"Dani you are not riding home today" I said firmly, as Julian got on the phone.

Between us, we slowly removed Dani's jacket. I could see straight away that the shoulder had dropped; it was dislocated, without a doubt. She could not move her arm, and so we sat there calmly, blood flow under control, waiting for the ambulance to come and take over. We were safe and out of harm's way, so there was no need to do anymore.

As soon as the beat-up old ambulance arrived, I started getting out of the way to let the paramedics do their job, but thought better of it as I saw two guys in scruffy shorts and flip-flops jump out and reach for Dani as if to pull her out of the ditch by her arms.

"STOP!" I yelled. "Don't touch her! Are you not going to examine her first?"

I turned to Julian once again, "Julian, please translate." I explained her injuries to them, and as the penny dropped, they decided to get a stretcher. Clearly, they were just the road sweepers; their job was to collect and deliver. I doubt they had any medical training at all.

With the (remarkably unscathed) hire bike safely in the hands of the police who had turned up, Julian and I got back on our bikes and told the ambulance we would follow. As we chased the kamikaze ambulance driver through blind bends and over broken tarmac, I found myself wishing I had thought to ask that all-important question, "Which hospital are you taking her to?" That old friend hindsight... all we could do was try to keep up.

What a ride! It's one I hope I never have the dubious pleasure of repeating. Dani's life was not in danger until she got into that ambulance. Looking back though, it *was* great fun when we got into the city and followed him with police waving us through traffic.

"How the hell did you two hooligans keep up?" were Dani's first words as the doors opened and we peered in.

"All right mate?" I asked. "Piece of cake," I lied, just as the driver came over and presented me with the bill for the ambulance service.

I'll never forget the doctor's face when he came to look at Dani's leg. As he removed the scarf and pulled the trouser leg up, he turned to me with a scrunched-up face and said in horror,

"Have you *seen* this?" I had kept my poker face well and truly locked down all this time, only to be let down by the doctor, who gave the game away in an instant.

While they took Dani into the operating theatre to stitch her up and put her shoulder back in place under a general anaesthetic, Julian and I rode two-up on his Enfield to the best bakery in town and picked up a feast for us all. As we arrived back at the hospital gates, we also picked up some sweet tea from the street vendors. We were hungry, and sure Dani would be, too. This hospital did not provide light refreshments.

Dani and I shared a room in the hospital that night. Her shoulder had been popped back in place, and her leg stitched up as good as new. She got the comfy bed. I got the chair.

At 6am the ambulance was ready to take Dani to catch her flight. I was left to settle the bill for our stay at the most expensive room in town. We thought it would cost a few hundred pounds. How naïve! I asked for some gas and air as they hit me with the bill for 427,526 Nepalese Rupees, including 13% VAT – about £2,900. Thank goodness for medical insurance!

Ambulance Service: 3,000 rupees (£20)

Repair under general anaesthesia: 110,770 rupees (£744)

Room for the night: 38,000 rupees (£255)

And various other charges, including all sorts of blood and other tests and X-rays of her shoulder, chest, spine, tibia, hand and pelvis.

The hospital certainly gave her a thorough check-up! As soon as Dani left, I set about preparing my bike for air-freighting to Bangkok, with a promise to myself that I would be back in Nepal one day with my roomie. We had unfinished business.

Chapter 12
The Bikers' Handshake

I could see three airport security guys gathered at the bottom of the steps as I disembarked in Medan, on the island of Sumatra, in Indonesia, and thought nothing of it until they addressed me,

"Miss Stephanie Je... Jee... Jeev"

"Jeavons. Yes?" *What the hell's going on?* Why were three security guys waiting for me? My mind went into overdrive as I desperately racked my brain to figure out what I might have done.

"Welcome to Sumatra. We are the members of a motorcycle club here at the airport. We got tipped off that you were coming by one of our brothers in Malaysia. We would like to take you to lunch." I really had to get rid of this first-instinct-guilty-conscience thing. I breathed a sigh of relief and tried to hide the transition in emotion.

"Wow. Well, thank you very much. I'd love to join you."

"Please miss. Your passport."

I handed it over and was whisked through the airport, past all the queues, through security and out the front gates before you could say "The Welsh girl has landed!"

So far, riding through Southeast Asia had been a breeze. The monsoon rain clouds somehow got snagged on my bike as I rode through Laos, and I'd had a rather unfortunate incident in Thailand with a Huntsman spider in the shower – who knew they could move so fast? Other than that, it was all rather pleasant; nothing complicated, cheap digs, and great food. It was also a break from the challenging stuff, and so, of course, after a couple of months I had to go in search of more excitement.

Sorting out my Indonesian visa in Penang, Malaysia, I'd put Rhonda on an onion boat and sailed her off to Indonesia. They wouldn't let me get on the boat with her, suggesting that it was not a good idea for a woman to be at sea for four days with an all-male crew. Having seen some of the crew on deck in their undies, I had to agree. Still, while I may have missed out on an adventure on the waves, I gained the experience of feeling like a rock star through airport security with my own personal biker gang chaperones. And it didn't end there.

Now that all the clubs in the area knew I was here, there was no chance I was going to be left alone – like it or not. That night, I was picked up from my hotel by a group of young bikers who were keen to bring me along to their regular Friday night meet. As Rhonda was still somewhere at sea on the onion boat, I jumped on the back of Omar's bike, a Honda CB150R. Later, I learned that in many parts of Indonesia, the term 'Honda' can be used for any brand of bike, just as 'Hoover' is used in the UK for any brand of vacuum cleaner.

The boys looked every bit the biker gang. Patched denim with leather waistcoats covered in patches, loud exhausts, and piss-pot helmets. With their Sumatran looks, all dark and mysterious, they looked far cooler than the British gangs of old – yet they were all on small bikes rather than big Hogs. It was a brilliant contrast.

I've never been a fan of the whole biker 'gang' thing. When I was in my own black leather and loud exhaust phase, our 'hang outs' were often disturbed by the arrival of the Outlaws, a worldwide biker gang (different from, but not unlike, the Hell's Angels) who loved to ruin a good party. These were the people who gave the rest of us a bad name back in the early 90s, and the reason many campsites or clubs would turn us away when they saw us turning up on our motorbikes. Don't get me wrong, we were rowdy enough, but always pretty respectful. Certainly, by comparison. There were no derogatory initiations; no initiations at all, in fact, although being able to down a pint of snakebite in under six seconds certainly gave you kudos.

"So where are we going? I shouted over Omar's shoulder.

"Sorry madam?" he replied, unable to hear over the buzz of the leather-clad swarm that surrounded us as we whizzed expertly through the busy streets.

"WHERE ARE WE GOING"

"Ah! To the mosque" he replied. *Hang on, did he say mosque?*

There must have been a hundred bikers standing around their steeds when we finally arrived outside a beautifully lit blue mosque in the centre of the city. Omar pulled up with our group, jumped off and politely took my helmet.

"You must meet everyone.." he said enthusiastically, "but first you must learn the handshake.

"Obviously" I replied with a giggle. "Bring it on Mr O". I'd learnt the bikers' three-stage handshakes in India and Malaysia, and I was about to learn the most complicated of them all, the Sumatran bikers' handshake.

I think I really nailed it by about intro number 23. This one had four moves and ended with a fist against the chest in a solute of solidarity. Most of the bikers didn't speak much English, but everyone was friendly and welcoming and clearly proud of their 'brotherhood' which, it seemed, was split into groups by brand of motorcycle. The most popular of all was the Yamaha RX King. An aggressive little two-stroke that could pull a wheelie with the merest tweak of the throttle.

Omar introduced me to handshake number 24,

"Welcome, I am Hassan" he said as we worked through the moves. I just about managed a "Hello" back as I tried to stay focused on the hands.

"You are a fighter rider" He continued "You are lady biker fighter rider". Bit of a mouthful, but I liked it.

Omar explained that I was a 'solo rider' or 'fighter rider'. This meant that I did not belong to a club and I chose to ride alone. A bit of a phenomenon around these parts it seemed, and I was rewarded that night, and for several more nights, with handshakes and motorcycle club T-shirts galore! As we walked around shaking hands, Omar told me that there were forty motorcycle clubs in Medan alone. They each had a leader, and there was also a 'leader of leaders'. Just like most modern-day UK clubs though, they were just a community-minded bunch who liked motorbikes. There were no initiation ceremonies, and women too were welcomed into the 'brotherhood'. They didn't even need to down a snakebite!

The following morning, I was summoned by this 'leader of leaders' for lunch. The guys seemed to think this was a big deal, so I made a special effort to brush my hair and went along. I was intrigued. He turned out to be of medium build and fairly average looking in a Sumatran biker gang leader kind of way. He spoke no English, so a translator explained,

"Our leader has a gift for you," before handing me a bag with a pair of trousers in it. "He was going to buy you a scarf, but thought you might prefer something practical instead,"

"Thank you," I laughed. "Yes, I am certainly a practical kind of girl. How thoughtful." The trousers were too big, just like all the T-shirts, which got sent home at the earliest opportunity. There was no room for all this on my tiny bike, but they would make great souvenirs some day.

The following day, Omar and his Honda gang took me out on a mission, the sole purpose of which, it seemed, was to make the white girl throw up. Or at least make her eyes water. I knew this game well. I had played it many times before, and I was up for trying anything. I used to have a saying that went, "Try everything twice. The first time you might throw up." In other words, some pleasures need to be acquired. I was sure I could handle anything they threw at me.

First, came the hottest chili I have ever tasted. My dish was called Bakso, and it consisted mostly of meatballs and noodles. I was instructed to mix in the chilies that came with it in a separate bag. The first mouthful was fine, and a slightly smug look may've crossed my face as I went for the next spoonful. The second hit the back of my throat with such force that I struggled to catch my breath. More water was ordered while my eyes streamed and my boys, as well as all the other locals, looked on and laughed. Round One to the Sumatrans.

After dinner, we rode out a little further and I could smell my next challenge before I saw it. We pulled up next to a wooden shack with a few tables inside and a massive pile of durian fruit outside. The smell was really quite pungent; reminiscent of rotting flesh, or maybe raw sewage. There was a sign on my hotel door that had a picture of one of these fruits with a line through it. Now, I could clearly smell why it was not welcome. The texture is quite unusual, too; it has a spiked, yellowy-brown husk with a fleshy inside that felt quite slimy as I put it in my mouth. The boys held their breath, poised and ready for the face of intense disgust. I disappointed them, because I didn't think it was all that bad. In fact, after a couple of mouthfuls, I found it quite pleasant.

Then came the TST; a drink made of tea, raw egg, and milk. It looked pretty revolting, though it turned out to be delicious. In fact, I ended up having many more as I travelled through Sumatra, south towards Jakarta, and beyond. It provided a nice bit of sustenance on the run.

Rhonda arrived at the port the next day. Despite the biker boys trying to persuade me to stay, it was time to go it alone again in true 'fighter rider' fashion.

My first stop from Medan was on the little island of Samosir, on Lake Toba – a large natural lake occupying the caldera of a super-volcano. It is one of the largest lakes in Southeast Asia and one of the deepest in the world. I knew as soon as I arrived that I would have to spend a few days getting to know the place. The first thing I noticed were the traditional Batak houses. The roofs were bright red and boat-shaped with intricately carved gables and up-sweeping roof ridges. The walls were mostly pastel-coloured or bare wood and featured little shuttered windows. They looked as if they'd come straight out of a fairy tale and I'd never seen anything like them.

The second thing to strike me was my top box. Literally. It sheered the bolts, broke free, and fell forward, hitting me on the back before falling off into the road. This was the second time this had happened; the first time was after the Himalayas. The bolts couldn't handle all the vibration and rough riding we were doing. If this riding could shear through metal, I wondered, what was it doing to my joints? No wonder I was tired a lot of the time.

"Must rest more," I promised myself as I picked up the top box and got back on the bike. I balanced it on the pillion seat, then held it awkwardly with one arm while steering with the other. Thankfully – with the help of a few locals – I found a workshop down the road. The young lads in there had me fixed up in no time for the Indonesian equivalent of about 50p. I shared my cigarettes with them before pottering off to check out more of the island.

It felt so relaxed here; a slow-paced way of life that seemed honest in some way. Cutting through the crap of modern-day stresses, it was simple and unadulterated. I stopped to take a picture of a pig that was grazing near the road. It was wearing a kind of wooden necklace held loosely with rope, almost like stocks. It hung down above its knees, and I guessed it was to stop it running while allowing it to graze untethered. I could hear some voices coming from one of the shacks nearby, and as I took my last picture, a man came out and made a welcoming gesture for me to join them. "Why not?" I thought. "No rush." I was intrigued and had long since learned not to let these opportunities pass me by.

Inside were seven or eight local guys of varying ages, drinking, smoking and playing a betting game with domino-like blocks which might have been QuiQui or Mahjong. The guitar player in the corner spoke a little English, and before I knew it, I was settled in nicely, drinking, smoking, and singing what sounded like Indonesian folk songs (I hummed along). I left the gambling to the experts, but it was a wonderful way to while away an afternoon.

The Sumatrans had a warmth to them that made me feel at home. The men held hands as they walked down the street, and people greeted each other with genuine hugs or lingering handshakes. I loved the fact that they ate with their hands, too. No cutlery. I followed suit. There is something really satisfying about scooping up a nasi goreng dish with your fingers or a piece of freshly baked flatbread. It has an earthy, no-fuss feel about it, like chopping logs or walking barefoot in the sand. These simple pleasures that we so often forget in our fast-moving and sanitised western world.

I rode down the middle of the country to Sungai Penuh, through the oldest tea plantation in the world. This area provides tea for the British royal household. It seemed an opportunity not to be missed, so I stopped in the middle to make a brew of my own on my little gas stove. Crossing the equator from north to south was completely uneventful as I totally missed the sign, but we carried on regardless, working our way through the swarms of killer scooters in the towns – often ridden by ten-year-old kids with their baby sister on their back – and towards the coast through some of the most spectacular rainforest I have ever had the pleasure of seeing.

Then there was the now-familiar wildlife and livestock to look out for – cows, goats, lizards, and bugs the size of your fist. The hornets and dragonflies seemed to stop and reverse as they realised they weren't going to make it across the road. Can bugs fly backwards? I could almost swear they did. The scorpions were so big I gave them the right-of-way and a salute at a respectful distance as they crossed, pincers held high in a proud boxer-style stance.

Despite a few minor hurdles, Rhonda and I covered a lot of miles in those two weeks. The last few hours were slow, as I was riding on slippery, hard-packed mud with many large potholes. As I worked my way around them, I spotted something moving in the road up ahead. The car in front swung left to avoid it. As it reared up to strike the wheel, I realised that it was an enormous king cobra – about three metres long with a distinctive hood that made it instantly recognisable as one of the most dangerous snakes in the world. I love snakes, and I knew that despite its fearsome reputation, the king cobra is generally a shy and reclusive animal, avoiding confrontation with people as much as possible, but clearly this one was distressed, having found itself in the path of the oncoming traffic, and now it was trying desperately to defend itself. I watched in awe as it came back up to its full height and bravely struck the next car. It was mesmerising, yet distressing to watch. I then realised it was my turn to pass. Since I had no metal box around me, I just went for it, giving it as much space as possible. It struck and missed.

"Wow!" I yelled, "That was amazing," as a rush of life shot through my veins and overwhelmed my mind with a jolt of exquisite ecstasy. A feeling no amount of DFS sofas or Opra-watching would ever give me, and better than any drug I'd ever tried. I looked in my mirror to get one last glimpse just as the car behind me took aim and drove straight over the top of it. Another two metres and he would have made it back into the jungle and away from people. A beautiful specimen of an endangered species, gone in a second. I got on the pegs, pulled back the throttle, and took my frustration out on the road.

The bungalows of the Lovina Krui Surf Resort appeared unexpectedly out of the dusk in the nick of time, quashing that first familiar fluttering of anxiety. I was riding in that crucial final hour of daylight, when I still had no clue where I was going to sleep and could vaguely hear the 'conundrum countdown' clock ticking in the background of my mind. I was in my 11th hour in the saddle and had eaten enough dust to give Dyson a run for their money. Hidden away on a little dirt track that sat in front of a long stretch of white beach teaming with turtles and coral, it was the perfect place to rest up for a couple of nights before boarding the ferry to Java.

****

The path of true adventure never runs smoothly, and after seven months of rough roads and mountain trails, Rhonda had a problem. As I jumped off at Bakauheni Ferry Port, I noticed one of my turn indicators was fried – completely melted. I had plenty of time before the ferry, so I took off my helmet and gloves to give it a closer inspection. My first thought was that I had an electrical problem, but Rhonda was a little out of shape! Her back end had drooped, and the indicator, exposed to the heat of the exhaust, had melted. I stripped off the luggage and removed the seat to find the sub-frame snapped in three places. Wow, I hadn't felt a thing! Shouldn't it have been more dramatic?

I wasn't particularly surprised, though. I considered strengthening the frame before I left the UK, but decided to leave it to fate and, at the same time, test the durability of the CRF250L. After all, at 250cc, she was designed for neither overland travel nor carrying luggage, so it was likely to happen, but I also knew it was likely to be an easy and cheap fix on the road. Mr. Honda had only trail riding and city commuting in mind with this one; a cheap and cheerful bike you could ride to work during the week and get muddy with on weekends.

Of course, I had other ideas. I could have chosen a bigger bike, generally considered at the time to be more suited to the job, but I wanted a bike that was light enough to pick up out of a ditch on my own. I wanted a bike that was agile and able to go over any terrain, while still being able to squeeze through heavy traffic. I wanted a bike that was simple and reliable; not flashy. I didn't care for flashy.

Fortune certainly favoured me that day – Rhonda snapped in a damn near perfect location. I was about to board the ferry over to Java, where I would find all manner of welders, and I was about to meet a local biker named Adi who was going to ride with me for a while. I didn't know much about Adi, except that he was a biker and probably a very handy guy to have around right now. Being a Sumatran, his language skills alone would be invaluable when it came to the repairs. He also had connections with the Borneo Orangutan Survival Foundation and had promised an introduction when I got to Jakarta. It was a long-held dream of mine to see the orangutans before we completely wipe out their habitat. For that reason alone, no snapped sub-frame was going to stop me from getting there. "He's cute," I thought as he pulled up and introduced himself. Was he really? Or was I getting lonely? Or Both? A Sumatran man about my age with a big smile and firm handshake – he'd do for sure!

What was I thinking? I looked like shit and probably smelt as bad. I always looked like shit lately; good hair days were a thing of the past and I'd mostly stopped caring. If male adventurers could look like shit, then so could I, goddamn it! That said, there were some days I wanted to feel feminine and sexy. I wanted to remember what flirting felt like. Where was Jenny, my girly advisor, when I needed her? I'd already used up her emergency girly pack; the stockings had been used as fuel filters, and the hairgrips to hold my hijab in place. I smiled at the thought. I don't think that's what she'd had in mind, but she would love it when I told her. I still had a couple of hairbands, but it was going to take more than that to make me feel sexy right now. I doubted I even had a hope of getting him drunk. He was probably a good Muslim boy, and I would likely be going to Jahannam (Islamic Hell) for even thinking such things.

"Did you know your sub-frame is snapped?" he asked, pulling me out of my daydreaming.

"Yes, I just spotted it," I replied, "Do you know any good welders?"

"Of course," he said. "We can get it patched here and then fixed properly in Jakarta. Let's go." I jumped on Rhonda and followed, noting that he also had a cute arse to match.

"Yup, straight to Hell," I giggled.

We raced through the traffic towards Jakarta. Adi had none of the previous over-protective gene I had discovered in many of his Malaysian and Indonesian counterparts, that was for sure. Much as I'd loved riding with those guys, I'd found their 'police escort' approach a little stifling at times and on more than one occasion I had insisted they back off. They seemed to overlook the fact that I had made it all the way there alone and had proved to be capable of looking after myself. They truly wanted to protect me while I was in their part of the world. It was a 'not on my watch' mentality that I both admired and detested in equal measures. Adi clearly believed I could look after myself and expected me to keep up. He was already a breath of fresh air.

It was hot in Jakarta, hot and humid; very humid. I started day-dreaming back to the Ice Bucket Challenge my son had nominated me for over Facebook while I was in Thailand; I'd kill for a bucket of ice over my head right now. Add to the heat, the devil-horned road users of Jakarta who take an inch – even if you don't offer – and you start to see a girl in serious need of an intravenous McFlurry.

While in Jakarta, I stayed with a biker named Dono, yet spent most of my time with Adi. Despite our cultural differences, we were the same, and, from the beginning, it was like being with an old friend. We had the same sense of humour and could 'take the mick' without the other getting offended. A true sign of a good friendship. His English was perfect, so I didn't need to slow down or simplify my sentences.

Riding in the city during the day was chaotic at best, so most of the time we rode around at night. The city was still surprisingly busy at 1am on a Friday night, though. Coming from the UK, I found it strange that I wasn't surrounded by drunk people. Instead, the bikers were out lining the streets, brandishing their club banners, or riding around enjoying the lack of traffic jams in the cooler air. Rhonda's subframe was now fixed up and stronger than ever, so Adi and I rode two-up and thrashed her around the highly unofficial racetrack on which bikers meet on Friday nights to run the two kilometres of tight, fast corners. It was highly entertaining to watch as young lads put their knees down and slid around the corners on everything from scooters to sports bikes – a place where everyone could take out their daily frustrations from the usual traffic jams. After that we went to an Irish bar, played pool, and had a couple of beers. So, he *did* drink. By now though, I knew he was married, so despite the very real connection we both felt, we kept it all about the bikes and the giggles. I taught him to play pool and by game number three, he was beating me. *Beginner's luck!*

The next day, he kept his promise to get me into the offices at the *Borneo Orangutan Survival Foundation*, and they agreed to let me work with them in Kalimantan, Borneo, for a week. But first, I had to have tests for STDs; standard procedure before you go hugging orangutans, apparently.

I was starting to worry about my timeline for Antarctica, so I left Rhonda with Dono and flew over to Borneo instead. If I didn't make my shipping for Australia in time, I would run the risk of missing my boat from South America to Antarctica further down the line. It was all suddenly a little tight and something had to give. *Where had the time gone?*

The charity sent a driver to pick me up at the airport. He spoke no English, but as we headed out of town and down onto a little dirt track into the trees he simply said,

"Welcome to the jungle." It was a line I was sure he'd used before, but I loved it anyway. Covering two thirds of the island of Borneo, Kalimantan is a wild province of Indonesia seldom visited by travellers. Shrouded in large tracts of rainforest, traditional Dayak villages are connected by a complex river system, but the most well-known inhabitants of this island can be found swinging through the trees, and I was about to fulfil a life-long dream.

I spent a wonderful week under the leafy canopy of this luscious rainforest. At first, they said I would not be 'hands-on' in any way but could watch how the staff worked from a distance. However, by Day Two I was walking and foraging all day with 'Jungle School One' – young orphan orangutans who were old enough to go out and start learning the lessons their mothers would have taught them. Normally, in the wild, they would stay close to their mother until the age of seven, but now it was down to their foster carers here at The BOS Foundation.

At first, they saw me as a toy; a new face with white skin and gadgets. These orangutans may've been young, but they knew every trick in the book when it came to dealing with rookies like me. They climbed in the trees ahead and waited for me to pass underneath, then grabbed my hair or jumped on me before playfully biting me. I'd been instructed to take off all jewellery, but was given special permission to bring in my GoPro video camera. The orangutans tried tricking me into letting down my guard, often with the art of distraction; a cute face here, a sweeping long arm there. Often, it would end in a gentle wrestling match. I learned very quickly that the best way to get them to let go or stop biting was to tickle them under the arms or on their potbellies. They reacted just like human toddlers, though their eyes gave away something deeper.

One of the young males decided he had fallen in love with me and clung to my leg wherever we went; his long arms difficult to untangle. I nicknamed him Velcro. Whenever we stopped walking, we foraged together or poked sticks in ant hills. Any fruitful pokes would see him licking those critters off the stick as if it were a great big lollypop. With his big brown eyes and his Prince Harry hair, I was already so very proud of him. When it rained, all their antics would cease, and they would come running for cuddles. We huddled together in a big soggy orange mass, holding the large leaves of the elephant ear plant over our heads for shelter. The smell was that of a damp dog in a well-planted greenhouse: the combined scent of vegetation, moisture, soil, decaying plants, wood and of course, well-saturated orangutan. It wasn't a bad smell. It was the smell of life, and I loved it. When the rain stopped, the play-fighting would begin again.

Much as this was a lifelong dream, I soon wished I was nowhere near these glorious and loving creatures. Many had harrowing stories, and what I really wished for was to see them back in the wild with no human contact at all; safe in an unthreatened forest, away from the threat of the illegal pet trade, and oblivious to our existence.

I flew back to Jakarta feeling a mixture of pride for my fellow human beings who worked so tirelessly to rescue these beautiful and intelligent creatures, and of despair, having seen first-hand the devastation we are causing with our insatiable appetite for convenience. We devour and destroy, much of the time not even realising it; by the time a product reaches our shelves, it's in a pretty package and a far cry from its ripped-up roots.

This was a journey of a thousand goodbyes and saying goodbye to Adi was almost heart-breaking. In such a short time, I had come to adore him as a friend, and I would miss him greatly. Forget the noise, the heat, and the permanent chaos; Jakarta was going to be a particularly hard city to leave behind.

"We will meet again," he said. "This is a Sumatran man's promise." I wished he could come with me.

"Yes, we will," I replied, "and that is my Welsh girl's promise." It had sounded so much cooler when he said it and I smiled at my own cheesiness. He took off his silver bracelet and put it on my wrist,

"This is to remember me." I wanted to cry. His sincerity was heart-wrenching. I gave him a big hug instead. I was going to be very lonely without him as I travelled down the islands, and to this day, bar once for an MRI scan, I have not taken that bracelet off.

Leaving the Ace Café, March 2014.

The Kurds washing Rhonda.

The heat getting to me in Iran.

My Dad and me in Iran.

Climbing up into the Himalayas, India.

The usual inquisitive
Indian crowds!

Considering a new mode of transport in India!

The Sikh horse I rode in Amritsar.

Some of the Sikh community in Amritsar.

Dani 'the leg' gets out of the
Kathmandu Hilton! Still smiling.

Playing QuiQui in Sumatra.

Inquisitive kids on Lake Toba, Sumatra.

A beautiful Batak house on Lake Toba.

Chapter 13
Life is Like a Cow

Nothing ever feels like those first few weeks of a long ride; that ostensible notion of escape, the trepidation, the energy and those nasty little butterflies in your stomach that won't leave you alone. Little else in life can evoke such passion as escaping one's self-imposed limitations… or those of others. Freedom. Liberation. It's a feeling like no other, albeit short-lived. A guy I once knew in Morocco, a nomad and a Berber, once said to me "Steph, life is like a cow. Some days you get milk. Some days you get shit." The same can be said for long-distance travel.

Journeys evolve. The longer you go, the more milk and shit you get. Just like a relationship, the first fluttering of lust-based passion gives way to something else. Something deeper, if we're lucky. It's really *not* for everyone and discovering that is just the beginning of another kind of journey, I suppose. It's a brave person who turns back to try another path. Perhaps those of us left wandering will always be searching for the buzz of that first hit; that first high of a new and empty open road; the first true love. It had felt so meaningful.

Crossing the border into a new country always gave me a buzz. A change was as good as a rest, if nothing else, but the thing that came really close, the thing that really got my juices flowing again, was a change of continent. Each one would see a dramatic enough difference in my environment that it would feed the flame that threatened to burn out, bring me back from the brink, and have me firing on all cylinders once more. Perhaps this is the reason I am yet to succeed in a long-term relationship? You work it out. I gave up trying a long time ago.

Jumping from Asia to Australia was about as dramatic as it got – certainly, at that point. From chaos to calm on so many levels. I kind of missed being the 'rock star' though, if I'm honest. I had got used to all the attention I drew in Asia, but in Australia, I was more likely to be just another white girl on a bike. There was no way my hooligan riding style would wash there, either. Still, it was fun while it lasted.

After leaving Adi, I'd ridden down to Bali, got nervous about my time constraints for Antarctica so flew directly from Bali to Darwin rather than continuing on to East Timor as I'd originally planned. With Australian customs being notoriously picky about the cleanliness of vehicles coming into their country, I had literally brushed Rhonda down with a toothbrush, scrubbing away eight months' worth of hard-earned dirt, before crating her up and flying her over. Unfortunately, I had missed a bit!

"Can you please remove the bash plate?" asked the keen young customs official, his clipboard highlighting his importance. *Shit! The bashplate! I'd forgotten that bit!* The dried-up lump of mud that came out may as well have been a kilo of heroin, such was the tension in the room as it fell to the floor. Our eyes locked. My heart raced. Mr. Clipboard made a purposeful scribble. If she failed the test, Rhonda would be quarantined for a week until a thorough cleaning could be scheduled, and a second inspection completed.

Mr Clipboard continued his inspection while I stood quietly watching. Holding my breath. After five long minutes he announced,

"OK! You have clearly gone to great lengths to clean this bike. Why don't we just clean this up together now and say no more about it?"

"I can go?". I couldn't quite believe it. Even a speck of dust was normally a fail from what I had heard. It seemed I'd got lucky with this guy.

"Yes, you can go" he confirmed with a smile, "You can breathe now".

The congestion of South East Asia now changed completely to empty roads; and 'no rules apply' became 'you can bet your ass *all* rules apply'. When it came to law and order, Australia took many prisoners. But before I worried about remembering to obey the rules of the road, I had to get through the Northern Territories. The Outback – the place where no one could hear you scream. Also, a place where no one could see you pee. That felt liberating, too. Especially for a Welsh girl with a walnut-sized bladder, like me.

It was almost impossible to believe I was really here in Oz, and yet, here I was, peeing next to a large and bizarrely-decorated termite mound, loudly singing "I Got You, Babe" to scare off the snakes and enjoying the faint smell of rotting kangaroo carcass. The termite mound was wearing sunglasses, a rather pretty pink bra, and a green baseball cap that, in my opinion, clashed. I briefly wondered what my 'girly advisor' Jenny would say about this.

There were dead kangaroos lining the long straight road ahead and for some reason, I still had my helmet on while peeing. This was bizarre. Either this was another one of my crazy vivid dreams or I really *was* here. The first seemed more likely, at this point. A split-second of panic washed over me as I processed this thought. *I might actually just be wetting the bed right now, tucked up in my old house back in Wales, and not actually squatting with my helmet on next to a bra-wearing termite hill in Australia at all.* Wouldn't that be an interesting end to this book?

I laughed out loud at the thought. Really loud. There was no one around for miles, so I made a point of increasing the volume because I could. I'd been riding this desert road for a couple of days now and had seen very few people. I finished my pee, took off my helmet, placed it on top of the baseball cap that was on top of the termite hill, took a deep breath, and screamed at the top of my voice in a mildly Scottish, mildly Australian, Mel Gibson-as-Braveheart-esque accent,

"Freeeeeeeeedoooooooooooooom!!!" until my breath ran out and I felt dizzy. Then, I took out the last of my Indonesian tobacco, made a roll-up, lit it, and stuck it in the termite hill's face. "You may as well have this," I said with a relinquishing sigh. "Too bloody expensive here. It's back on the no-smoking wagon for me."

I jumped on Rhonda and opened her up. I had lots of time to look around as there were no corners on this road for miles and it was a good surface. I saw several more of the fancily-dressed termite hills dotted around. Perhaps one in every thirty had some kind of get-up. Some had T-shirts on, some were dressed only in underwear, others in hats and scarfs. Each one was a piece of art in its own right and there was often a hint of irreverent or obscene humour behind them. It seemed this long stretch of highway had become a canvas for road-tripping artists. I could only assume that one person had started it in a random moment of inspiration to entertain himself on a long, boring road trip one day and, over the years, others had followed suit, as people do. I wondered at the minds of the people behind each one. I considered doing one of my own, but I couldn't spare the underwear. Instead, I got a picture of one wearing my helmet and gloves with Rhonda parked next to it.

I made up my own entertainment as I travelled through this far-reaching mass of deep red soil and pale blue sky. This, despite what people had told me, was not a boring road; at least, not for me. I saw it as an opportunity to play. I counted dead kangaroos and worked out the average number of kills per road train – big trucks with multiple trailers attached and 'roo bars' to clear the kangaroos and debris as they went. I based my calculations on the number of road trains I imagined came through at night, when the 'roos are mostly on the road, and the amount of time I imagined it would take each one to decompose in this dry heat. Utter nonsense of course, but it came to about four per road train if you're interested.

After that I decided to duct tape my GoPro to my handlebars and record myself dancing to *Bohemian Rhapsody* as I rode along. Later, I would post it on my blog and get a sarcastic message saying, "Riding with no hands? New South Wales Police are impressed!" To which I replied, "I wasn't in NSW," and concluded with a smiley face. I assumed it was written with a sense of humour, but who knows? I'd been warned about Australian police. Of course I had. So far, I'd been warned about pretty much everyone.

Deserts encourage imagination and playfulness in me, and I love them for it. It took me a while to realise what it was about them that I loved so much: where else do you have the time and the space to be this silly? To allow yourself to be crazy and childlike with no fear of judgement? To see the horizon all around you, uninterrupted by people or buildings?

I also used to think that one desert was much like another until I actually visited one after another and realised that each one had its own personality; its own colour and flavour. The one thing they do have in common is that they all command respect, no matter how play-inducing or beautiful they are. The situation can change in an instant with a foolish mistake or a sudden and dramatic change of weather. They can swallow any man, woman, bush or beast that does not follow a few basic rules of survival. The desert is a leveller, and in that, I find a certain comfort. There are no labels in the desert.

It feels honest. Deserts cut through the bullshit and take you back to basics.

I played along the highway for several hours before pulling up near an unlikely river and set up camp. There had been no point in stopping too early, as it would be too hot to sit around. This looked like the perfect spot – a little green oasis clinging dependently to the riverbanks. It had a couple of picnic tables and a handful of eucalyptus trees for shade. There was a croc gate in the river where it flowed into a shaded marshy area. If I wanted to swim, then I was more-or-less safe if I stayed behind that gate and around the spring.

There were a lot of fallen leaves under one of the trees, so after checking it for snakes with a big stick, I set my tent up right on top of it, giving me extra cushioning and shady branches above. Perfect. It was still a couple of hours before the sun clocked off, but it was cooler now. The smell of eucalyptus wafted through the air as the faintest breeze jangled the leaves above. I unpacked my kitchen, placed my top-box on the ground for a stool, and began preparing dinner; a delectable dish of pasta with a tin of tuna mixed in and a triangle of Laughing Cow for decadence. Laughing Cow, the soft cheese that travels well, never melts, never goes off, and is widely available all over the world. You will find it in the furthest, hardest-to-reach, and hottest corners of the world. If there was a nuclear fall-out, the only things to survive would be cockroaches and foil-wrapped triangles of Laughing Cow cheese – probably!

As daylight gave way to a rosy dusk, I decided to call it a night. In the morning, I would go for a run around the shaded spring area, and then maybe bathe and refresh before hitting the road early. It was difficult to find exactly the right time to start a ride here. If you left too early, you risked being taken out by a kamikaze kangaroo who was still grazing in the cool early light of day. They always seemed to run alongside, then turn sharply, and dive in front of me. The threat was real, as they would have you off your bike and swallowed up by the waiting desert, given half a chance. Another carcass for the body count. I smiled as I realised I'd made it onto the next level of the platform game. Now, instead of rickshaws and psycho holy cows, I had kamikaze kangaroos. If I left it too late though, the sun would suck me dry of energy before I'd even begun.

Lazily putting the 'kitchen' in a bag to wash in the morning, I got into my tent and zipped her up for the night, as ever with the forced hope of an eternal optimist that I wouldn't need to open it again until morning. I knew I was lying to myself. Peeing in the night was a pain in the neck, and the more I tried not to let the urge win, the more it invariably did. I placed my head torch and Leatherman in their usual spot of the little inner tent pocket next to my head, fluffed up the clothing that I used as my pillow, checked the air pressure on my wafer-thin blow-up roll matt, and slipped into my three-seasons North Face sleeping bag.

As I lay comfortably in my little green nylon bubble, I allowed my mind to wander over the last few months. I was on the other side of the world, for crying out loud. I'd put 20,000 miles on Rhonda, over two continents and was now onto my third. I'd been living on the road for eight months and it would soon be Christmas. I'd always wondered what Christmas in Oz was like. Did they decorate with fake snow and reindeer?

I'd left Darwin only a couple of days earlier, and now I was heading steadily south and east across the outback some two thousand miles via Mount Isa and Longreach towards the Sunshine Coast and Brisbane, where I would meet up in a couple of weeks with a guy named Shane. A Kiwi living in Australia, and fellow overland biker, we'd met online via a web-based community called Horizons Unlimited. We were travelling in opposite directions and should have crossed paths in Bangkok. He had been a few days away, though and I'd grown impatient. I hit the road and headed north for the Golden Triangle instead of waiting for him, and we missed each other. We were still in touch, and now he was back in Australia to arrange his visa for Pakistan (you have to get these in your home country), having left his bike in India and flown home in time for Christmas. It looked as if we were going to get a second chance to meet in person. We were then going to meet up with Jeremy, another overland biker, from Canada. We had shared much route info with each other, and he seemed nice enough. Shane though, seemed nicer!

As I lay there wondering what would happen when you put three solo riders together and whether there was any chance we might spontaneously combust on contact, I heard a rustle of leaves right outside the tent.

"What the…" I paused and held my breath. *What's that? A mouse?*

More likely a snake out here. I was zipped in. It was fine. *Don't stress.* I allowed myself to breathe out just as the creature moved again. *Shit!* It sounded bigger than a mouse or a snake. *What else could be out there in this part of the world? A dingo maybe? Shit. Are dingoes dangerous? Damn it. Why didn't I do more research? Think woman, could it be a croc from the river? Don't be ridiculous. Relax. Keep still. It'll go away. Count and relax.* By the time I got to six, there was another shift. This time, it moved the side of the tent. I grabbed my torch and Leatherman out of the pocket, put the torch on my head, and opened up the Leatherman. I could now see an indent in the tent. Whatever was out there was now lying against it. *Had I stolen its bed? A dingo. It must be a dingo.* I couldn't think of anything else at that moment. Fixed on the image of a small dog-like creature with big fangs snoozing *right* there. I just laid there, too scared to move, too scared to go and check it out, just waiting for whatever it was to make its next move. It didn't, and so eventually I drifted off, Leatherman in hand and the light from my torch casting a psychotic shadow on the ceiling – blade and grasping hands silhouetted beautifully as I drifted into sweet oblivion. I woke briefly at one point and remember turning over, knocking the 'thing' with my elbow and saying out loud, "Oh, sorry mate" before drifting back to sleep again, too sleepy to be fully aware of the ridiculousness of this statement.

Morning came, and with the new light came the confidence to investigate. The warm indent was gone. Whatever it was must have buggered off as the sun came up. I guessed I'd never know. As I crawled bleary-eyed out of my tent, I heard something move behind me. I jumped up and spun around only to find myself staring straight into the eyes of a swamp wallaby.

"Oh jeez. You little shit! You frightened the life out of me!" I laughed with utter relief and then harder with the realisation that this was my little warm indent. A cute little wallaby, knee high and almost fluffy enough to cuddle. He didn't look impressed though. "Sorry, mate. Did I steal your bed?" I asked and crouched down offering a hand of friendship. I expected him to run, but he hopped forward to sniff my hand. "Hello mate. You're a cute little thing, aren't you?" Just then he snorted, rose up to full height, and kicked me in the shin before hopping off in a huff. Okay, apology *not* accepted then, I take it.

A few days later, I found myself on the coast and couch-surfing my way south towards Shane in Brisbane. I got drunk with many an Aussie on my way down, never failing to keep up with their beer drinking; seeing it as a matter of Welsh pride.

At one such overnighter, I stumbled upon a religion I had never encountered before. I'd stayed with followers of many creeds along the way but this particular night, I found myself staying with worshippers of The Flying Spaghetti Monster, otherwise known as Pastafarians. According to adherents, Pastafarianism is "a real, legitimate religion, as much as any other." Although "the only dogma allowed in The Church of the Flying Spaghetti Monster is the rejection of dogma," some general beliefs are held by Pastafarians. For example, they believe that the universe was created by the FSM while very drunk, the effects of which can be seen in the resulting imperfections and contradictions in the universe. Pastafarianism was founded in 2005 when a physics student sent a letter to a Kansas school board satirically critiquing the theory of intelligent design by citing "evidence that a Flying Spaghetti Monster created the universe." The joke grew and for some reason really took off in Australia and even more so in New Zealand, where ordained Pastafarian Ministers can actually legally marry people. While staying with my two new twenty-something-year-old friends, I discovered that they also occasionally wear colanders on their heads and always drink lots of Captain Morgan's rum. I also discovered drive-through off-licences for the first time in Australia. Coincidence? I think not.

Shane took a day's ride to come out and meet me, rocking up in stained, cut-off jeans, steel toe-capped boots, and riding a big old ugly-as-hell Kawasaki VN800 cruiser.

"G'day! Sorry about the bike," he said as he cut the engine and dismounted. He must have read my mind, "It's my step-dad's. It will have to do with Donkey still being in India". 'Donkey' was his beloved Suzuki DR650 trail bike. He looked a little embarrassed. Slightly awkward. Shy. Nowhere near as confident as he had come across over text or Skype.

Our plan was to meet up with Jeremy the next day, and ride together for a few days before Jeremy continued north. Shane and I would go back to his parents' and spend Christmas there. But first, we found a cheap hotel on the outskirts of Brisbane and spent the night getting to know each other a little better. We had been flirting online for months, and we both knew that this night was not going to be all about talking. We had done enough of that.

"How was last night?" asked Jeremy with a wry smile as we sat around a bar table in the wonderful town of Nimbin the next day; a colourful little town where the bus shelters were painted with rainbows and the locals all had long grey hair and beards to match. A community of hippies who had moved there in the seventies.

"Best night of my life" replied Shane. Jeremy and I took one look at each other and erupted into fits of laughter. "No, no, I didn't mean that... I meant..." But we weren't listening, and his explanation got lost in our hysteria. We were never going to let him live this one down. It's probably worth mentioning that we were quite nicely stoned at this stage. I'd say just marginally to the left. Nimbin is famous for it. Everyone there grows their own food, they have community craft fairs, and sell weed in the local shop. It's not (or wasn't at the time) strictly legal, but the local police leave them all to it as they're clearly not bothering anyone. Quite the opposite, in fact. If only every town were so community-minded, happy, and industrious. They even have a 'Mardi-GRASS' every year. A whole festival devoted to the marijuana leaf.

Over the next couple of days, the three of us had a great time riding around and checking out new places and nice roads. No one spontaneously combusted, but it did take us a while to decide who was going to lead. Being independent riders who each like to be in control, you would probably expect that we all fought for the job. Far from it. I think we all wanted someone else to be in charge for a while. It made a refreshing change not to have to make all the decisions. One of the best things about riding solo is that you get to make all your own decisions. No compromising. One of the worst things is that you *have* to make all your own decisions. No matter how tired or fed up or scared you are. This was a holiday for all of us and we enjoyed the break with like-minded people who understood without having to explain. People who knew why you were doing it and so didn't need to ask the question; whom we could relentlessly rib around a bar table without fear of offending. A bond was formed between the three of us in a very short space of time.

Jeremy moved further south on his own mission to conquer the world. I would see him again a year or so later in Canada. Shane and I spent some time at his mum's on the Sunshine Coast, and Christmas was, of course, spent mostly on the beach.

It seemed Shane and I were becoming a bit of a thing. That wasn't meant to happen. This was fun. No emotion. That's what we agreed. But we had ignored the fact that a bond developed way before we ever met in person. We had already shared our deepest fears. We had laughed and cried together from separate filthy hotel rooms in different parts of the world. We had even been taken ill at the same time and offered a virtual hug and mop of the brow when the places we found ourselves in offered no solace, no comfort. We were the same. We understood. Either way, I knew I was going to have to leave him. I knew my narcissistic journey had to continue to its eventual conclusion, and I knew that conclusion would only be recognised alone, as I had started. This is something he would later resent in me, but no more than I resented it in myself. I decided to leave earlier than I needed to rather than face any deeper dilemma.

I spent New Year's Day getting lost in a web of tangled trails in a labyrinth of trees somewhere en route to Sydney. The map just showed a mass of green and a load of lines that, to me, spelt good times. Just what I needed. I blasted along the shaded narrow dirt tracks and over streams, completely regardless of direction. I took no points of reference in case I got lost. One tree looked much like the next, anyway. At one point, I dropped the bike with a rookie mistake and had to remove all my luggage to get her upright again. Ten minutes later I did it again!

"Arghhhhhhhhh. Stupid cow. What are you doing? Get a grip!"

I squeezed my leg out from under the bike and was grateful to find I was not stuck. Close call. Then I took my jacket off and threw it on the ground. I looked around. I had no idea where I was. I had been in there some time though, and had seen neither man nor beast. I was thirsty and sweating and a little bit agitated. What was the point of this 'mission' anyway? I was confused at leaving Shane. He had rocked the boat with the power of his easy, lopsided smile. I took my top-box off and grabbed the water bottle out of it. Then, I sat down on my jacket and lay back on the bank looking up at the flickering light through the trees. The green was so pale and delicate against the blue sky. Rhonda lay next to me where she had fallen.

Was I right to move on in this blinkered single-minded fashion? For me, this whole thing had begun with the notion of freedom. The definition of freedom, generally, is having an ability to act or change without constraint. Was I now not bound by my own self-imposed rules? I stared at the leaves dancing with the light and allowed my eyes to become unfocused; dreamy. This canopy collage combined with the smell of earth and bark always brought me to a calm place. My favourite place. I was pushing so hard in search of freedom that I never stopped to ask, what happens when I get it? Was this it? Was this where I got off? Maybe I should change direction and go wandering with Shane. At what point do I actually stop pushing? I wasn't sure if I would recognise the signs.

I wasn't sure what was important now. But I knew I wanted freedom with a purpose, a project, a vision to pursue. It's great, at first, to wake up and do whatever you want; go wherever you want. If you don't like it, move on. Romantic, like the notion of riding off into the sunset. After a while, though, this kind of freedom can itself feel stifling. No anchor. No meaning. No challenge. How can you have reward if you have no challenge? Nope, I did have a choice, and I chose the mission. It was measurable; defined. Anchored, yet transient. I had chosen those measures. After that? Who knew?

"Maybe in my next chapter I can have the little house in the hills with the dogs, chickens and three-legged donkey," I thought, as I got up and put my jacket on again. "My search is over. Right now, though, I'd best search for an exit before it gets dark."

It took some getting out of that colossal forest. I wondered if I had got myself lost deliberately. Probably. I often did strange things when my mind was preoccupied. It sounds self-defeating, but it helped to clear my head, and the trails were delicious for my little dirt bike, which devoured them with ease. The little Honda named Rhonda that apparently had no chance of making it around the world.

I arrived in Sydney a couple of days later, just in time for my visit to the Morning Sunrise TV studios. Remember them? The Aussies for whom I had promised 'to do handstands' upon my arrival in Sydney, when speaking to them from London, just before my departure. Unfortunately, my airtime was cancelled at the last minute due to a terrorist attack in France. With it went my opportunity to win another £50 bet with Pete (my unlikely word this time being 'Techno Tractor'. I'd already planned my sentence too). Instead, I set about organising Rhonda's shipment to Buenos Aires, in Argentina. On my way to the airport the following morning, I got my first flat tyre of the trip. A great big rusty nail in my back tyre that saw me snaking down the road, but fortunately coming to a stop still upright. I had to get Rhonda to her flight on time, damn it. I'd have to work fast to get her fixed. I had all my tools and a spare inner tube, so before long I had pushed her to the garage a few metres away and got under way as quickly as possible. It was around 7 am, and I would've killed for a coffee. A few minutes later the garage attendant came with exactly that. No charge. He'd read my mind.

We were done in no time, and Rhonda made her flight. Onwards to Continent Number Four, South America, and then of course the big one – Antarctica…but not yet! When I went to book my own flight, I discovered it was the same price to go via New Zealand, and so, on a whim, that is exactly what I did. Shane was already over there visiting his Dad by now and I thought, "What the hell. One last time." So I flew to Auckland and spent a week roaming around the North Island in a borrowed camper van with Shane and a bag of weed while Rhonda waited patiently at Buenos Aires airport. We parked up in various spots, shared our playlists, enjoyed the view, and fed scraps to the kookaburras who made us paranoid with their laugh-like chatter.

I left feeling I hadn't quite finished with Oz *or* New Zealand. Those seven weeks had felt rushed. While Oz wasn't my favourite place, I marked it on my mental 'to do' list as a place that needed further investigation, and New Zealand needed a *lot* more time. Antarctica was my only priority right now though. *That* would surely feel like that first unforgettable hit of the wanderlust drug.

Chapter 14
The Landing

I found the 'Ice Bird' moored-up at the docks in Ushuaia, nestled amongst maybe thirty other vessels of similar ilk. They all looked the same to me. Floaty things with big poles and people wandering around them carrying vegetables for the hold or sitting on deck putting knots in ropes for no apparent reason. These people are generally known as 'grotty yachties'; a breed I had only briefly encountered once before, in Turkey, when I had stayed with Fuat.

Getting Rhonda out of the airport in Buenos Aires had been no big drama, and I'd quickly found myself on the most direct route (Ruta 3) on the east coast of Argentina to Ushuaia.

The famous Patagonian wind had grown stronger as I headed further south towards Comadoro Rivadavia, 1,700 kilometres from BA. Entering the area settled by Welsh emigrants in Victorian times, en route, I was interested to see if I could find any Welsh speakers in the little town of Trelew, or any red dragons to make me feel at home. The Welsh settlers first arrived in Patagonia in 1865. They had migrated to protect their native Welsh culture and language, which they considered to be threatened in their homeland. I'm sad to say I couldn't find any Welsh speakers, though many of the road signs were written *yn Gymraeg* (in Welsh). I did also find lots of dragons (painted murals), learnt a little of their history, and stayed in a place called *Ffem Taid* (Grandad's Farm).

Other than this little gem of Welsh history, this was the most boring stretch of road yet, and I found it extremely difficult to keep myself entertained, though it wasn't without drama. The concentration and physical demands needed to fight the constant side-winds made it possibly the least enjoyable part of my journey so far. The town had nothing to offer either, aside from overpriced hotels and annoying one-way streets. I stayed a full day because I needed to rest before tackling more of the same on the way to Puerto San Julian, 400 kilometres further south.

I found a hostel where the owner spoke English. That was enough for me, I took the windowless 'cell' on offer and parked Rhonda in an indoor car park two blocks away. The next day, the owner told me the wind was going to be bad and that I should not attempt the road ahead as it took out cars and trucks regularly in those conditions.

"You will have no chance on a bike" he said firmly.

Now you can say I was foolish to ignore local advice, but if I'd taken everyone's advice, I would not have made it that far in the first place.

"Oh no Madam, you cannot take that road. Men with knives will get you."

"Oh no Madam, you must not go there on your own. You will be raped."

"Oh no..."

You get the picture. I decided to go for it the next morning.

"Think positively," I told myself as I left the hotel, but before I even left the shelter of the buildings, I was fighting the wind; and the wind was pretty damn determined that day.

I checked my map and saw the next town was just 70 kilometres away. If I couldn't handle it, I could at least crawl there and then push on to San Julian the next day.

As the buildings ended, it was a constant struggle to keep the front wheel moving forward. (Looking back, I believe part of my problem was the location of my tent on the bike – which I later moved. I had it strapped to the rack in front of the handlebar-mounted screen and I think it was acting almost as a sail!). It felt as if the front wheel was being swept from underneath us. I kept to a slow speed and moved on. This was not good. I wondered how long it would take to do the full 70k at this rate. *Not to worry; knuckle down and think positive.*

As I reached the headland and rounded the corner, the cliffs that protected me on one side ended, and an almighty gust of wind took hold of me. I could not hold Rhonda, and we were thrown into the path of an oncoming lorry, and slid to a stop, laying on our sides. Fortunately, the truck stopped.

I jumped to my feet and started dealing with the situation in hand. Check for further hazards (traffic), switch engine off, and see who the hell is going to give me a hand to lift this beast off the ground. Amazingly, the first two cars slowed, gawped, then continued on their way; the third stopped and its driver ran over to help. By this time, the lorry driver was down from his cab and, after checking I was okay, at my side ready to drag Rhonda back to the upright position. I was fine, and the three of us got Rhonda over to the side of the road.

The driver of the car got on his way. The lorry driver went back to his cab but did not leave. He refused to move until he saw I was able to ride again. This was when the fun started.

The wind was coming in gusts, and at times so strong that I could barely stand. I knew I had to get back behind the shelter of the cliffs, but I couldn't get my leg over the bike, let alone hold her up. I lifted the sidestand in an attempt to face her in the right direction, but nearly lost her again, managing to get it back down just in time before we both went over. I clung on to her for support as I tried to figure out what the hell I was going to do next. The truck driver came back, and between us we got her facing in the right direction. He held her in place as I got on. After a few attempts, I was moving again, albeit slowly and unsteadily. It was just a couple of hundred yards to the cliffs; I could do it. And I did!

I parked up again and assessed the damage. Aside from a broken gear lever (I had a spare for just such an occasion) and a clutch lever that was now pressed against the handguard, all appeared to be fine. The truck driver sailed past, waving, and I was left sitting on the side of the road. That's when I started to laugh. It wasn't hysteria; it was probably relief. I giggled as I wondered what I would do next. *No rush; just relax for a minute.* I got my phone out to check for a signal. Present. My neck and back were hurting by now, but I could easily ride. I was stuck between a rock and a hard place. I didn't want to face the road back into town, and there was no way I was going to poke my head out from behind the relative shelter of the cliffs in the other direction, either.

What were the chances of hitching a lift with a motorbike? I giggled again as I pictured films where the good-looking girl puts her thumb and leg out as the guy hid in the bushes until someone stopped. Perhaps I could hide and Rhonda could be my lure? I decided to just try waving down vans instead. It was worth a try. I had nothing much to lose.

Miraculously, it worked. A man in a white van stopped, and between us, we managed to get Rhonda into the back. He spoke a little English and was happy to help a biker in distress. It probably helped that I offered to pay him.

We strapped her down as best we could, but it wasn't enough. The van was blown all over the place, and as the driver fought to keep her in a straight line, we heard an almighty crash and knew without looking that Rhonda had fallen over. We stopped and managed to get her back up again, as the driver constantly apologised. It wasn't his fault, but he clearly felt responsible. The right-hand mirror was hanging off now, but there was no serious damage. I felt I got off lightly after the near-miss with the truck, so I was still smiling – much to the driver's surprise.

After what felt like the longest 3,400 kilometres of my life along a mixture of deep gravel, scarily windswept, and just downright boring roads, I had made it to Ushuaia. The end of the road. The end of the world, for most. But for Rhonda and me, this was just the beginning of a whole *new* adventure.

It was the first week in February, the air was cool, and I was early. Cathy, the skipper, had invited me to stay on board and help prepare the sixty-foot sailing yacht for our voyage across the Drake Passage in a few days' time. I had no idea what that entailed, but I was happy to be the general dogsbody as required, if it meant saving money on hotel rooms. It would also give me time to familiarise myself with my new environment before the rest of the crew arrived. I was about to step into a whole new world and was determined to adapt quickly; even get ahead of the game if I could.

The air was quiet and still, with the odd tinkling of cables against masts, and my feet tapping out news of my arrival on the well-trodden wooden slats of the jetty. As I approached the boat, a blonde head popped out of the pilot house and turned to greet me with a smile, "Hey, you must be Steph. I'm Cathy."

She had a strong Kiwi accent; slim, mid-fifties, I'd guess; hair loosely tied back to frame a gently weathered face sculpted from a healthy life balance of hard work and pleasure. In her navy-blue quilted waistcoat, she oozed 'yachty' in the way that horse people ooze 'horsey'.

"Hi Cathy, nice to meet you, at last." I extended my hand with the aim of a confident approach; I pulled it off.

"Come aboard," she said and turned to head back inside, "I'll get us a G&T."

"My kinda woman," I replied cheerfully as I awkwardly navigated the gap between the jetty and the boat.

Cathy had two boats here at the dock. Newly separated after her husband was found to be having an affair, Cathy now ran one boat and her husband the other. She gave me the full run-down of these recent events as we sat around the small table in the sunlit pilot house. I nodded and looked appropriately appalled between sips, as she regaled me with the details of the split, which was clearly still raw to her. My face may have been sympathetic, but in truth, my mind was wandering and I was keen to change the subject.

"So, can I take a look around?" I asked, when she took a breath. "I just can't imagine what this will be like with nine of us on board."

"Sure," she replied. Her face and demeanour suddenly changing as she snapped out of her personal nightmare. "It will be cosy, but you'll get used to it." She smiled now – a genuine smile. "Make the most of the space before the rest of the crew arrives."

I spent my first full day on board cleaning and getting acquainted with the boat. Once everyone was on board, we would all eat and take our turns on watch in the pilot house; there was just enough room for all of us to sit around the table. Down a few steps, there was a narrow galley kitchen where we would take turns cooking, and a small office space with barometers, maps, and radio equipment. Off that, was a doorway that led into one of the bedrooms, and then a tiny bathroom with shower and toilet. On the other side of the kitchen, there was another door which to led to another, even smaller, bedroom. I couldn't imagine all of us in here at the same time, but I decided I might actually enjoy sleeping in the little bunks, with a lee cloth to stop us falling out and no headroom, if only for the novelty factor. There would clearly be no privacy once we set sail. These confined spaces alone would be a test, with few secrets left regarding one's personal habits. We'd have to get on, come hell or high water. Probably both.

The next to arrive was Ollie, another Brit; clearly a well-educated young man of twenty-nine who had already sailed around the world. He told me his next mission was to ride a motorcycle around the world, and then to fly around. He seemed a strong and dependable type, with his tall build and neat ginger hair, but perhaps a little green around the gills. I had no doubt he could do anything he set his mind to, and I got the impression he had made the most of what was probably a very privileged upbringing. Ollie had never sailed to Antarctica, though.

This was also the first time Cath had gone across without her husband – as a team. Like me, the rest of the crew was either new to the boat or completely new to sailing. I remembered the words of Fuat, in Turkey "Crossing the Drake Passage is, to sailors, what Everest is to climbers." There was a lot to teach us in a very short space of time. I hoped that I was capable of grasping it all, and that we were up to the job. This was going to be a true baptism of fire.

Cath expertly cooked dinner that night, and the three of us sat at the table discussing the plan for the bike. Cath and Ollie would figure out a way of getting her on board, while I figured out how I was going to protect her from the elements once we were at sea.

"We can use the boom and rig up a winch, then lower her down and put her on the hull until we get going," Ollie said confidently, "as long as she doesn't block the anchor hatch. We'll have to take some measurements, but I think it will work. She'll get fewer waves there than on the bow, at least."

"Great," I said. "Once she is in place, I'll prepare her as best I can for the onslaught of salt water. This is not the kind of thing I can Google. I don't think anyone has ever done this before." With a bit of common sense, I was fairly confident that we could keep her safe. "I think the important thing is that we can wash her down regularly with clean water," I concluded.

Cath replied,

"There is a water maker on board, so as soon as we get into calmer waters on the other side, we can do that." Ollie and Cath seemed to be taking

Rhonda's safety as seriously as me, and not just as an afterthought. That was reassuring.

One of my biggest concerns was that Cath was picking up a group of kayakers who were flying into Antarctica after our trip. She had to drop me off on King George Island, from where I would hopefully be able to fly back to mainland Chile with the other 'guests' while she continued on her mission for another twenty days *with* Rhonda. Meanwhile, I would have to find my way back to Ushuaia before being reunited with her mid-March. I wasn't keen on this one bit, but I would go along with whatever was required to make the trip happen.

"We'll work it out," Cath said reassuringly. "You can probably hitch a ride back on one of the ships." It wouldn't be the first time I'd hitched a lift with my motorbike, but this? This was going to be interesting, if nothing else.

Over the coming days, the rest of the team arrived and final preparations were under way. The hold was rammed with enough food for a month and perhaps enough alcohol for a year. Fresh vegetables, pasta, rice, spices – the works – got packed in one-by-one. Then the alcohol. Gin seemed to be the preferred beverage and outweighed anything else by far, but we also had wine and a few 24-can packs of beer.

Rhonda was unceremoniously winched on board with a complicated web of ropes and pullies. It was a tight fit, but we could just get the anchor hatch up and work around her. I then removed her battery, taped up the air filter, oiled the chain, sprayed her all over with WD40 before wrapping her with cling film, and duct-taped a bin bag to what now looked like her head. I almost felt guilty for this act, but it had to be done – and it would have to do. Time would tell whether it was enough to avoid mechanical problems later.

There were now nine of us aboard. Pete was an Aussie who would soon prove himself to be an accomplished sailor and friend to all. He had a suitably rugged beard and wore dungarees over a big belly, which all added to his dependable, seafaring profile. Ronnie, too, was a sailor of old from the 'States and wore a roll neck Arran jumper just like my grandad used to. They busied around helping Ollie get the rest of us 'ship-shape' with lessons on ropes, knots, and buoys. The rest we would learn as we went. The general chores were assigned in pairs via a rotation system; our cooking and cleaning partners would remain the same, but our watch duty buddies would regularly change, or we would do it alone.

For the cooking, I was paired up with Yvette, a softly-spoken, middle-aged vegetarian from Australia. Like me, she was a total novice to the sea, and not that confident in the kitchen either; not when it came to cooking in a tiny rocking galley for nine people anyway, and certainly not when it came to dealing with meat. We vowed to figure it out together when it was our turn. The remaining three were Lesley, who was Ollie's mum, and a young couple from Australia, Sally and Xavi. None of them had signed up for the additional challenge of landing a motorbike on Antarctica, but they were in. Rhonda only added to the buzz, and everyone was keen to see us make it.

We finally set sail from Ushuaia on a gloriously sunny Sunday afternoon. The hope was that we would get across in five days, but conditions would decide. The forecast threatened a storm near the tail end of our journey, but based on the weather patterns, this was probably the biggest opening we had for another five days. Once we were out past Cape Horn, we would be at the mercy of the Drake, so it was a tough call. Cath made the decision and we went for it.

The Beagle Channel offered us calm waters as we proceeded through the breath-taking scenery of Tierra Del Fuego, the 'Land of Fire', on the port side, and a series of small islands to starboard. The jagged mountains that rose up from its banks offered shelter as well as an opportunity for Ollie to continue his lessons on rigging and key-word communication, which he did from the stern with great authority. A school of dolphins played in our bow-wave as we got to grips with terms like 'aft' and 'leeward'.

"You really need to remember these terms, as they will get shouted at you a lot," Ollie said with a firmness that emphasised the seriousness of his words. "When the sea is rough, we need to be able to communicate and work quickly and effectively."

A few hours later, we moored up at Puerto Williams, a small Chilean village and naval base on the island of Navarino, just south of Tierra Del Fuego. It is actually the southern-most inhabited place on earth (although Ushuaia claims to be). This would be our one and only stop before we tackled the thousand-kilometre ride across the Drake Passage. If anyone wanted to jump ship, now was the time. Ronnie did just that. Ronnie and Cathy were already struggling to get along, and after a heated debate at the port, Ronnie left the expedition.

Cathy was proving to be quite a bossy character, perhaps more than her title required. Her monumental mood swings were already apparent, with one of her outbursts leading to a full-blown screaming match with her husband across the port. When in these moods, nothing we did was right, and it was clearly something we were going to have to learn to manage. Ronnie, too, had a habit of not listening and assuming he knew best. He had no time for moody women and she had no time for know-all Americans. It was inevitable that they would lock horns, and better now than later. There would be nowhere to run to once at sea.

This left us down one experienced sailor and with a skipper who might just be on the verge of a nervous breakdown. Time would tell. We were all strangers now and would have to settle down to our jobs quickly and learn to work as a team, regardless of personal problems or conflicting characters. Living as closely as this in an unforgiving environment would soon either build strong bonds or create enemies of even the most imperturbable.

The rest of us decided to jump ship as well, two boats down, to an ancient-looking wooden vessel named The Macelby – a floating bar sitting on an angle that seemed to get more acute, the more alcohol we drank. This was our last chance for a piss-up as we'd all have to remain sober until we reached Antarctica. It was also a great opportunity for a good old-fashioned team-building exercise. Our poison for the night was pisco sour, a drink I first became acquainted with in Peru back in 2013, when a taxi driver warned me never to drink more than three in one night. This became the first team challenge. The second was navigating our way across the decks and back onto the Ice Bird at 3am, having drunk at least four each. What did that taxi driver know, anyway?

The following morning, I was supposed to drop all my riding gear and luggage off with someone at the port. This was Cathy's plan to save space on the boat and keep it safe while we were at sea. I would have to come back this way to reunite with Rhonda, whatever happened. Cathy seemed

147

impatient to leave and suggested we just leave it where it was.

"Will it be safe in the hold?" I asked nervously.

"Yes, it will be perfectly dry and not in our way," she confirmed.

With that, after downing our Stugeron sea sickness tablets and making sure everything was tied down, we set sail for Cape Horn and the open waters of the Drake Passage. These rough seas could dislodge furniture, footsteps, and most likely our stomachs, from their moorings. The experience is often described as like being thrown into a washing machine. I made one final check of Rhonda's moorings before going below deck to check that the rest of my belongings were safely stored away.

"Good luck, old girl," I said quietly. "See you on the other side."

It wasn't long before rough waters started a tsunami of sea sickness. I felt annoyed to be the first to go down with it, and wondered if my Stugeron had even had time to dissolve. I was in the pilot house with Pete when it began. I grabbed my little 'sick jar' and threw up what smelt and looked just like last night's pisco sours.

"Ah!" said Pete, more matter-of-fact than surprised. "All part of the adventure," as he reached for the tissues that were tucked behind the wheel. He'd clearly seen it all before.

"Yup," I said, then retched and added a little more to the contents of the jar, "All part of the adventure."

Within a couple of hours, there were five of us in the pilot house throwing up into our sick jars. When one threw up, we all followed suit, each with our own personal retching sounds. We'd then put the lids on the jars and lie or lean back again. It reminded me of The Frog Chorus from Paul McCartney's song, "We All Stand Together". Some of us were worse than others; for the first twenty-four hours, I couldn't move at all without being sick. If I needed the loo, I had to run there, run back, throw up, and lie down again. As the boat rocked, our jars got thrown around and mixed up on the floor. My stomach lurched with the new wave and I grabbed the pot nearest to me.

"Oh, God!" I said between retches, "This isn't my jar!" The smell of your own vomit was bad enough, but someone else's? That was just wrong. This started a whole new wave of retching, but this time with giggling in between. Then Sally let out an almighty noise along with the contents of her stomach.

"How can such a noise come from such a dainty woman?" asked Ollie, laughing. We all laughed again and took the piss out of each other's sick noises as we continued in our chorus. Eventually the waters calmed and our stomachs settled, one by one.

It remained fairly calm over the next couple of days. We kept rotating on our three-hour watch shifts, making notes in the log every hour. Distance travelled since last reading, direction, speed, depth, weather conditions, and a space for noting any additional observations. For the most part, there was little to report, as all we saw was the ever-moving horizon, with the occasional treat of a soaring albatross. Three hours on, six hours off. Most of us used the time between shifts to sleep, as activities like reading or writing only made us feel sick again. Sleeping was easy and numbed the monotony.

The first clues that we were getting closer to land came four days into our voyage with sightings of petrels, brown skuas, and the odd seal head bobbing about in the waves. Then, in the last hundred miles, the promised storm hit us. It was impossible to do anything on board without getting thrown around.

"One hand for the boat and one hand for you," Cathy reminded us as we struggled to move around. Some jobs were harder than others with one hand. Carrying cups of tea was fun. Cooking on a pivoting hob was entertaining, but having a pee was often the most difficult. Especially the bit when you need to pull your pants back up with one hand whilst holding on for dear life with the other as the motion threatens to launch you into the not-so-far wall in front of you. With just sixty nautical miles to go, the waves reached uncomfortable levels, so we brought in the sail to avoid over-power from the gale.

Late that night, I was in the pilot house with Pete and Cathy. It was getting harder to see in the dark as we kept an eye out for hazards in the angry water. A thick fog had now descended too, and we were entering iceberg territory. Cathy seemed nervous and decided to go and wake Ollie for an extra pair of eyes. Between us, we watched and crept forward as best we could, all our senses on high alert, trying to make out whether it was a 'berg we could see ahead or just a little light coming through the clouds. Our eyes were playing tricks on us.

After three hours, I went off watch and squeezed into my bunk, pulling up the lee cloth behind me; but sleep was impossible. I was still on high alert, particularly as the boat was thrown around by the bigger waves crashing against us like concrete blocks. Drawers broke their bindings and flew open, the contents banging and clattering about as they took their opportunity to escape. I wondered how Rhonda was doing on the aft deck. She was on her own though; there was nothing I could do for her now. I braced myself as the swollen sea gave another angry punch. Each one felt as if we'd hit a solid object, and each had me wondering if we had found an iceberg. It was a long and lonely night that had me hoping that whichever way I went, it wasn't by 'Davey Jones' locker'. The sea could be so cold and hostile – not a place I wanted my life to end. Given the choice, I would take a bed of moss, a canopy of green and one last taste of heroin, any day. I would choose anywhere right now, rather than this. I closed my eyes and silently hoped I would get to see my fortieth birthday – on dry land.

Morning broke, along with the storm. The waters calmed again, and the sun came out just as we spotted the Melchiors. These low, ice-covered islands sparkled in the new light and lit the runway to our Antarctic adventure. We all came on deck and sat quietly, admiring the islands and the icebergs, as we slowly and silently navigated our way through them. The air was fresh, the water was calm, and we had just enough breeze to fill our sail. Wrapped up in our thick down jackets, our faces warmed by the sun, it all suddenly seemed worth it, and I hugged my knees as I tried to take it all in. It was still a few hours to the peninsula, but already we were being honoured by the presence of humpback whales and gentoo penguins that slid off the ice into the water nearby. It was the most beautiful desert I'd seen yet. We'd made it.

We arrived at our first anchor spot before noon. Paradise Harbor is a wide embayment that was first named by whalers operating in the vicinity in the 1920s. This day, it looked like the perfect spot for landing Rhonda; sheltered, no penguin colonies to disturb that we could see, and a nice stretch of rocky beach fringing around the ice. Ollie and Sally dropped the anchor, unpacked the Zodiac inflatable dinghy, and set off in search of 5,000-year-old ice for the G&Ts. Pete lit the barbie, and I set about untying the lamb we'd strapped to the aft before we left. 'Lambie' as we affectionately named him, had been given some frilly knickers to wear for the crossing. Feeling a little sordid, I reluctantly removed the offending item so I could set about my butchering. This of course, did not go unnoticed,

"Do you Welsh normally bother taking the knickers off first, then?" asked Cathy, as she came out of the pilot house with gin and a plate of sliced lemon in hand.

"That's rich, coming from a Kiwi," I laughed as I picked up the knife and stabbed it between Lambie's ribs.

Butchering was a skill I'd learnt during my time as a zookeeper. Local farmers would bring in stillborn calves and we would compete at who could butcher them the fastest. I held the record for a while and made short work of Lambie. His ribs were on the gas before the Zodiac was back with the ice.

The food and drink tasted all the better for our efforts, surrounded by the most spectacular scenery I had seen or was ever likely to see again. Our restaurant on the edge of the world.

"I'm in Antarctica!" I whisper, as I wipe the porthole next to my bunk the following morning. Not even in the earliest seconds of my waking day could I forget *this* time. I clamber down from my bunk before my consciousness has fully reached all parts of my body, and accidentally kick Yvette in the head on the bunk below. "Shit, sorry," I whisper. Yvette grumbles, turns over, and begins snoring again. No movement from Sally and Xavi in the bed opposite. Okay, good. It is the morning of the 14th February 2015, and still early, but I am already ridiculously excited. Today is more than just Valentine's Day. Today is V-Day. Today is the day I realise a goal that has for so long seemed like one of my B-movie ideas; low budget and completely unrealistic. But it is happening right now. Today, I will land Rhonda on Antarctica.

I grab my towel and slip quietly into the bathroom. I will treat myself to a shower today. Sod the grotty yachty rules of one shower a week. It has been at least five days, anyway, and today I want to feel fresh for my big day. I catch my reflection in the mirror as I hang up the towel. The reflection is smiling. I smile back and whisper,

"Not so far-fetched now, is it?"

After my shower, I go into the galley and make brews for everyone as loudly as I can. Cups clatter, teaspoons clink, and doors bang; my wake-up call to the crew.

"Tea's up," I shout before taking my own brew on deck to drink it in the morning sun. It is a glorious morning. Perfect. Not a breath of wind or a cloud in the sky. Rhonda is still strapped to the aft covered in cling film and duct tape, just as I had left her. "Mornin', girl," I say as I sit next to her. "Today's the day. I hope you're going to start for me."

Lesley appears on deck for her morning smoke. Brew in one hand, and a fresh pack of Marlboro Reds in the other. She always starts the day with a fresh pack.

"Want one?" she asks, as she sits next to me offering the smokes.

"No, thanks. Save it for the celebrations afterwards," I reply. I know she only has a limited supply for the journey, and I hadn't brought any for myself. I have a feeling I might want one later, though.

I set about unwrapping Rhonda and washing her down with fresh water. I'm pleased to see that the WD40 and wrapping has kept her rust-free and fairly dry, considering what she's just been through. I fetch the battery from below deck and connect her up just as Sally and Yvette appear on deck.

"The moment of truth," I say with a smile. Sally rummages in the deep pockets of her jacket.

"Hang on, let me film this," she says and pulls out her GoPro. I press the button and Rhonda starts on the first turn. Not even the tiniest hesitation.

"Woo-hooooo!" everyone cries all at once.

"That's my girl," I laugh. Rhonda has crossed the Drake Passage and not even missed a beat. Now all we have to do is get her off the boat and balance her in the tiny Zodiac across the deathly cold water to shore, and we are home and dry.

My stomach lurches at the thought. It's a particularly small dinghy, or 'tender', designed for six people, and a long way down to it in the first place. We will need five people in the boat to balance her for the half mile to the shore. If we get it wrong, Rhonda will not survive a dip in this icy cold sea. There are not enough 'dry suits' for all of us, so if one of us falls in, he or she will only have minutes before their body starts shutting down and the situation will become life threatening very quickly. If we make it, then what? How will we get her ashore? I guess the goddess of brute force and ignorance will be showing her face today.

And so, it begins. The team is ready to go and there is nothing else to do. As with the loading, we use the boom as a crane and winch her up and over the handrails before slowly lowering her down into the open arms of Ollie, who gently guides her into position in the centre of the Zodiac. It all looks so precarious, but so far, so good. My confidence grows a little as the first part is completed.

"Okay, come on down," Ollie calls, as he releases the rope and sits in position between the Ice Bird and Rhonda. "Just climb around me, slowly." He is holding Rhonda up now, and if one of us causes too much imbalance, he'll probably be over backwards and in the water with Rhonda on top of him. He doesn't look phased one bit, though.

"This is ridiculous," I whisper, as one-by-one we climb down to join Ollie. First Pete, then Sally, then Cath. I grab the GoPro I'd set up to witness this whole crazy day, and climb down to join them. Cath takes control of the outboard motor, while the rest of us hold the bike.

The crossing goes without a hitch. It's difficult to see the ice chunks, as Rhonda is obscuring Cath's view, but she navigates through them well with a little guidance from Sally and Pete, up front. The beauty of this place hits me once again,

"Wow, look at that one," I say, pointing towards one of the icebergs. It looks like an electric blue diamond, shaped like a work of art. These extra-blue 'bergs form after the ice above the water melts, causing the 'berg to overturn so that the water-smoothed portion formerly below the surface now is above. The air bubbles have been squeezed out, so on a sunny day like today, the ice refracts the light, colouring it in spectacular shades of blue. It seems unreal. Nothing could possibly be this beautiful in real life. I laugh out loud with sheer delight. Here I am, living my dream, and yet it's *better* than my dream. I could never have imagined this.

We make it ashore easily, and with brute force, lift Rhonda out of the boat. Touchdown! The Honda has landed! I throw my arms in the air and unashamedly whoop. There is nothing left to do. Then, I just stand there silently; a smile on my face. I just want to savour this moment. Soak it all in. Remember every detail.

"Come on then," laughs Cathy, "Are you going to ride this thing, or not?"

"Damn right, I am," I laugh, and grab my helmet.

I turn the key, and away we go, riding steadily along the sand. As I reach the headland and prepare to take the corner, I stop dead. The ice cave in front of me is amazing.

"GUYS!" I shout, "You've got to come and see this." They are already on my tail.

Before long, I ride off to find a quiet corner of my own, while the others do their own things; taking photos, checking out the rocks, or writing their names in the snow. Dismounting, I pat my old friend on the tank and say quietly, "Thanks mate. We did it." Then let out a louder, "Ha! Check us out," as a feeling of pride washes over me.

I find a suitably-sized rock on which to sit and give this time to sink in. Soon, memories of the last few years begin trickling through to the forefront of my mind, bringing with them a boatload of emotions: the effort; the self-doubt; the joy; the fear; the friendships; the regrets; the gratitude; the ups and downs and every little thing before, during, and in between. Then the floodgates open, and tears flow freely down my flushed cheeks. I take in a deep breath of sea air. The sigh that comes back out is unmistakably that of a happy soul; an enlightened soul; free from past or future. I look around, and in the serenity of this glistening desert, I am overwhelmed by a peace that I have never before encountered. "We did it," I repeat, as the tears reach my smiling lips. "We did it."

Everything and anything seems possible, now. I feel a weight lifting from my shoulders that I hadn't quite realised I'd been carrying. The memories evaporate. I am truly 'in the now', and life feels as clear and as beautiful as that electric blue iceberg.

As I sit on the rock next to Rhonda breathing it all in, I notice a curious penguin has come ashore and is checking us out from a few feet away,

"Hello," I say quietly, "Nice to make your acquaintance."

We both sit there for some time, just observing each other. Sharing a tiny space on the planet and a peaceful moment together, before he turns and waddles back into the water and away. I smile, as I imagine him running home and telling tales of strange red dragons on the shores. Of course, he won't. He's a penguin. Like most animals, he is naturally unburdened by events of the past, or the future. I am of no consequence to his 'now'. For this brief moment, and for the first time in my life, I know just how he feels.

Chapter 15
Smoke Me a Kipper

The peace I felt on that rock stayed with me into the next day as we pulled up the anchor and set sail for Waterboat Point. Something had changed in me. It was as palpable as the cold air that tingled on my face as we sailed away from Paradise Harbor. I hoped the feeling didn't melt, once back in warmer climes, but deep down I knew this had been an experience that would last a lifetime.

Cathy's voice pierced my thoughts as I wound up the last of the anchor line. "I just managed to get hold of the base on the radio. They've invited us to come ashore." She was referring to González Videla, a Chilean Naval base. It sits at the historic site of Waterboat Point, which has been used for research since the British Imperial Antarctic Expedition of 1920–1922 when two stalwart men, Thomas Bagshawe and Maxime Lester, stayed an entire year there doing scientific research and living in the abandoned whaling boat after which they named the spot. Coincidentally, the legendary explorer Ernest Shackleton died of a heart attack on South Georgia in January 1922, at the very start of his final Antarctic expedition, just as Bagshawe and Lester's ordeal was ending.

Cathy explained that today it was home to thirteen men and a large colony of gentoo penguins.

"Sounds like fun," I said. "Shall we take them some beers?"

"We can do better than that," she said. "Let's bake them a cake."

I wasn't sure which they would prefer, but as they were nearing the end of their six-month shift, I imagined their supplies would be low and they'd be grateful for anything.

The overwhelming stench of penguin shit was the first thing to hit us as we were helped off the dinghy by the waiting Chilean officer. The island was crowded with gentoos. Several pale-faced sheathbills, or 'shit birds' as they are affectionately known, scurried around amongst them in the mud, scavenging for eggs and fresh faeces to eat. Not an aesthetically pleasing bird, they reminded me of a dirty white gutter-foraging pigeon. Their movements, too, were that of a skittish character up to no good.

The penguins themselves were notably unconcerned as we wandered amongst them. They had no fear of us and seemed more interested in a large brown skua circling above. The officer came and stood next to me, then pointed over at a ghostly white lone bird standing on an outcrop of weathered rocks nearby. At first, I couldn't make out what I was seeing.

"What is that?" I asked, unsure as to whether he would understand my English.

"Albino," he replied. I was looking at a rare albino gentoo penguin, and this realisation was nothing short of thrilling.

"Wow, it's beautiful," I said, as the officer smiled the smile of a proud father.

"Señor Bustos," said the officer next to me, offering his hand. I shook it and replied, "Pleased to meet you. I'm Steph. Just Steph."

"Welcome, Just Steph," he replied with an expression that gave away no signs of intended humour. *Was that a joke or a language thing?* I was left wondering as he led us through the throng of penguins towards the base, and past the sign that told us that Cartagena, Columbia was 9,189 kilometres in one direction, and the base was a hundred metres in the other.

From the outside, the base looked like a cosy farmhouse nestled amongst the rocks. The inside consisted of a lounge, a well-equipped gym, and a well-stocked bar with its very own karaoke machine. *Note to self: must get on that later!* After the grand tour, we were shown into the dining area where they had prepared a lunch for us all. They had even baked *us* a cake.

"I knew we should have gone for the beer," I giggled under my breath to Pete as they brought it in and added it to the table which was already piled with enough food to feed a small country. On the other side of me, was a naval officer who spoke a little English.

"So, what do you do here?" I asked, genuinely intrigued. I managed to decipher that they were made up of navy and air force personnel. Their job was to conduct scientific research and help with any rescue situations.

"You like Wi-Fi?" he asked, now probably tired of trying to find words I would understand.

"You have Wi-Fi as well?" I asked. Was he pulling my leg? Out came the password. That made me laugh. *Why have a password? Who was going to steal their Wi-Fi out here? Damn those pesky porn-watching penguins!*

After such a great welcome, we invited a few of the men on board that night to join us for beers. Of course, they jumped at the chance for a change of scene. A night out at last! Before we could say 'Club Ice Bird', we were squeezing more bodies into that pilot house than I ever thought humanly possible. The alcohol flowed, and the drinking games began. I did not always understand the rules, but I did seem to be good at them; winning not only one Chilean Navy officer's name tag, but two; not to mention a Chilean flag. All of which were unpicked with a knife by me whilst still being worn by said officers! I just hoped I didn't get stopped on my way back into Chile and have to explain my souvenirs. By the time the 'South Pole dancing' started on the mast, our skipper decided we'd all had enough fun for one night and sent the rowdy sailors back on the Zodiac with Ollie.

As dawn broke, the sun cracked my eyes open with no apology. *Why did these portholes not have curtains?* I had a monumental hangover. Coffee was ineffective, and by 8.30am I was in desperate need of a distraction. I found it in the form of a kayak. It was another glorious morning, and after climbing into our dry suits, Cath and I took off for a spin. Within ten minutes of entering the water, we were graced with the presence a couple of humpback whales not far from us. Now, *this* was a hangover cure! We watched in silent awe as they continued on their path, apparently oblivious to our presence.

Then the icebergs caught our attention again. There was a field of them; all shapes and sizes, but mostly quite small on the surface by comparison to some that we had seen. The light reflecting off these magical ice sculptures created blues and greens that were out of this world. Their combined shapes and curves went beyond beauty. Beyond words. Beyond reality as I knew it.

We got as near as we could without the threat of getting caught up in an overturn and just sat there, floating in a daydream – silenced by the world at its most alluring. The only sound was that of the water gently lapping against our kayaks. It was dreamy all right; it seduced the senses, and I had to fight the urge to draw closer. It was almost calling to me, inviting me to take my chances for a momentary taste of Utopia; hypnotising me. I'd been hypnotised by promises of Utopia before and fallen for this trap many times, but not even my wildest dreams could have conjured up this vision before me. Resisting temptation is not my forte, yet somehow I avoided being drawn too dangerously close to these bewitching 'bergs.

The thing I loved about sailing was that there was no rush and no noise. We travelled through calm waters at a modest four miles an hour. It forced us to stop and 'smell the flowers'. Not that there were any flowers to smell, of course. Antarctica was mostly devoid of colour, like a perfect monochrome photograph. We examined every 'berg, enjoyed every whale encounter, and never got tired of the penguins, stopping at almost every colony we saw. In time, we would put our cameras down and just enjoy; appreciate. Rhonda was now sitting on the bow, and often I would sit on her as we rode the waves together – like a child on her rocking horse.

I didn't need a motorbike to enjoy Antarctica, but there was something special about having my travelling companion with me as we experienced this continent together. She had come to be more than a machine. She was almost an extension of me now, part of my identity. Together, we were Steph and Rhonda. In truth, she was probably my most successful relationship to date.

As we came close to the Ukrainian science base of Vernadsky, we were intercepted by three guys in a large Zodiac. They'd spotted us approaching, and came over to greet us. After a very short exchange of pleasantries, they mentioned the elephant in the room: "You have a *motorbike* on board. We love motorbikes. Please bring it ashore."

Cathy agreed immediately. I already felt we'd pushed our luck and was not keen to go again. Rhonda had had her moment, but it seemed I had no choice but to go with it. The Ukrainians were excited, and who could blame them? It's not often you see a motorcycle sailing through the icebergs, and it seemed everyone wanted to meet her. It made a change from counting penguins, I guess.

The landing went well the following morning, and this time I wasn't so nervous; although I don't think I would ever get totally comfortable watching my bike being dangled over icy salt water and lowered into a small Zodiac. Fortunately, by now we were well rehearsed. We took her around to the boat ramp where the strapping scientists were waiting with a crane. They carefully winched Rhonda out and placed her gently on the jetty. Wasting no time, I jumped on and rode up the wooden walkway and around the side of the building. Here, we had to get her through a narrow passage between the deep snow and the building. It was a very narrow, icy walkway, though, with a little engine and a lot of brute force between us, we got her through. On the other side was a pile of rocks, which was easy to ride over, then more snow and a frozen stream, which we navigated with no trouble. The tyres weren't gripping at all, so we just pushed and shoved her up the hill to a point where I thought we could get some good photos. One of the guys was into motocross back home and really keen to meet Rhonda. He was so happy to have a bike on his island that I suggested he ride her back. The smile said it all, and off he went. When he got off he came running over and shook my hand.

"I am so excited, my legs are shaking," he said. I laughed and said,

"I know the feeling."

From here, the boys asked if they could take Rhonda inside the building for a photograph.

"Why not?" I said, and so we set about getting her up the wooden stairs at the back and through the narrow doors. Once inside, we had a quick shot of their home-brew in celebration. Then we took a team photograph that they explained they would frame and put on the wall along with other historical photos in the base. The photos went all the way back to the 1960s when the base was first set up and run by the British. It was handed over to the Ukrainians in the mid-1990s. There was a photo for every year. It was a very proud moment to think that I would soon be joining their line as the photo of 2015. Rhonda and I are forever now part of Antarctic history, in our own little way.

That night, we had a wonderful spread and enjoyed some great hospitality in their cosy bar area. The home-brewed spirit of almond snaps, sugar, 5,000-year-old glacier water, and some other unidentified liquid helped to keep the conversation going and more than made up for any minor language barriers. The meteorologist got his guitar out mid-evening too, and of course, I had to sing along. I think I may have even sung in Ukrainian at one point! It was as close to karaoke on Antarctica as I was ever likely to get, having totally missed my chance on the Chilean naval base.

Ollie took some of our crew back to the boat at around 12:30am, and that's when the more extreme party games began. One involved sitting on a chair with a beer bottle on the floor next to you. Starting from a normal sitting position, the aim of the game, was to clamber right around the back of the chair and to return to a sitting position, all without touching the floor, having picked up a beer bottle with your teeth en route.

Needless to say, I failed that particular challenge (although I came very close) as did everyone else bar one young man who became the hero of the night. Most landed on the floor with the chair on top of them. Thankfully, we had the base doctor as adjudicator, so he was always on hand should any bloodshed result. It didn't.

Later that night, I found myself in the base's workshop raising the temperature of the meteorologist who'd been leading the pack all day in the motorbike escapades – not to mention the drinking games and the karaoke. It was a drunken fumble; not my finest hour (or five minutes in this case), but the real turn-on was the realisation that this sexual congress was taking place in the remotest place on earth. *Wow! Best not put THIS in the blog!* It felt devilishly liberating. Half an hour later we went back up to the main room and carried on partying – with a big smile and zero regrets.

I clumsily boarded the Zodiac with the remaining crew at around 3am and somehow, thankfully, we managed to navigate our way back through the icebergs to the boat without incident.

We set sail much later that morning, and as we passed the base, we were saluted from the balcony – led of course, by the Ukrainian meteorologist. I saluted back and shouted,

"Smoke me a kipper. I'll be back for breakfast." Never had there been a more appropriate time to use that phrase, famously coined by the dashing Captain Ace Rimmer, of the TV comedy series 'Red Dwarf', who would say it whenever he was off on another heroic death-defying mission.

I sat and watched as the meteorologist, the base, and its water tower with Vernadsky written in large letters, disappeared forever. Little did we know then, that our Antarctic parties were over. The Lamaire Channel had other plans for us.

****

Antarctica's mood took a turn for the worse as we headed north into the Lamaire Channel on the evening of 22nd February 2015. I guessed she wanted to remind us who was in charge here, and so, with the utmost respect, we found a sheltered bay to hunker down in for the night.

By the early hours, we'd lost an anchor line to a fast-moving iceberg, and all hands were on deck trying to replace it as the boat swung wildly in the strong winds. We'd already had a couple of smaller 'bergs hit us but, using the Zodiac, we'd managed to redirect them without too much fuss. Antarctica really was irritated that night, and I wasn't sure I liked her tempestuous side.

As daylight came, the crew worked efficiently, fighting the wind to free the boat from her overnight anchor spot. Xavi and I were on the aft pulling in the lines; it was imperative we pulled them in quickly to avoid getting them caught up in the ship's propellers as she swung in the wind. Ollie was at the wheel ready to engage engines; once free, we would have to drive quickly forward into deeper water to avoid being blown onto the rocks. Cath and Sally were in the Zodiac releasing the shorelines; and the rest were at the bow.

Suddenly, there was a really loud bang, and we grabbed the rails nearest us as the *Ice Bird* rolled to about a 45-degree angle. Then, as quickly as she had leaned, she righted herself. For a moment, everything went quiet. Quiet and still – as if someone had pressed the pause button. Still gripping the rails, Xavi and I looked at each other with questioning eyes. *Are we damaged? Are we stuck? What the...?* As rookies, we didn't have much to go on! Then someone pressed 'play', and we looked to see everyone was okay before figuring out what had just happened. The depth meter had read eight metres, and we only needed three and a half. There must have been a big rock sticking up which hit the keel and pushed us over. Thankfully, the boat's steel hull could take a few scrapes, and it appeared we had got away with it. There were no injuries, aside from Yvette and Lesley, who had fallen on top of each other and sustained minor bruising.

We pressed on, and by mid-morning we were into our usual routine. I hooked up to a line, connecting myself securely to the boat, and ventured outside, where I set about butchering a little more of Lambie for a tagine I was making for dinner. It wasn't easy on a rolling deck in the rain, but it was a strangely satisfying job, none-the-less. As I put the tagine in the oven later, I entered a minor state of anxiety. *Is there enough meat? Did I put too much cumin in? What if it all goes horribly wrong?* I laughed at myself and muttered,

"So… you've survived battling with an iceberg and now you're stressing over a *tagine*?'

As we travelled further north, the weather gradually got worse and less predictable, and we were beginning to get a little less tolerant of each other. There was little surprise when the bickering started. With all-night anchor watches now required at every stop, people were getting tired and perhaps a little nervous, too. Dealing with tears and tantrums from various members was now part of the daily routine, just as much as iceberg slaying and running aground. These came as a shock to the skipper as much as it did to us rookies, and on one such occasion she shouted at me,

"What do we do?". This was unnerving as you can imagine!

"I don't know" I replied, "I'm the biker. *You're* the bloody sailor!".
We were *all* on a massive learning curve in uncharted waters; at the mercy of this frozen desert, as brutal as she is beautiful. The parties were now a distant memory.

Finally, the weather broke, and as quickly as she had turned, Antarctica went back to being a sparkly and magical little kitten again. We made the most of our respite, sitting on deck, taking in the scenery, and watching out for whales. We weren't disappointed either, meeting a family of orcas on one day, as well as the now common sightings of humpbacks and their trademark dramatic tale swishes.

After a lovely evening of wine and steak, I squeezed myself into my bunk and snuggled in with another episode of *Breaking Bad* on my laptop, raising the lee cloth for just a tiny bit of privacy. I was half-an-hour into a particularly exciting episode when there was a sudden nudge to the boat and a slow grinding noise like metal-on-metal or a bad Freddie Krueger movie. I looked out of my porthole and saw nothing but white.

"Iceberg!" I shouted as I reached for the wool socks I had been given by the guys at Vernadsky and, in doing so, fell out of my bunk. The others were up, and we were all on deck within seconds. The 'berg was a big one, and it pushed up against us with some force. As it slid along us, we tried to push it away, but there was no moving it. As it reached aft, it looked as if it was going to rip off the Zodiac. It *just* missed. Then we realised the aft line to the shore was going to get in the way. With the bow held by the anchor and the aft held by the rope, something was going to have to give.

"We need to cut the rope," I said, and immediately wished I'd thought it through before saying it out loud. There was no need to cut it, as it was attached to a winch. All we had to do was let it out and down in to the water and the 'berg should be able to float over it, and be on its merry way. My foggy brain had let me down, but at least we were safe from this particular ice monster. There could be more though, and they move quickly, so we had to start up anchor watch again. Our respite was over. I went back to my bunk and placed my wool socks in a more convenient location – right next to my life jacket. Then I went back to bed to finish Episode Ten of *Breaking Bad*.

The next day's plan was to get across the open water from Enterprise Island to Deception Island. There could be no stopping en route once we got going as the waters were too deep to anchor in. The problem was that the sea was still rough, and we would have to travel at double our recent speed to make it before dark. The chances of managing that were slim to zero so, predictably, we ended up having to take a detour into uncharted waters looking for a safe mooring spot for the night.

Closely monitoring the depth, we sailed along Trinity Island in search of shelter. We had been sailing since 9am; it was now 7.30pm and we were no more than halfway to our planned destination. Thankfully, we found a spot and anchored up for the night. I went down to the hold to get the last of the wine to have with dinner and decided to check on my luggage whilst there. It had been moved to a lower spot for some reason, and as a result, everything was sopping wet. The kit is waterproof up to a point, but it wasn't designed for *that* much water. Luckily, I had most of my electrical stuff and my important paperwork in my bunk. However, the rest of the paperwork was ruined and worst of all, so was my Garmin GPS. The device, along with all the maps and my routes thus far, was destroyed; the entire history of the trip to date, gone. I was pissed off.

My desire to be back on the road and alone was now palpable. My patience was wearing thin and the lack of privacy was stifling, but the lack of control over my situation was the worst thing. I was unable to make the important decisions, or even the basic ones, when it came to my safety and that of my equipment. I had promised myself to stay cool throughout this part of the journey despite my hatred of being caged in – being trapped. Unlike many of the crew, up till now I had managed to keep my emotions in check, outwardly at least, but I would be glad when we finally got off the boat.

Later in the evening a blizzard arrived, so we would all be on anchor watch again. When anchoring, the rule is that you let out four times as much chain as the depth of the water. This often keeps the anchor from dragging. However, it also allows swing, which is fine as long as the wind is blowing in the right direction. Our main concern now was that we were anchored fairly close to the shore to make use of the shallow water, leaving us open to the possibility that a shifting wind would leave us grounded. The only sure thing about the weather in Antarctica is that it changes *extremely* quickly.

Yvette and I did the first watch while everyone else went to bed. An hour in, the conditions suddenly worsened. The monitors showed us swinging in the opposite direction and further out than before. The depth, previously steady at 30 metres, came up quickly. By the time it reached 18, we were ready to call the skipper. Then, the automatic alarm sounded; it seemed the system was of the same mind. Ollie got up instead of Cath, and agreed we needed to act quickly. The wind changed and we were now in imminent danger of beaching. Jumping into action, we put our wet weather gear on and went out on deck. I jumped into the anchor locker and fed the chain into place as Pete pulled up the anchor before Ollie quickly drove us forward into deeper water.

Now, we faced a dilemma: we could not anchor where we were and had only one navigational light to guide us. Cath decided to move on and try to find another anchorage a few miles away. There wasn't much choice, but this meant navigating in the pitch dark through challenging waters, relying only on the radar – and it clearly wasn't picking up everything. Eventually, I went to bed and left the second watch keeping an eye on things. It made sense to get some sleep while I could.

At 3.30am, Cath came running into the forward cabin,

"Steph, Sally, Xavi – we need you on deck." I was amazed at how quickly I responded from a deep sleep; instantly, I replied,

"OK," jumped out of my bunk, got my wet weather gear on, and was out of the door. Sally and Xavi were close behind and awaiting further instructions. I was expecting to pull ropes in or feed them out, but the next instruction sent a shiver down my spine…

"We need you to climb onto the boom, guys." We'd actually done this before, during the day, and in calmer conditions, so I knew what it meant: we were stuck on rocks again and needed to get the weight over to one side to see if that freed us.

The deck was covered in snow, and Xavi and I could only find our Crocs deck clogs. It was a real blizzard now, and the snow was accumulating thick and fast. We were sliding all over the place, but there was no time to find our boots. Together, the three of us managed to climb up onto the roof of the pilot house, and then up onto the boom. We huddled together, held on to the ropes, and waited. Ollie swung us out at a 90-degree angle to the boat. We sat there and waited for the jolt as we heard,

"Okay, hang on guys".

We dangled over deathly cold waters as they drove forward trying to free us. There was a real risk of the boat lurching sideways and – as we weren't tied on – spilling us into the sea. Nothing was happening though; no movement. The only thing left to try was pushing the aft over with the Zodiac. The water was rough, but Cath climbed in and got to work.

It's funny the things you think as you are shivering on a boom in Antarctica wondering if you are about to fall into an icy sea: *I haven't cleaned my teeth. My mouth feels furry! I'm wearing pink Crocs. I don't want to die in pink Crocs!* I wasn't even sure I was awake. We had gone from deep sleep, to sitting on a boom in a blizzard, shivering uncontrollably, in ten minutes flat. Looking around, disoriented, I tried to assess how far we were from shore.

We were utterly alone and off course in uncharted waters. This was a tidal spot, and the tide was working against us. If we didn't get off these rocks soon, then there would be no chance of getting out at all.

"We are definitely having a hot chocolate after this!" I said as we huddled closer together enveloped in darkness. My companions agreed.

It seemed everything was against us; the tide, the time, the dark, the roughness of the sea, and the weather, but after lots of pushing and pivoting and repositioning, we finally freed the boat! As Ollie tried to get us into deeper water, we were swung back in and we carefully climbed down and went inside, huddling together under a sleeping blanket. Cathy came over and said,

"Well done, guys. Good job," as she handed us our well-earned hot chocolate.

The snow kept falling and all we could do was keep moving a little at a time until the weather cleared enough to give it one last push over the open water to King George Island.

****

King George Island has to be the ugliest part of Antarctica, with nine countries holding bases there. Chile, Peru, Brazil, Argentina and Uruguay, plus China, Russia, South Korea and Poland. Their work ranges from geographic research to biological and ecological studies. The bases are all interconnected by a maze of roads and tracks that stretch for more than twenty kilometres. There's nowhere else on earth where you can literally walk between Russia and Chile...with a side trip to China and Uruguay along the way. It was not a pretty sight in this vast and beautiful wilderness, but it was a useful gateway to get out of here, and I think we were all glad to see it. The *Ice Bird* certainly took a beating on this expedition and would go home with some pretty substantial battle scars.

A couple of days earlier, we had contacted the crew of a Russian ice breaker named *Polar Pioneer*. They had agreed to collect Rhonda in the morning and take her to Navarino Island off the coast of Chile. The rest of us had a charted plane coming to take us back to mainland Chile. I would have to find my own way back to Navarino Island and await Rhonda's arrival there. Spirits were high amongst the crew as we squeezed ourselves into our bunks and pulled up the lee cloth for the last time. I think by then we were *all* looking forward to being back on dry land.

The radio woke us with a crackled message at 7am, announcing the plane was delayed. Conditions were unsafe for landing, as the strip had no lights and visibility was poor. They would try again tomorrow. I wasn't keen on this new plan, but there was nothing I could do about it, so I busied myself prepping Rhonda for her send-off, while making a mental note to try to get ashore to one of the bases later in the day, to let people know I was safe.

The Zodiac arrived mid-morning. The crew quickly introduced themselves as Peter, a large Russian chap whom you just wouldn't mess with, and Martin, whom I later discovered was an ex-US Navy Seal. We were in good hands. Conditions were worsening by the minute though, so we quickly dispensed with the pleasantries and got to work.

Using the tried and tested method of employing the boom as a winch, we lowered Rhonda in. The receiving Zodiac was much bigger this time, but the wind was making things a little trickier.

"The weather is changing fast," said Peter in his strong Russian accent, "We must get her on board quickly before it is too late." *Before it's too late? What did he mean by that?* It all sounded a bit dramatic, so I asked Ollie to come with me to help with the delivery.

We drove through the water; crashing off the waves and getting drenched in sea water with every bounce. Rhonda was lying down. She was my biggest concern, but the guys assured me they would wash her down and stow here away once we were safely on board. Conditions had worsened by the time we arrived, and the Zodiac crashed up against the port side of the ship.

"We must get out of the water quickly," yelled Peter, "Or we could be pulled under." *Pulled under? Really? What had I got myself into this time?* Damn, I hated the sea. What crazy fool would ever put themselves through this for fun? I smiled through the chaos as it occurred to me that *I* was the only crazy fool here. I certainly didn't see anyone else in Antarctica trying to board an ice breaker in a storm with a motorbike. My good friend 'brute force and ignorance' still reigned supreme, and I could credit many of my successes to her. The god of not-thinking-too-much and hoping-for-the-best.

Ollie worked hard to get the luggage up quickly into the hands of the waiting Russian crew, falling over several times in doing so. The plan was to get the luggage on board and then get ourselves off, before winching the Zodiac up with the bike still inside. Ollie fell again, and Martin grabbed his life jacket just in time to stop him going over and getting crushed between the two vessels. The weather was worsening faster than we could work, so they made the call to winch us all up together. The crane hook came swinging down over our heads, and it took Peter's every effort to grab it and get us hooked on. We were then lifted to safety – it felt wonderful to be on board.

They gave Ollie and me a tour of the ship, including the captain's bridge. The last time I'd been anywhere like it, I was five years old and my grandad (the captain) was offering money to the kid who saw land first. We were on his cargo ship and on our way back to live in the UK from Canada. I smiled, feeling close to him. Proud. I could almost smell the smoke from his pipe. The water was calmer now, innocent.

Our last outing in Antarctica promised little excitement as Ollie, Xavi, Sally and I took a trip to the base in search of good coffee and Wi-Fi. As we got halfway across the two-hundred-metre run, something popped up next to the Zodiac. A big head. No. An *enormous* ancient-looking head with deep, threatening eyes; gateways to a dark and distant past, promising death to all those who peered inside. Xavi was first to shout,

"*Leopaaard seeaaal*!"

Leopard seals are fierce predators known to bite into Zodiacs with their powerful jaws and long teeth. We'd been warned to look out for them when kayaking, as one had recently killed a diver on a scientific research trip. We'd come across one earlier in the trip, but from the safety of the *Ice Bird*. Now, we were like sitting ducks – all fluffy and prey-like with our warm blood and down jackets; nicely packaged and ready to eat.

"Shit!" I squealed, jumping to the other side of the boat. The sea leopard lurched forward as if he was about to haul his ten-foot-long, 900lbs of blubber right in there with us. His head came over the side and rested on the boat right where I'd been sitting a moment ago. "We need to get out of here," I continued with some urgency now, "Ollie, let's go!"

The seal pulled back and swam around to the bubbles generated by the outboard.

"That could be a threat to him," Ollie said, killing the engine. I suspected Ollie, not oblivious to the danger, just wanted to stay longer and enjoy the moment. It actually *was* an amazing moment; I couldn't deny him that. Few come that close to a leopard seal and live to tell the tale. Part of me was with him; it was the experience of a lifetime, and we were honoured to be in the presence of such a magnificent beast. The other part of me wanted get the hell out of there and ensure that *this* lifetime lasted beyond our encounter.

The seal was intrigued, though it was hard to tell if he wanted to play or kill; but its eyes were unmistakably those of a remorseless predator, so I was betting on the latter, although I'm not sure his 'play' would have been acceptable either. His nostrils flared, and with a snort, he turned and sank back into the darkness from whence he'd come – leaving us ecstatic from our close encounter.

The 'plane got clearance to land the following day. It was time to get the hell out of Dodge, or rather, Antarctica!

Chapter 16
Navarino Island

Upon landing in Punta Arenas, Patagonia, in the southernmost region of Chile, I rented a room in the cheapest hostel I could find, threw my bag on the floor, and flopped on the bed. It felt so spacious. I rolled around giggling and making like a snow angel before remembering I needed to find the Wi-Fi password. I sighed light-heartedly, got up again, and rummaged around, eventually spotting it on the back of the door. Grabbing my MacBook from my bag, I began searching for flights to Navarino Island the next day, where I would meet Rhonda at Puerto Williams.

I called my parents first.

"So glad you are OK" was the first thing my Dad said, the relief evident in his voice. "We had expected you back a couple of days ago. We were worried". I hated worrying my parents. I'd given them enough of that in my life.

"Sorry. I was actually only due back yesterday, but the plane was delayed due to bad weather". There had been some miscommunication somewhere along the way. My parents had supported the trip from the very beginning. They had never tried to put me off. Both of them understood my love for motorcycles as they both rode themselves, and anyway, they had given up trying to change my mind on *anything* a long time ago; choosing to remain stoic in the face of my crazy ideas; a long line of boyfriends; sudden job changes. The list goes on! They had offered support rather than criticism or naysaying. Dad had helped me to prepare Rhonda before I left, and they had both agreed to take care of my dog, who was now being treated like the favourite grandchild, by all accounts. They wore my fund-raising T-shirts with pride, and they seemed happy that I was pursuing my chosen version of life rather than being a passive victim of it. I was certainly a talking point at Mum's badminton club!

Next, I called Tim, my friend and Antarctica sponsor, to let him know I'd made it back to the mainland. He was overjoyed to hear my voice. As we chatted, I felt myself glowing; happy I'd achieved what he always believed I could. His praise and congratulations meant as much to me as those of my own parents, and as I talked him through the whole Antarctic experience, I relived every emotion in glorious detail.

"I'm so proud of you," he said finally. The tears rolled easily, then. He was such a kind man. I still had no idea why, but he had taken me under his wing. Tim was someone I looked up to and respected, and those words meant so much.

That evening, as I lay on my bed writing my blog, I heard British voices on the landing outside. I decided to nip out and say hello. A quick chat revealed they were British military who had just finished a kayaking trip.

"Nice" I said "When do you fly home?"

"Tomorrow morning" said the man who had introduced himself as their leader before he'd even told me his name. "Why don't you come out to dinner with us tonight? We can share adventure stories".

"I can't think of one reason why not to" I replied.

Dinner became a club and one drink became several. The tables melted away into a dance floor and the dodgy dancing became a lesson from the locals in salsa. It was turning into a great night, but after the *blue* drinks came out, it was never going to end well.

The big guy, maybe six foot four and quite wide, struggled to stand. After a couple of wobbles, he fell over and landed heavily on the dance floor.

The military boys kicked into action and began running around like headless chickens! Somehow, I was the one who ended up with him draped over my shoulder (all five foot four and nine stone of me) walking him down the stairs. With sober hindsight, this was not a great plan. Nevertheless, it worked, much to the amazement of the squaddies who were now flanking me. I used my firm voice again (the one I normally reserved for dogs, drunk Turks, and Iranian taxi drivers) to talk him, step-by-step, to the pavement where I sat him down and kept telling him not to fall asleep. His head bowed, and his full weight fell into me as he lost the fight against the blue drink. Mr Team Leader barked orders as the taxi arrived and two of the men got in, dragging the now practically unconscious heavyweight lump with them. The rest of us walked back to the hostel.

As we turned the corner and walked into the courtyard, we saw him lying at the bottom of the stairs; silent and motionless. A bright crimson pool of blood had already formed around his head – in stark contrast to the grey of the concrete that his skull had smashed into. His two escorts stood over him, also motionless; their wild, frightened eyes fixed on their friend and not even registering our arrival.

We ran to him, and the younger escort finally spoke,

"He just fell down the stairs." He looked frozen, as if awaiting orders, "We couldn't hold him".

The orders came from Mr Team Leader, but it was the last thing I expected him say.

"You," he said, pointing at the frightened young squaddie, "Go upstairs and get a notebook. Write everything down that happened. *Don't* mention alcohol." The young man nodded and was back in a flash. He sat on the stairs and started writing a narration of the events as they unfolded.

The man had fallen into what resembled the recovery position. On closer inspection, I found him to be breathing without obstruction. His pulse was good, but he was cold.

"Fuck the notes. Get a blanket," I yelled at the scribe. He stared at me, unsure what to do. I was not his superior. I turned my attention to the other guy who was still just standing there, "Wake the owners. Get them to call an ambulance. We need their Spanish. Hurry." The young scribe was still on the stairs. The only part of him moving was his hand furiously writing out the scene. I briefly wondered if he was using quotation marks and quoting us verbatim.

The team leader shouted at the other guy, who had come in behind us and was now crouching down next to us,

"Put your fingers in his mouth to stop him swallowing his tongue. Keep them there." He was almost screaming now, hysterical at seeing his comrade on the floor, but he was not thinking. I dug deep for my gentlest, calmest voice so as not to startle him further, and said,

"You know that could be a bad idea. He is already in the recovery position and his airways are clear. He has a head injury, though, and could start convulsing. If he does, you will lose your fingers." I'd seen enough seizures to know this was a real possibility.

"I don't care about my fingers," he barked, "I just care about my friend."

"Yes, but…" I trailed off. There was no point. I was an interloper in a badly written play and these guys would only follow orders from their boss. The last thing we needed was two casualties, but I let it lie and went to see what was happening with the ambulance. I was completely sober now, and completely calm. Almost too calm, as if it wasn't a big deal. Emotionless, while everyone else was running around trying to remember their training through a fuzzy, alcohol-induced haze and a good sprinkling of sheer panic. It was their friend, after all, not mine. But I did know my place, and that was to let them get on with it, regardless. Too much meddling might see Mr Team Leader really lose it, and that wouldn't help anyone.

Much of the hour waiting for the ambulance to come was actually spent just trying to get through in the first place as the 'phone line had been engaged. As soon as the paramedics jumped out, Mr T began barking orders at them, too. He was far from calm, and getting in the way, but he would not hear my words as I tried to calm him. As they put the injured man in the ambulance, he shouted at two of the guys to follow in a taxi. As soon as they left, his demeanour instantly changed and he turned his attention to me.

"I'm sorry you had to see that," he said, in a cool, film star hero style. I scored him eight out of ten on the 'patronising male' scale. The only things missing for the final two points were a raised eyebrow and the words "little missy" at the end. "Are you OK?" he asked, still in his ridiculous action hero voice.

"I'm fine. It wasn't me that fell. How about you? Will you follow on to the hospital?" He turned to me then, and stepped closer, all panic gone, and his focus completely on me,

"I'd rather go back to your room" he said seductively. I'm pretty sure a single eyebrow raised that time.

"Are you serious?" I asked, stepping away from him. "Goodnight. I hope your friend is okay."

"Please don't mention this on your blog," he called after me as I stepped over the pool of his friend's blood and climbed the stairs to my room. "At least, don't mention the alcohol."

"What a slimy bastard!" I muttered as I locked the door behind me.

I'd had enough action for one night. In fact, this last few months felt like enough action to last me a lifetime. Now, I wanted calm. I lay on the bed and entered the sleep of the intoxicated. I found out the next morning that the injured guy was doing well after an emergency operation to relieve the pressure on his swollen brain. He would not make his flight, but he would be okay. It wasn't a great day for the British Army.

The short flight to Navarino was uneventful and I spent my first few days on the island enjoying an easy solitude. The locals were preparing to settle in for the winter, piling logs up against their very basic, shack-like houses, and nailing things down that didn't need to move for a while. When the wind blew, the wood of the houses creaked as though it had a good mind to snap, and the corrugated iron roofs threatened to fly away from this bleak, yet beautiful place. I stayed with a local family who gave me food, shelter, and a chair at their sturdy family table. There was always a log on the fire, and they did their best to make me feel welcome, despite the obvious communication barriers. In fact, no one spoke a word of English. As the foreign visitor, the linguistic responsibility lay firmly with me, but I had failed to learn anything beyond *'gracias', 'sí', 'por favor'* and *'tienes un cenicero?'* (Do you have an ashtray?). No one seemed to mind the long silences between us, though, least of all me.

On the sunny days, I wandered around the coastline overlooking the Beagle Channel. I would stop and sit in the windswept grass near the grazing horses and enjoy the feel of the ground. I loved how dependably solid it felt. At first, it felt strange, and on the first day after landing in Punta Arenas, I felt 'land-sick' for a time. Who knew that was a thing? But the lack of movement had definitely caused some nausea.

This place had an authentic end-of-the-earth ambience. Everything about it made me feel as if I couldn't possibly be any further from home, and yet I didn't mind. In fact, I realised I felt very little; not quite numb, but mildly depleted, displaced from my emotions.

During my second day on the island, I made friends with an Alsatian dog. He was suddenly just there next to me as I wandered around the village, and he never left my side until I got on the ferry to leave. He waited for me outside the house and walked with me wherever I went after that. I offered no food, and little in the way of conversation, but he hung around anyway. I was glad of his silent, dependable company.

Rhonda arrived at Puerto Williams a few days later. It was good to see her. I joined the crew on the *Polar Explorer* for a coffee and catch-up before they winched her onto dry land.

"Good to see you girl," I said as I pushed the button to fire her up, amazed that she started first time. I set off to the customs office to explain where she had been and why she had no stamp out of Argentina. No one spoke any English, so after a few confused looks from them, and a well-rehearsed look of innocence from me, we got through without delay.

We still had a few days to wait for the ferry back to Punta Arenas, which came only once-a-week. It looked as if the locals were preparing for a long, hard winter, and I hoped to be out before the snow came, although the idea of getting stuck there for a few months was not entirely displeasing. What drama could I possibly find on this sleepy little island? The more I thought about it, the more it sounded like heaven.

Landing Rhonda
for the first time
at Paradise Harbor on
the 14th of February
2015. An amazing day!

The crew of the Ice Bird enjoying a well earned and extremely chilled glass of Champagne after successfully landing Rhonda at Paradise Harbor.

Riding Rhonda for the second time on Antarctica after a successful landing at the Ukrainian science base.

Ollie in the Zodiac pushing mini 'bergs away from the Ice Bird.

A view from Vernadsky science base. The Ukrainian base where we landed Rhonda for the second time and partied until the early hours!

Ollie attempting the drinking game at Vernadsky – and struggling!

The Ice Bird moored off the peninsula. Rhonda takes pride of place on the bow.

Our local inquisitive leopard seal!

A view from the Ice Bird.

Dropping Rhonda at Paradise Harbor. My fourth continent fall!

Ice caves near Paradise Harbor, Antarctica.

Chapter 17
The Atacama

After a good cleaning and service back in Punta Arenas, Rhonda was gleaming. I took out my marker pen and proudly added 'Antarctica' to the list of countries on her windscreen – a ritual I had started after my first border crossing. Then I counted down the list. It confirmed that I had now ridden through twenty-five countries and on five continents. It was all there in black and white, written by my own hand. So it must be true.

We were also coming up to our first anniversary on the road. I guess I could call myself an adventurer now. And why not? I had already tried pretty much every other 'hat' and none had really suited me. The problem was, my helmet wasn't quite fitting now, either, and Rhonda was the only one gleaming. An undefined sadness seemed to have descended on me like a cloud.

I'd been pushing for months with Antarctica as my main focus – it had to be, to make it happen. But with my first real goal accomplished, I wasn't quite sure how I was supposed to feel. I felt depleted; dissatisfied, alone, numb. It had been creeping in since I'd landed back on the mainland. Perhaps it was psychological; was I suffering some kind of post-event blues? I did feel a vague sense of loss. Perhaps it was the anti-climax. Maybe it was the fear of life never feeling that good again; never being able to return to that ultimate peace that I had felt on that rock in Antarctica. Then again, maybe I was just tired. Yes, that was it. Just good old-fashioned tired. But this fatigue, both mental and physical, stayed with me for some time.

For the next couple of weeks, I went through the motions of a daily routine. Get up early. Pack the bike. Ride until the sun shouted last orders. Find a spot. Unpack the bike. Set up camp. Too tired to cook; make coffee instead. Climb into tent, curl up, and fall into a deep sleep, no longer worried by the sounds of the world around me. It was just another day. Eat, sleep, ride, repeat. That was my thing now. The bike kept going, and I just kept her pointed north towards the sun.

My next few weeks consisted of riding steadily northwards overall, but with several easterly and westerly excursions over and through the Andes as I switched between Chile and Argentina as I went. I maintained momentum, but I felt nothing; no joy at the sight of a volcano or a towering granite peak; no shot of 'life' as I observed the llamas effortlessly springing over the roadside fences; no buzz as I looked skyward at the infinite array of hawks and eagles above me; and not even a respectful salute to the armadillo that crossed the gravel road ahead. That dark, impenetrable cloud of sadness was obscuring my view, changing my perspective, and along with it, the world around me.

Those long-stretching gravel roads were harder than they should have been. The Patagonian winds that swept across this remote and pristine land felt cruel and bitter; the gusts felt like a game of cat-and-mouse. A jab from the left, then another from the right, before SWIPE! Keeping the bike upright was hard work, and my hands would freeze, forcing me to stop every twenty minutes to defrost them. (My heated grips had proved inadequate when they worked, and they'd long since stopped working; and I have trouble keeping my hands warm even in Wales!). What was once a challenge, was now an affliction. Even on the warm days, the vast empty spaces between towns no longer held any playfulness. I was tired. It was as if my senses had been saturated, filled up to the point that they couldn't take any more beautiful views or new experiences and I was scared witless that this was it for me. Should I just ride home really fast and be done with it? I was thousands of miles away though and riding a 250cc bike! How fast did I really think I could go? No, that wasn't the answer.

That's when I decided to find myself a cheap hotel room. A dark one, the darker the better. Somewhere I could 'hole up' for a few days. I didn't want a view of the mountains or a balcony. I just needed to lock myself away for a few days and shake it all off. No stimulation. I needed a 'system reset' – like an overloaded hard drive, I needed to switch off and then on again.

I crisscrossed between Chile and Argentina on my way north, and headed for the town of Perito Moreno, Argentina – not to be confused with the Perito Moreno glacier, which is further south (they are both named after the 19ᵗʰ Century Argentinian explorer Francisco 'Perito' Moreno). It turned out to be a great choice, as there was nothing of interest in this town; no significant landmarks, no cute little coffee shops, no impressive cathedrals or ancient monuments. There were no leaflets full of suggestions of things to see, or travellers telling me what I had missed. It was the perfect place to hole up for a few days and do absolutely nothing.

I called Shane a couple of times. He was now back in India. We held onto each other as much as we could, the same way our relationship had started: across continents, piecing together conversations through text, WhatsApp, and sometimes Skype video calls if the Wi-Fi was good enough. Other times, long silences. It was good to have a friend who understood what each of us was going through without saying those dreaded words, "You're living the dream" or worse, "It's all part of the adventure," although we often did, sarcastically, just to get a giggle.

After a week of doing very little other than talking to Shane and watching movies from my hard-drive, a little bit of my happy-self started to emerge again, crawling out tentatively from under the rock where she had hidden. My energy was returning. Not fully, but it was enough to consider moving on again.

Then I received a message from a Facebook friend, Claire, a Brit who lives in Australia. She was riding around South America on a Honda XR250 Tornado, currently en route to San Carlos de Bariloche. I checked my map; it was 800 kilometres north of me, and I could be there in two days – the day before my 40ᵗʰ birthday. I told her I'd see her there, and she agreed to have a cold beer waiting for me when I arrived. This plan was enough to shake off my melancholy, and I left the hotel the next morning looking forward to the road ahead. I was also looking forward to a beer.

Claire and I hit it off straightaway and spent the next few days riding the iconic Ruta 40, a highway stretching 5224 kilometres up the east side of the Andes. Known to the Argentinians as *La Cuarenta ('The Forty')*, it has a reputation as one of the world's wildest and least-travelled roads. Che Guevara travelled along much of it in his famous *Motorcycle Diaries*, engraving it even further into international legend. It was the perfect (and most spookily appropriate) way to celebrate my 40th birthday. We stopped at every Ruta 40 sign to take momentous photographs of each other doing handstands against them.

We played like big kids enjoying a summer holiday, only now, instead of a hose in the garden, we had motorcycles on *La Cuarenta:* riding the empty gravel highway that wound its way past snow-capped mountains, alpine lakes, and forests while finding perfect little camping spots along the way. There were no time constraints, no rules, and I could not have wished for a better way to celebrate.

We stopped for lunch that day in a trendy vegan restaurant just outside San Martin de los Andes. It looked expensive for my budget. *What the hell. It's my birthday!* As Claire was a vegan, it seemed only right that we made the most of this find. The meal was delicious, and just as I finished and put my knife and fork on the plate, the rather camp and quirky little waiter came out singing "Happy Birthday", holding a fruit crumble with a normal 'house' candle sticking out the top. I was so touched that she had made the effort to mark the occasion, but as I thanked her, she said,

"It's not from me. It's from your Mum." I laughed, a little confused. Perhaps she meant in spirit. After all, how could it *actually* be from my mum? I had only just met Claire myself, so my mum definitely didn't know her, or indeed where we would be stopping.

"No, seriously," she said. "Your mum contacted me a few days ago; you must have told her you were going to meet up with me. She found me on Facebook and asked me to get you a cake. The crumble was the best they could do here, but yeah, happy birthday, mate."

I wanted to cry. I'm not normally that emotional but it was such a lovely gesture to have gone to the effort, and after a tough couple of weeks, it meant the world to know I was not alone, no matter where I was.

About 60 kilometres north of San Martin, we turned off Ruta 40 and headed back through the Andes into Chile, to a little town called Pucon in the centre of Chile's Lake District. It lies on the edge of Lake Villarrica and sits in the shadow of the Villarrica volcano. The town is known as a hub for adventure tourists who come for the watersports and the hiking. We were there for the coffee shops that come with tourist spots, and the promise of some great off-road riding.

I can't recall how we met Matt, a twenty-something American from Boston. Blonde hair, blue eyes, and a great sense of humour. He was one of those bubbly characters who always had something funny to say; often inappropriate, but always funny. He was working in Pucon for three months as a mechanic for a motorcycle tour company. The three of us hit it off immediately and decided to attempt a ride up the great Villarica volcano herself. This was, perhaps, not a great idea, as she was particularly active at the time. She hadn't erupted since the 1980s, but in the last few days something had upset her. She was smoking and grumbling like a grumpy bear coming out of hibernation. At night, we watched from our hostel balcony as red plumes of molten rock shot up into the starlit sky in a spectacular display of strength and foreboding.

In true 'nothing can happen to us; we're just observers' fashion, we went for it anyway. But why not? In reality, it was a good time. If she was going to erupt, then we'd have no more chance in town, with the coffee-drinking North Face-wearing hikers, than we would on the now empty trails that ascended her slopes. We'd have the place to ourselves – Hikers 0, Bikers 1. But the police called 'foul' on our game. Part-way along the trail, just before the ascent, we encountered a barricade, two green and white police cars, three police officers in high-viz vests, and a few 'volcano voyeurs' with long lenses and flasks of coffee, all at the foot of the billowing Goddess Villarrica. There was clearly some unease about the situation, but the fact that no one had been evacuated yet told us the danger was not necessarily imminent.

We stopped for a while to admire the view,

"Quick, get a picture of me," giggled Matt as he ran over to the police car, and leaned over with splayed legs and arms as if he had just been arrested. I got the shot just as the policeman came over to see what he was doing. Thankfully, the officer was smiling and happily posed with his handcuffs as well to make the next photo more authentic. We took the opportunity to ask if there were any chance we could sneak past the barrier, but his goodwill did not extend that far.

"Don't worry. I know some trails that will take us around it instead," said Matt, who'd already been here a few weeks. We jumped back on our bikes and followed his BMW F800GS. Turning off the main gravel trunk route, we found ourselves on a tight and twisty single track that flowed like a sweet caramel river, carrying us deep into an enchanting forest. Our knobbly tyres were made for this and gripped the loose dirt with ease as we stood up, gripped with our knees, and let our suspension soak up the whoops. We ducked under branches, dived into gullies, and played under a canopy of green that smelt of fresh pine and sulphuric ash. After the long stretches in Patagonia, it was delicious to get off the gravel and do some more technical trails. With no luggage and the company of other riders, it was playtime, and it felt oh, so sweet. We didn't get any closer to the peak of the volcano, but it was great fun.

I stayed a day longer than planned just to hang out with Matt and Claire, but my chain was now practically hanging off the well-worn sprockets. Although I'd tightened it twice in the last few days, it was already sagging again. There was nothing for it, but to take a steady ride to Mendoza, 1200 kilometres north, to see if I could get the parts there. We allowed ourselves the luxury of one last 'posh coffee' together in one of the cute little coffee shops before I hit the road again.

"See you on the road" I said, twisting to grab my jacket from the back of the chair. I hadn't realised a waiter was coming up behind me, and as I put my arm up to get it in the sleeve, I knocked the heavily-laden tray of drinks straight out of his hand and sent it flying through the air before smashing to the floor; glass, milkshake and possibly a skinny latte or two – I couldn't be sure. Everyone paused for a second, motionless, reconstructing what had actually just happened.

"Oh no! I'm so sorry!" I said, as the waiter just stood there looking at me in disbelief – hands still in the position of holding a tray. I tried to help as he jumped into action, but he politely suggested I leave it to him. With everyone still watching me, I flushed crimson, grabbed my bags, smiled at the guys and made for the exit.

"Be careful out there," shouted Claire, still laughing behind me,

"And if you can't be careful" I replied over my shoulder, halfway out now, "have fun". It was something my dad used to say, having long given up on any serious parental advice. And with that, I was gone.

Crooning happily in my helmet, I took the quieter roads east to a lazy border crossing back into Argentina, and back onto Ruta 40. At this pace, it would take three days to get to Mendoza, and I was happy with that. In total contrast to just a few days ago, I was happy with everything. The sun shone and I was free from burden again, despite the fact that Rhonda now had a few minor problems. The biggest of these was the chain, which, potentially, could snap in this condition, but worrying wouldn't change anything. It would either make it, or it wouldn't.

Mendoza was a flop. The Honda dealership finally opened after a two-day holiday only to reveal that they didn't have the parts I needed and that it would take weeks to get them in. By now, my chain was in desperate need of replacement; my headlight was working only intermittently and I needed a new back tyre. Also, the top box bolts had broken for the third time from the punishment of all the corrugated roads and my head bearings needed a clean and grease, or maybe replacement. That meant Rhonda's steering was really stiff and she was feeling very awkward at slow speeds. On top of all that, the clock was now flashing on and off for no good reason. My options were to try to keep heading north to Salta, a two-day ride away, in the right direction, or to detour west to Santiago, capital of Chile, where I knew for sure there were lots of bike shops, and the parts would be cheaper. Decision made: I headed west, back into Chile.

Santiago is a colourful city, and it was easy to hang out there, not least because I had found Claire at the hostel. There was also Mathieu – a KTM rider from Belgium whom I had already bumped into twice on the road, and Sheldon – an Aussie with whom I had spoken to previously, online, riding a BMW R1200GS. We spent most of our days sorting out our own personal bike problems, and most of our nights in the bars – often laughing at Sheldon's Tinder escapades. He was a true red-blooded male Aussie all right.

News had come in that the worst floods in Chilean and Bolivian history had hit a very large area just north of us. This area was on my route, and the reports had said that many of the roads were closed. I wasn't sure if I could make it out the way I'd planned, or indeed, if there was a way around, but on the eighth day – with Rhonda all fixed up – I pointed my wheels north again and set off towards the Chilean coast. Claire and Sheldon chose to stay in Santiago a little longer, while Mathieu was heading for the MotoGP race meeting at Termas de Rio Hondo in northern Argentina.

I soon came across the devastating results of the flood in an area normally considered to be one of the driest places in the world. The towns of Chañaral and Taltal seemed to have suffered the worst of it, with many lives lost. Both towns had been cleaning up for over a week, and there was clearly still so much still to do. Chañaral looked as if something had clawed a line straight down the middle, slicing entire buildings in half as easily as a hot knife through butter, taking homes and businesses down to the sea and dumping them there. The bay was now a pile of floating trash; the village destroyed. Trucks and cars stood abandoned in the mud that had enveloped them, and the people who remained looked tired and beaten.

The roads in between were mostly damaged and washed away, so I carefully picked my way through, along with many truckers, all of whom seemed to keep an eye on me. They moved out of the way to let me through, and they beeped and waved every time I stopped, or they stopped. We were all sharing the road and watching out for each other with a real sense of camaraderie in the face of this natural destruction around us. It had been a sobering sight.

By the time I got to Antofagasta, Mathieu had caught up with me again; his plans for the MotoGP had been scuppered, as many of the passes into Argentina were closed due to the flooding. With little choice remaining, he too found himself travelling up the west coast of Chile and into the Atacama. At first, I wasn't sure I wanted the company. I'd enjoyed the last few days of solitude on these long desert roads enveloped by a never-ending red, arid terrain that suggested I might be on Mars, not Earth. There was no life to be seen between towns; neither bush nor beast for hundreds of miles, and I remembered now that I liked being alone in the wilderness. Would company spoil it? But Mathieu had an easy way about him. Maybe in his thirties, he was smart, funny, and easy on the eye. He was an engineer, so he was also a useful guy to have around. On top of that, he was doing far better than I was with the Spanish. I decided I was happy to ride into San Pedro with him, at least, and we could go our own way from there.

San Pedro de Atacama is a town set on an arid high plateau in the Andes mountains of north-eastern Chile. Its dramatic surrounding landscape served as the perfect playground for two hobo bikers looking for some fun! And by fun, I mean dunes. We dumped our luggage as soon as we arrived and ventured into the sand. Matt's KTM Adventure 990 V-twin was a great bike, but a heavy beast to heave out of a rut. Even a small 250cc like Rhonda can feel pretty heavy when you're battling soft sand and desert heat. Together though, we could take on anything, and laugh at each other as we took it in turns to fall off or bury the back wheel in the sand.

After two days of motorcycle mayhem, we ventured higher into the hills to check out a geyser field. El Tatio is one of the highest geyser fields in the world and the largest in the southern hemisphere; its eight distinct geysers sit high in the Andes at 4,320 metres (just over 14,000 feet). We ventured out at first light, while it was still cold as the difference in temperature between the air and the geyser water creates a most spectacular display. Or so we were told, and we were willing to bet our extremities on it. We were *freezing*.

Mathieu was in control of our ride for the day. Rhonda's headlight had gone on strike again, and this mission definitely called for one. Riding two-up on the big KTM for 110 kilometres of dark, cold, and winding roads, we hoped for a spectacular sight. It was strange, at first, not being in control, but after twenty minutes on the back, I was glad to have him take the brunt of the icy wind while I tucked in behind him. I couldn't feel my hands. My feet, too, were going numb and my legs were tingling. I stole a glance at the wonderful display of stars above me in the clear, pre-dawn sky, before tucking my head back down and resealing the draughty spots. It was 5:30 a.m.

The scene that greeted us was straight out of a post-apocalyptic movie, with snow-covered volcanoes surrounding jets of steam rising like a white fog from bubbling holes in the crusty surface. The sun rose like a goddess over this gurgling volcanic plateau and we worshipped her as she warmed our faces and painted the surrounding volcanoes with a gradual shifting of hues. It was truly breathtaking. We had not been disappointed.

Our plan was to stick together again today, head towards the Argentinean border and find a nice spot to camp en route. The plan failed.

I took the lead out of town on Ruta 27, but before long, Rhonda was struggling to maintain our speed. The temperature was dropping again and I realised that we were actually riding at an altitude of more than 4,500 metres – nearly 15,000 feet. The climb up was so gradual that it hadn't occurred to us until our lungs started feeling tight and the bikes lost power. Both humans and machines were working harder to get the required amount of air. We stopped at the snowline and got as many layers of clothing as we could back on again; from one extreme to the next, twice, in just a few hours. The wind was bitterly cold, but the riding was phenomenal. Every corner we turned led to another breathtaking view; miles of unadulterated red and white desert; a contrasting mixture of sand and salt guarded by rock formations standing tall over the windswept plateau. Silent. Immense.

We considered a couple of places for a campsite, but it was just too cold. With nothing to burn, we would have *endured* rather than enjoyed. Once the sun went down, we would be left with the icy wind and sub-zero temperatures. Instead, we decided to carry on to the border and see if we could get a bit closer to sea level before the day was out. Darkness loomed, and Rhonda still had no lights. By now, the cold was deep into my bones, compounded by my tiredness, and sniggered at by the falling sun. We had not stopped since 5am, and I was now shaking. I needed a hot shower or a fire to get warm again. We were still in good spirits though, and rode on in the hope that we could make it to the town of Susques before nightfall, where we might just find a hostel. Something. Maybe. Anything, as long as it had heat. As we rode on, the light changed to a dusky pink. The soil and the rock faces changed to beautiful green and pink pastel colours. Even the cold and the threat of darkness could not stop me from enjoying this other-worldly scene.

Darkness, real darkness, fell as fast as a guillotine blade. There would be no encore to our beautiful view; it was gone. With no headlight, I had to rely on Mathieu completely and I was so happy not to be alone. Riding as close as I could, all I could see was his taillight.

When he went around a corner, the light went out completely. I stuck to him like glue after that. We had no choice now, but to ride on. Luckily, the road was extremely quiet; we saw maybe two cars in all. We rode in total darkness for around thirty kilometres before reaching Susques. Then my headlight decided to start working again, and I had light for the last three hundred metres. Typical!

Susques was actually a tiny little village, made up mostly of mud huts. My heart sank until, miraculously, we found a very basic room for the night in a tiny house. The room had a couple of beds and nothing else; not even heat. I was amazed to find that it had a shower, of sorts, and hot water! Total bliss. That was all we needed. We used our sleeping bags as blankets and fell asleep listening to a podcast about the British education system. This is how Mathieu learnt English, and it turned out we both loved to go to sleep to a voice.

We woke at dawn with dreadful headaches and stuffy noses. I felt sick and threw up straight after our homemade breakfast of avocado, bread, and cheese. The previous day had been a long one, with extreme temperatures and altitude changes. We were still at 3,800 metres, and so put it down to these factors.

Although our plan was to ride together to Salta, that morning we both felt we wanted to do something different. I saw no reason to head for a big city, so I turned Rhonda towards Tilcara, while Mathieu decided there was a road to be ridden south of Salta. We split the rations of water and said our goodbyes once more.

"Adios, Amigo. See you in Bolivia." But I never saw him again.

The Atacama Desert had to be high up on my list of most beautiful places in the world; right up there with Antarctica. Two contrasting deserts. I'd travelled from the coldest place on earth to the driest place on earth, and *both* had made me throw up.

Chapter 18
Broke in Bolivia

As the speedo struck 65 mph, I shut my eyes and counted,

"one thousand, two thousand three thousand..." I managed fifteen seconds before opening them again. My focus came back in a rush of white, as if someone had stuck a bright light over my eyes. I checked the speedo again. I'd dropped to 35mph and not even felt the difference. *Has my direction changed, too?* I couldn't tell. There were no points of reference. I had challenged myself to thirty seconds of blind riding at speed on Uyuni Salt Flats. The largest salt flat in the world, it stretches for 11,000 square kilometres and is surely a gift to bikers. (Although, ironically, only a few weeks earlier, in January 2015, after heavy rain, it had been more like hell on earth to the motorcycle competitors in the Dakar Rally who had been decimated by the wet salt and water of Uyuni attacking the electrics and alloys of their machines). But for me, now, it was a playground. A blank sheet. No, not blank. An even-weaved canvas of polygons, shaped by thermal contractions, just waiting for a needle to cross-stitch the hell out of it. My bike was the needle; my imagination, the thread.

I got back up to speed, said,

"Have faith," then shut my eyes and tried again, this time managing seventeen seconds. What a rush! Bizarrely disorientating. "That will do," I giggled and pinned the throttle back once more for a long blast for the sheer hell of it. This time eyes wide open. All my senses fully focused now, and Rhonda's engine thumping a beat to our creative dance.

My ride into Bolivia had been fairly straightforward as far as my rides went these days. The border crossing at Villazón took a couple of hours and nearly half a packet of my emergency biscuit rations. I followed that with about three hundred kilometres of dirt, sand, and corrugated tracks through the cactus-covered mountains into the town of Uyuni.

I was less than proud of my riding skills that day – I lacked the courage required to ride the sandy stretches and planned for the falls rather than the successes. Speed and conviction really *are* your friends when it comes to sand, yet despite having lots of experience, I didn't do what I knew I should. All I could think about was how heavy my bike and luggage would be to pick up should I 'bin it' and made it twice as hard for myself by taking it slowly and paddling through the deeper sections. I wimped out and then placated my self-flagellation by pointing out that I had, in fact, made it through and that there were no prizes for style.

Rhonda's suspension was no match for the miles of corrugation that followed the sand. My vision blurred as my eyeballs rattled in their sockets and I wondered whether the bolts holding my top box would break again.

Rhonda and I arrived in the dusty little town of Uyuni later that day, none the worse for wear. The unpaved streets were lined with stick-and-mud houses and patrolled by a distressing number of stray dogs. Everywhere I looked, there were fornicating dogs; many unable to get free after their moment of pleasure – a sight that always reminded me of the Push-Me-Pull-You character from Dr Doolittle because they always seemed to end up facing in opposite directions, but with their genitals still locked together. It was something I never got used to, though. Not the fornication, necessarily– at least they were getting that – but the number of unloved animals that littered this country and so many others. Scrawny and mange-infested dogs and cats left to fend for themselves and procreate until there were so many the rubbish bins could no longer provide for them all. Many had twisted, broken limbs; some had gaping wounds; all were malnourished, let down, and abandoned by the humans who once loved them. But then I guessed life wasn't easy for anyone up here on this high plateau where nothing ever grows. Since Turkey I'd carried dog biscuits and saved any food scraps in my panniers for the worst cases I came across, but it was a mere drop in the ocean. I told myself that if I could make a difference that day to one dog or cat, then it was worth it. My heart never hardened, but my wheels kept turning. They had to. There was always worse to come, and there were too many lives to save.

After the salt flats, I continued vaguely north to the city of Potosí, where I sat for a day or two crunching numbers. I was in grave danger of starving myself if I didn't figure out a way to make some money soon. Despite sticking to my measly budget, my bank balance was, unsurprisingly, low. I had planned to be on the road for eighteen months, yet my overly optimistic budget only stretched – and I do mean stretched – to sixteen at a real push. I was now thirteen months in, and with the best will in the world, I only had enough to see me through to Colombia. From there, I had to figure out how I was going to get across the Darian Gap into Panama. However I did it, I knew it would cost more than I had right now. There were the small matters of Peru and Ecuador to cover before that, not to mention the thousands of miles afterwards to see me home. *How the hell am I going to make this work?* There had to be a way, but I was damned if I could see it. Brute force and ignorance would only get me so far. I had to come up with a plan before I was obliged to eat those dog biscuits myself. I decided to ration myself to just one simple meal a day. Thankfully, a plate of chicken and rice was less than the price of a Starbucks coffee back in the UK. It was all relative though; I wasn't in the UK and I had no income. I pushed on to La Paz. I knew the answer to my money worries would come to me somewhere in the midst of my helmet time.

The highest capital city in the world, La Paz rests on the Andean Altiplano at 3,650 metres, three times higher than the tallest mountain in Wales. I was getting used to the altitude by now, having been on the Altiplano – the 'High Plain' – since northern Argentina. My body still felt as if I were pushing through deep water though, and unless I was sitting, my breathing was like that of an old woman with a sixty-a-day smoking habit. It did get easier; I guessed that was partly due to my internal mechanics adjusting and compensating for the thin air, and partly just getting used to the feeling. Still, I felt a little foggy today, as I worked my way towards the capital.

La Paz means 'Peace'. "How ironic," I thought, as I joined the waterfall of traffic that ran down into an enormous stadium-like bowl and joined the cavalcade of honking-horns, angry drivers, and chasing dogs. And boy, were they chasers here! *Funny how the dogs behave collectively differently, depending on the country.* By the time I found a hostel, I had been chased up and down these awkwardly steep and cobbled roads no less than four times. One such occasion resulted in my size-6 Sidi motorcycle boot connecting with an oncoming canine jaw as it launched itself at my leg, an action I would have preferred to save for the clearly blind taxi driver who very nearly took me out just a few minutes later. I kicked his wing mirror instead and swore at him in Welsh; a language that lends itself beautifully to obscenities. Swearing, singing, and occasionally rugby – that's what the Welsh are best at.

My body strained under the pressure as I climbed three flights of narrow stairs to my room; helmet tucked under one arm, tank bag under the other, and a roll-top bag in each hand. Dropping one of the bags at the door, I took the key from between my teeth, opened the door, and bundled everything in before dumping it all on the bed. The hotel was in an elevated position, and my room was at the top of the building. The parking was down the road, the stairs were a pain, and the room was tiny, but the panoramic view was to die for. As the natural light faded, the city lights took over, working their way up the side of the natural bowl before merging seamlessly with the stars. There was no horizon. I was inside a ball of twinkling lights.

That night, I came up with a plan. I would try to set up an online raffle. How hard could it be? First, I would contact my friend Richard, who runs a marketing company back in the UK, and see if he could help set up a website. Then, I would contact all the sponsors who had helped me with my kit and see if they would be up for offering prizes. People would pay to enter and 'hey, presto', everyone's a winner! My sponsors would get some promotion; the entrants would obviously be in with the chance of winning some prizes and I would make some money to keep going a little further. That would buy me some miles and some time to come up with the next plan.

I sat typing out emails furiously for a couple of hours at least, then, feeling weary, I got up to go and find some coffee. I got about one step before my head spun out and my body gave way beneath me. I fell on the bed and hit hard, like a fighter in a UFC knockout. *What the hell?* I tried to get up, but my body was too heavy, so I lay there not trying to move for a while, telling myself not to panic and to just focus on the pattern on the wall; it would soon pass. And it did – but not before I'd drooled all over my nice floral sheet. Within a few minutes, I was back in a sitting position trying to figure out what had just happened. I decided it was probably a combination of lack of food and high altitude. I decided that the next morning I should hang out and find a cheap breakfast for the energy, and some coca leaves for the altitude. Death Road could wait.

As luck would have it, my hostel was just around the corner from La Paz Witches' Market. Of course, it was. Isn't there a Witches' Market on every street corner these days? This market has been run for generations by the Yatiri, or 'witch doctors', who are recognised by their tall black hats and coca pouches. Although we might call them witch doctors, the Yatiri are medical practitioners and community healers among the Aymara of Bolivia, Chile and Peru. They use both symbols and materials such as coca leaves in their practices. I roamed the stalls for a while, checking out the merchandise that ranged from medicinal plants to potions, to native herbs, to dried llama foetuses. I also spotted black penis candles, armadillos, various parts of frogs, naked couple figurines, aphrodisiac formulas, owl feathers, dried turtles, starfish, and snakes. Need to punish a cheating lover? They had something for that. The dried llama foetus was for prosperity and good luck. I just wanted something for the altitude, so I bought a bunch of coca leaves to chew on my journey. I'd leave the rest to chance.

Coca leaf is the raw material for the manufacture of cocaine but the transformation requires several solvents and a chemical process. The native people use coca leaf as a stimulant, like coffee. It is also said to help with altitude and stomach problems. I'd had it last time I was in Peru, and found it was a bit like having half a pint of beer on a summer's afternoon – a very mild, uplifting sensation. I stuffed the carrier bag full of leaves into my bag and went in search of breakfast. I'd seen enough of the market. While it was interesting, I suspected that it was more for tourists these days, and possibly far removed from its original roots. Anyway, if you've seen one llama foetus you've seen them all, right?

Back at the hotel, I found a few emails from sponsors who'd replied to my requests for prizes. They were in, and Honda was in, in a big way – they'd offered a brand new motorcycle as a prize! The same model as Rhonda, a CRF250L I could set it up as 'Win your very own Rhonda'. This was amazing. I also got luggage, some motorbike clothing, a weekend in France and several other prizes. Enough to make it a great raffle. This was going to save the day.

The next morning, I set off for Peru with a smile on my face and a pocket full of coca leaves.

Chapter 19
Purging for Pacha Mama

"The first two stages are going to be very hard," explained the stout little Peruvian man with a weathered face and soothing voice "Once you swallow the plant, you will feel it merging with every part of your body. Stay in control. Be strong, but let it do its work." He had an indefinable resemblance to a goat, but I knew that focusing on such details would not stand me in good stead later on in the proceedings,

"Okay," I thought, "I can do that. No big deal."

"Then you will feel like you are going to die." *Ah, okay. Big deal!*

I was in a little round hut somewhere in Peru, about to try a hallucinogenic plant called ayahuasca in an ancient Incan ceremony. "You will never have felt anything like it before," he continued. *"We'll see about that,"* I thought, but said nothing.

I looked around and saw that the nervous faces of my fellow rookie dimension-travellers had turned to horror, while all I could think was, "Is this for real, or just another witches' market?" Still, I was now in, and wasn't about to back out. After all, I had just spent the previous day abstaining from food, alcohol, and caffeine for this. Not only that, but I'd drunk eleven glasses of 'purgative medicinal volcanic water' until my shit ran clear, like a watery enema, and I could be pronounced 'clean' by the nurse. All part of the service.

"Close your eyes and breathe deeply. Remain positive," he continued. "After around thirty minutes, your body will begin to purge. This is the plant getting rid of all the negative forces in your body. It is good; let it happen; don't fight it." By 'purge', he meant throw up, and it didn't sound too positive to me. The little man then got on his hands and knees and demonstrated the 'purging position'. I wondered when we were going to get to the good part.

These ceremonies were started by the Incas centuries ago, and over the past ten years, spiritual healing clinics have been growing in number, catering for the thousands of people who travel to Peru every year to take a journey into self-discovery. It is believed that this plant extract, helped by guidance from your shaman, opens the doors of perception and allows you to see into another world where the lucky ones will get to meet the Inca deity Pacha Mama – Mother Nature. "Sounds interesting," I thought. I had no money to visit Machu Picchu anyway. In fact, I didn't have the $250 required to do this 'Inca trip', either, so I contacted the organisers to see if I could do it in return for writing an article on them for my blog. After an intense medical questionnaire, they agreed.

Set in the quiet little town of Písac, the retreat did feel pretty damn chilled, I'd give it that. I did want to take this on with an open mind, but in truth, I expected nothing more than a simple drug-induced high with a shaman who, I imagined, probably stripped off his robes and nipped out in his jeans for a cheeky cigarette straight after the show. I find it hard to take anything seriously in which the organisers put their spiritual endeavours into a glossy brochure and profit from it. That makes it something else. That makes it a job. *I know, I know. So cynical, for one so young.*

Písac sits in Sacred Valley, about forty minutes northwest of Cusco, the ancient capital of the whole vast Inca empire, and it was hard not to feel just a little moved towards the spiritual side by the landscape of lush terraces and ancient Incan ruins. The garden in the retreat was beautiful, too, and the only sound was the birds chattering as they flew amongst the flowers in full bloom.

I was introduced to my fellow 'trippers' before we entered; a professional couple from Belgium, an Australian student and two young guys from New York who were just about to start their university degrees – one in law, and the other in business. Once we were inside, we had to adhere to the code of noble silence – no talking; only meditation. There were dogs there and dogs always calm me, so I found that bit very easy.

The ayahuasca ceremony involved many preparation rituals before the big event. First, we stood in the garden and asked permission from Pacha Mama by holding three coca leaves between our fingers and listening to the chanting of our shaman. Then the shaman cleansed us further by placing us in a circle in the garden and hitting us with flowers as he chanted.

Then it was time for the coca leaf reading. The shaman called us in to the house one by one. I went first, and through an interpreter, he read my future by throwing down the leaves. He looked a little shocked at the overturned leaf set apart from the rest.

"Well that's ominous," I thought with an internal, and slightly sarcastic, eyebrow raise. Does everyone get the same look?

"You are in good health," he said, "But you have a great fear of something." He then asked what it was.

My mind drew a blank. In an attempt to be helpful, I said the first thing that came to my mind,

"Heights?" I asked, and half expected a buzzer to sound confirming my ineptness. It seemed he was looking for something deeper, but I had nothing to give.

"You have split up with a partner," he continued

"Oh, yes. Several," I said happily.

"...but you are going to fall in love with someone you are already in touch with. The leaves are very clear."

Try as I might I found it all hard to take seriously. This was for people desperate for answers. I might like some answers, wouldn't everyone? But I was pretty sure this man in front of me had no short cuts to them. Later that evening, we were shown to a little round thatched hut, decorated in soft tactile materials and rich earthy colours. They led us inside and instructed us to sit in a semi-circle on the cushions provided. Next to each cushion, was a bucket. First, we were all given water and earth that we mixed together in our hands and rubbed all over our bodies. This was followed by perfume. Then the shaman slowly and deliberately poured glasses of thick, brown liquid. He blew pure tobacco smoke into the top of each one before the guide handed them around one by one. We sat cross-legged in the candlelight, holding our elixir, and waited for everyone to be served.

"Now you must drink," said the funny little goat man. "Drink it all at once, then lie back, close your eyes, and breathe."

"Here goes nothing," I thought, and downed my potion without hesitation. It tasted foul – like a mixture of dirt and Marmite.

The guide blew out the candles and lit a tiny fire in front of the shaman. I watched his glowing face in fascination as he sat in his white robe and began his shamanic chanting. He was now the only thing visible in the dark room. Shamans say they were taught their chants, or icaros, by the plants themselves. Since they're taught directly by the plants, each shaman has his own distinct set of icaros. Just as shamans believe that every animal, tree, and stone is its own being with a spirit, they believe that every being also has an icaro unique to them. This icaro is provided to help guide us onto the spiritual plane as the ayahuasca takes effect. I closed my eyes and listened. All I knew was that it sounded like it was coming from an ancient place, its voice deep, soothing, and melodic; sometimes mimicking natural sounds like birds and maybe the rustling of leaves. It was like nothing I had ever heard before.

They put out the fire, and the singing continued in complete darkness. It sounded as if the shaman were moving around the tent, but he stayed right where he was, invoking the spirits. I began to feel like I was floating just a little way off the pillows. I pulled at the warm blanket I had been given and snuggled into it. Soon, the purging started. Not me, the others; more than one began moaning and retching. I could feel the liquid coursing through my veins, but I didn't feel particularly bad. I saw lights in the darkness – swirls and patterns of colour – like flowers and leaves flickering in a warm light. One of the guys cried out, and I was suddenly aware of him being helped up; to the bathroom, I think. He was crying.

Then, voices all around me. The young American soon-to-be lawyer spoke the loudest,

"I feel so much love. I love you all. Hey, Australian guy next to me." He didn't answer. "I love you." He continued unperturbed, "Steph?"

"Yes?" I replied, knowing the answer.

"I love you."

"I love you too, mate," I managed, and I found I was now struggling to speak, as if someone had disconnected my lips from my brain. People were talking all around me now, but not to each other. It was as if they had just met up with old friends and were catching up. Some later said they'd met Pacha Mama and experienced something life-changing. I did not. It was a while before anything much happened beyond the floaty feeling for me, which was beautiful in itself, but not life-changing. Maybe one hour in, I started shaking uncontrollably. I was so cold, despite my layers of blankets. Goat Man came over and put another blanket over me. I chose that moment to sit up. He took my hand and bleated in my ear,

"Leeet's breeeath togetheeer", which we did. Slowly and calmly as one. I had moments of relaxed and beautiful floating, followed by uncontrollable shaking. I opened my eyes to a colourful darkness; where the hand in mine was now a cloven hoof, albeit with thumbs, and Goat Man's ears were now horns. He was a cartoon in a cartoon world. Everything around me was moving; shapes dancing; breathing as though I was tapped into the pulse of the world.

"You must take more," he whispered again. "Some people need a bigger dose." He was concerned I had not yet purged. I couldn't face drinking more, despite knowing this may be the only solution to reaching Pacha Mama. My mind and body completely rejected the idea, and by doing so, I sat up and purged. Then, I continued to happily float around the room in a swirl of lights to the gentle rhythm of the earth's pulse.

After about three, maybe four hours, the feelings went away, and I landed back on my pillows; relaxed and sleepy. Soon, I was checked again by the nurse and helped to my bed. It was over.

In the morning, we had breakfast together before meeting the shaman and his translator again to discuss our visions. One at a time, each person got their turn to speak and share their experience. Most told a tale that related to their life in some way and that they had been guided along the path by old friends or strangers. Those who said they met Pacha Mama told us she had shown them the most amazing things. They were given advice and guidance for their future, and now felt so much love for both nature and people. The majority felt they had had a life-changing experience and were positive their future was brighter for their new-found knowledge.

Listening to them telling their stories one after the other was very emotional; many cried with joy. I cried with them, and as anyone who has tried hallucinogens of any kind knows, this is a common side-effect. They were probably going to feel like shit in a couple of days during the stage commonly known back home as 'come down Tuesday'.

I walked away feeling slightly disappointed that I had not felt more. Perhaps, in truth, I had been foraging for that 'Antarctica' feeling again or searching for that buzz of my first few days on the road. They had been real; earned, induced by actions not drugs; yet they were feelings I feared I would never find again once all this was over. I had a wonderful evening; another story to tell for sure, but not life-changing by any means. Having tried many drugs by this point in my life; including LSD and good old Welsh magic mushrooms, I'd certainly had better mind trips. Perhaps I'd grown out of it. Perhaps my life was just crazy enough for now. Perhaps Pacha Mama had nothing she wanted to say to me. She had, after all, already shown me some pretty amazing things. Maybe it was someone else's turn. With that, I went back to Cusco to gorge. I'd had enough purging for one day. My true path to happiness that night was in a large pepperoni pizza with extra cheese and an ice-cold beer.

Having failed in my mission to meet Pacha Mama via the ancient Inca methods of ceremonies and hallucinogenic plant forms, I decided it was time to revert to my old ways of getting close to Mother Nature. I was going to have a night of camping in the Amazonian jungle. "Not a bad second best," I thought, as I jumped on my bike and headed out of the city on a dirt road that took me back over the Andes towards Kosnipata Valley and Manu National Park.

You won't find Chontachaca on the map. Its twenty-two inhabitants are blissfully hidden in the misty Amazonian cloud forest about a six-hour ride out of Cuzco. I found it quite by chance, having chased a mountain track off the tarmac and onto the dirt for miles. Twisting through the glorious Peruvian mountains I easily forgot the chaos of the Cusco traffic and the hordes of holidaymakers that frequented the city. This was more like it; the crowds and diesel fumes wonderfully replaced by fresh country air – like wine – and mischievous livestock. I found my way to Paucartambo first, a small town on the edge of the Andes. There, after some debate with the locals, they directed me up another mountain track. The track got narrower, and the obvious signs of landslides more frequent, but all was well, and I continued on my way, safe in the knowledge that there was a town nearby. If

I got lost, I could always head back there and find shelter for the night.

About half an hour in, I found a landslide that covered most of the narrow track. There was just enough room for Rhonda and me to sneak through – if I didn't look down. We made it, but there were more obstacles ahead. Most of the dirt road was torn away from the side of the mountain, leaving a gaping hole. There was even less room here, and I didn't fancy my chances, although I bet the locals did it without blinking an eye. I turned back and found another track to follow, and it was the best thing that could have happened that day. I worked my way along the trail until I came across a clearing in the trees and stopped to take in the view. Sprawled out before me in a relentless blanket of green, filling the valley and the surrounding mountains below, was the Amazon. It was magnificent.

I began my descent down into the valley, with no real idea of where I was going, but I couldn't resist finding out. My lights were still not working, so if I got lost and night came, I could be monumentally stuffed. The drop-offs along the road were no joke, but again, I didn't care; I just kept riding. I was on a roll. The road had to go somewhere, and if not, I had my tent and if I had to, I could always camp next to the road, except that the road was narrow, with a drop on one side and a cliff on the other. There was no side to this road. Still, I didn't care, continuing through streams, under waterfalls, and into the start of the rainforest itself. For two hours, I went deeper and deeper, until I knew it was too late to turn back and get out before dark. I saw no one; not even the obligatory man with a donkey that you see just about everywhere when you think there could not possibly be anyone else around. This road was far more spectacular than Bolivia's Death Road, which I had had to do before I left the country, if only to say I'd done it. I knew with a name like 'Death Road' (changed from the original 'Yungus Road') and the fact that *Top Gear* had featured it in one of their stunts, it would be a tourist trap. I hoped they didn't find *this* place and turn it into another gimmick. I felt as if I were riding into a lost world, and the solitude alone was epic.

I came down through several climates before the road flattened out and finally, I saw my first building. Standing outside, as if expecting me, was a little man with white hair, maybe in his late sixties, and smart, in a scruffy way, with an old shirt neatly pressed and an older silk scarf around his neck. He waved, and I stopped to say hello, perplexed at the site of him here, in what appeared to be the middle of nowhere, standing at the gate of what looked like his house. It was a wooden house, surrounded by fencing that kept out the encroaching jungle and framed his garden with a host of shrubs and flowers. It turned out he *was* expecting me, at least for the last few minutes. He'd heard me coming down the valley, and had come out to greet me, so infrequent was the passing traffic.

"Come in for tea," he said, in an accent I couldn't place. He didn't even ask my name.

Inside, the house was even more eclectic than the garden, and it might explain what he was doing here. It was filled with insects from all over the world, all carefully framed and displayed in glasshouses or frames. Beetles, butterflies, stick insects, and moths of all shapes and sizes, each beautifully preserved and catalogued. José introduced himself as he fiddled with the gas hob for the kettle; he was a Majorcan who'd settled in this place about six years earlier, having spent most of his life travelling the world, studying insects.

"I just fell in love with this place" he offered, "My garden is designed to attract the insects. Come, let us take our tea. I will show you around".

"Wow, you really are the insect man of the Amazon," I said, as we pottered around the garden, with José explaining each plant and pointing out hidden insects along the way.

"You must stay the night," he said, "and tomorrow my friend Paulo will be here. He will take you to sleep on our platform, if you like."

"Sounds good, thank you," I replied. I had no idea what he meant by 'platform' or who Paulo was, but I instinctively knew that if José was suggesting it, I probably *would* like it.

Paulo and his six-month-old sausage dog turned up the next day in a beat-up old 4x4. Thirty-something, from Barcelona, Paulo had found himself down this road a couple of years before, and he, too, found it impossible to leave, having fallen head-over-heels in love with the place.

We parked the truck a hundred yards away and waded across the river to get to the platform built in the trees. There were a couple of beds covered in mosquito nets, a couple of bean bags, and a basic kitchen area with a compost loo outside. It was only four kilometres from José's place, but, due to the dense jungle and rocky terrain, it took us about an hour to get there. It was probably quicker to walk, but the ride was a lot of fun.

"You sleep out here alone?" I asked, as I dumped my rucksack on the floor and flopped on a beanbag.

"Me and my puppy," he confirmed. "I love it here."

"What's his name?" I asked, curious to know every detail.

"Just Puppy,"

"Hello, Just Puppy" I said, and smiled as a memory of the Chilean officer in Antarctica popped into my mind – the one who thought my name was 'Just Steph'.

I could see why he loved it here. On his own in the trees, he must have felt like Tarzan – but it couldn't be easy. Aside from biting insects, there were many other things to consider. The jungle has pumas and jaguars, poisonous frogs and snakes, plus the indigenous tribes and drug runners who also frequent the jungle. It may be a big jungle, but I would hate to meet any of them in the dead of night, on my own, with only a small sausage-shaped puppy to protect me. I was impressed, and all at once that 'impostor' feeling came rushing back to me. *This* was an adventurer. *This* was commitment. I was not a patch on Paulo.

That night, after a few rounds of Blackjack by candlelight, I climbed into my sleeping bag, tucked in the mosquito net, and lay there enjoying the surrounding sounds of the insect chorus accompanied by the gentle flow of the nearby river. It was a peaceful night and I slept like a stick insect.

I woke early, as usual, and took my toothbrush for a wander up-river to bathe in the fresh water while Paulo and Puppy slept in. I hadn't brought a towel, but it was warm, and I would soon dry off. Finding a suitable rock pool, I stripped off, got in and lay there naked; liberated.

Hearing a squawk above, I looked up to watch a pair of macaws flying over. Then a rustle in the trees drew my attention to a pair of eyes watching me with intense curiosity and suspicion. A squirrel monkey, I think. I then realised that there were maybe three or four of them watching me from their vantage points; probably thinking, "What is this thing, and what is it doing here?"

They were both valid questions. Who *was* I, and what *was* I doing here? I'd been so many things in my life, it was hard to know which was the *real* me, if any. I knew what I definitely was *not,* and that was a human resources manager. I wasn't even that person who had ridden through Europe just over a year ago. She was more frightened than she cared to admit. Was it really only a year? That seemed like a lifetime ago now, and yet as I lay there pondering the meaning of life and my reasons for being there, an age-old worry crept in. Maybe I *was* still a human resources manager. Maybe I still worked silly hours and battled the rush hours to get there every day. Maybe all this was another crazy dream, and soon I would wake up to find the river was actually just a wet dream and the squawk of the macaw was my 6am alarm. Then I smiled at the thought.

Nope. This was real, this moment right here. This moment in time. *This* was why I travelled. These moments were what I was looking for. It wasn't so much an ultimate moment or an end goal, but a collection of moments. These were what made it all worthwhile. They were priceless; soul-repairing. I didn't need ayahuasca or any other drug for that. I needed to feel something real. The good and the bad combined were what made me feel alive. There was no one without the other. No light without shade.

I realised then how little I was worrying lately. Money; the road ahead; the noises outside the tent; the strangers I met along the way; the fact that I was officially homeless and had no real plan for when I finally got home. *Home!* It seemed so far away. Back in a land where people knew me; longer than a day, I mean. Where people showered and brushed their hair with a regularity and frequency that now blew my mind. A place where clothes smelt of soap and were carefully selected from a whole wardrobe of other clothes. A place where they put their biscuits in tins – not tank bags – and picked food from their fridges, fresh and cool, without a second thought. A fridge. Now *that* I missed. I would never again take one for granted. I also missed the British sense of humour, soft pillows, and my dog. I'd really missed my dog!

How would it feel to get home though? Would I ever fit in again? I could barely imagine how it would feel or, indeed, how I would cope in that strange land that now seemed so far away; so organised, so foreign.

I splashed my face, pulled myself out of the river, looked up at the monkey and said,

"One thing I *do* know, is that it's high time I bought some new underwear." With that, I picked up my frayed black knickers, held them in front of me, and addressed them directly, "Your days are numbered, my friend. I don't know about me, but you? You definitely have no place back home."

Chapter 20
From Peruvian Prostitutes
to a Colombian Crash

Lima, the Peruvian capital, was my first brothel experience. I'd gone for the cheapest 'digs' I could find. The only thing that gave away its perfectly legal, if slightly disreputable operation, was the constant wailing and banging from the surrounding rooms. Still, it was clean, it was cheap, and they offered beer on room service – nothing else – just beer. If I turned my music right up, I could barely hear the raucous counterfeit chorus of ecstasy through those flimsy walls.

I'd left Rhonda with a local Honda dealer for a couple of days to see if they could figure out my intermittent headlight problem. The biggest problem was the 'intermittency'. When I picked her up, they were confident the problem was fixed. The headlight certainly worked, so what else could I do but pack her up and get going? *But wait. Where are my keys?* Unpack. Find keys. Repack (more tidily this time). Hit the road. Headlight goes out again. *Damn it!*

Despite the glitches, I could not resist one last ride into the majestic Cordillera Blanca mountain range for some hiking and biking before leaving this fascinating country. I'd ridden in Peru back in 2013 with a group of guys on an off-roading trip and so I knew what treats were in store for a biker with a taste for rough roads and rugged trails. Not far from the Equator, this region forms part of the Northern Andes – the spine of South America – and is considered to be the world's highest and most glaciated tropical mountain range. Leaving my luggage behind in Huaraz, I headed a further twenty kilometres up the trails, and then a further three-hour climb on foot to Cherub Glacier Lake at an altitude of 4,600 metres before heading further north the following morning along curvaceous mountain passes to Cañón del Pato and onwards.

I waved to the Quechua women in their beautiful red woven dresses and traditional hats, or *monteras* – all an indication of which village or region they are from – as they went about their daily business of fetching and carrying. Many of these mountain roads were built centuries ago by the native people of the region (pre-Inca), although the forty tunnels through the Cañón del Pato were originally part of a railway built in 1952 but closed after a major earthquake in 1970. There were no electric lights, and many had bends which obscured any sunshine from the other end. *What I wouldn't do for a working headlight right now!* Without it, all I could do was aim for the pitch-black hole and hope for the best as I beeped my horn and rode on through. The horn was less of an attempt at echolocation and more a warning to any other potentially lurking no-headlight life forms that I was coming through, ready or not. It went against all my instincts, and I definitely let out a few bat-like squeaks – but what a rush. Endorphins? *Check!*

It was a delight to ride this area and I felt the total unification of mind, body and motorcycle as the Andes once more attacked my crusty outer layer and soothed my very soul with their primordial elegance. Never before had my 'helmet time' felt more meditative – almost spiritual – as I was propelled upon my steed, gliding along the dirt, to the regular rhythm of her single-cylinder 250cc four-stroke engine.

In total contrast, I spent the next three days in the fishy-smelling town of Chimbote on the coast, sitting on the toilet and throwing up in the bath after my stomach finally met its match in a badly cooked chicken dinner. I couldn't leave my room at all, and I wished I had someone to fetch and carry and mop my brow, bring me soup and administer sympathy. But I didn't. I endured and I drank water and I threw up until my body felt weak and spent and empty. Feeling less meditative, more bilious now, I got back on my bike and rode the path of least resistance along the Pan American Highway to the border with Ecuador at Macará.

Ecuador had cheaper fuel and quieter roads – definitely a bonus. I also found, to my surprise, that more people spoke English than anywhere else I'd been so far in South America.

"Eggs. What the hell is eggs in Spanish again?" I muttered absentmindedly as I searched the shelves of the first convenience store I'd come to. The shopkeeper turned around and offered,

"*Huevos.*" I laughed and replied,

"*Sí, huevos.*" We then had a pleasant chat in Spanglish while he helped me with the rest of my shopping. He even threw in some complimentary Ecuadorian sweets to try later with my roadside breakfast of eggs, cheese and avocado – a safer bet than stick-shack chicken.

Ecuador was friendly, and the roads were smooth, new, and growing all the time. China had arrived and made an agreement, probably along the lines of "You give us oil, and we build you roads." But I didn't want smooth, polished roads; those were for people wanting to get from A to B as fast as possible. It was going to take some planning to get off them and find something a little more stimulating.

I started by poring over my freshly-bought map at breakfast before deciding to contact Freedom Bikes, a motorcycle company in Quito who offer trail riding all over this area. Without hesitation, Court Rand, the owner, very kindly offered me a selection of off-road routes that would take me cross-country through the hills, valleys and dense cloud forests all the way to the capital.

The trail was a little tricky at points; steep and muddy with ruts, rocks, and hairpin bends all at the same time. I dropped Rhonda somewhere just past the little village of El Corazón on the western side of the Avenue of Volcanos – aptly named because it's flanked by eight dizzyingly-high snow-mantled mountains that form part of the Andean spine. I think tiredness from my recent bout of food poisoning played a part, with the heat and altitude compounding my clumsiness. Of course, the cavalry – a little old man in a cowboy hat leading two horses carrying barrels of water – arrived just *after* I'd unloaded her and picked her up. He looked completely bemused at the sight of me holding my bike precariously, red-faced and sweating, with luggage strewn across the path. We had a conversation, but I have no idea what he said to me, or vice versa! He didn't seem to mind, but once again, I felt an opportunity go by, even though, by now, I could pick out a few bits of meaning from the Spanish; common questions like, "Where are you from?"

and "Where are you going?" Unfortunately, the real colour of the experience is in the details that still escaped me after all these months in South America.

After a couple of days camping out somewhere between Zumbahua and Chugchilán, overlooking a misty valley and giving my body a little time to catch up, I left the beautiful Ecuadorian countryside with its colourful indigenous settlements, and hit Quito, the capital, to rent an apartment for the equivalent of £10 a day.

As soon as I got in, I cooked some *huevos*, made a cuppa, and pulled my laptop out to check on my online raffle. It was going very well, with hundreds of tickets sold. There were just two weeks before the competition closed, and it looked as if I was going to hit my target. If I made it, I would have enough to see me through Colombia and Mexico, and I was already three-quarters of the way there. I decided to call Shane with the good news. Since I had good Wi-Fi, I made a video call so I could see his face. I missed his face.

"I've been thinking," he said, once I'd finished raving about my successful campaign. He clearly had something on his mind and was patiently waiting for me to take a breath.

"Go on," I said, his tone alerting me to the fact that this could be big.

"I thought I might come and meet you in Mexico and ride with you for a while."

Shane was now back in Australia. He had left his bike in Bulgaria and gone home to earn some more money as a trucker before going back to continue his trip. He'd been dithering over his plans for a while, though, and couldn't seem to make his mind up where he wanted to go from there. He was more of a drifter than me, though lately it had seemed that he was searching for some direction in his life. I paused to consider this new information.

"OK, you hate the idea. It's fine. It was just an idea," his voice slightly sulky now.

"No, no," I said quickly trying to figure out what this meant exactly. Did he mean keep riding with me all the way back to the UK? I wanted to see him, but I wasn't sure I wanted *that*. We had already gone through this a thousand times. "I like the idea, but how will you do it? Ship your bike from Bulgaria? And what about…" He interrupted,

"I'm not sure, but maybe I could buy a bike in Mexico and then sell it when I'm done."

"Oh, so you would just be joining me for the Mexican leg?"

"Yes, don't worry," he said. "I wouldn't want to cramp your style for too long." Again, with a little hint of sulky, "and anyway, you know what I'm like. It's best we keep it just as fun, like we said." But I didn't actually know what he was like. He'd never displayed to me any of the nasty jealous streaks that he feared so much in himself and spoke of often.

Well, if it worked for him, it worked for me. We agreed we'd figure it out; a pair of misfits taking it one country-sized chunk at a time, always with an end in sight: the border. We didn't need to worry about the future, that was a pastime best left to other people; normal people.

After two condensed weeks in Ecuador, it was time to head for Colombia. I had one overnight stop en route to the border but misjudged my timing to the town where I'd planned to stay. At the hour I had expected to be rocking up at some cheap hotel, I was still riding down the bleakest, windiest, mountain pass with big drop-offs on every switchback, the sun and temperature both going down, and no headlight. I went as fast as I dared, stealing glances at the sun, which seemed to be enjoying the race. Who would get to their finishing line first? I had no chance. As the sun said its final farewell and kissed the horizon, I had no choice but to stop.

"What the hell am I going to do now?" I asked aloud. There was no way I could make it down this mountain road in this darkness, and it was just a bit too windy, exposed and, quite frankly, creepy, up here to camp. The mist swirled across the barren landscape. All I needed now was for a hirsute shapeshifter to appear, and we had ourselves a werewolf movie! Nope! Camping was my last resort. I pulled out my head torch and stretched out the band to fit it over my helmet. *Would this do?* It was a poor light at best, and once on the bike, I would see very little. I could easily make a mistake, and if I missed a turn there was certain death waiting for me just a metre or two off the path.

"Death by werewolf?" I muttered, "or death by precipice? Which would you prefer, Rhonda?" I didn't like my options. Just as I was erring on the side of precipice, I heard a rumble in the distance. *What was that?* I took my helmet off to listen carefully. *Was it thunder?* Then I heard it again; distant and getting closer. It was a diesel engine; no doubt in my mind now. A truck was working its way down the mountain and it had nowhere to go but right past me. If I waited long enough, I could jump in behind and follow its lights all the way down to the town, where I imagined werewolves had long since been ostracised. I pulled Rhonda off the road a little and waited. It was at least five minutes before the truck arrived, but it did arrive, like a gallant mythical basilisk charging through the night. My saviour.

As soon as he passed, I jumped on Rhonda, fired her up, and tucked in right behind him like a hidden parasite feeding on his diesel fumes. He probably had no idea I was there. Half an hour later we were safely in the town and I peeled off to find a welcome bed for the night.

****

Colombia. Best known for coffee and cocaine; it once supplied 80% of the world's cocaine habit. At one point, the government effectively lost control of the country to the FARC guerrillas due to the lucrative return from kidnapping and drugs. The fight went on for decades, with people like 'drug lord' and terrorist Pablo Escobar becoming worldwide household names. He was a *really* bad man that made our Kray twins look like Simon and Garfunkel. Back then, Colombia was home to some of the most violent and sophisticated drug traffickers in the world. Today, it is still a major supplier of cocaine, cannabis, heroin and yes – very nice coffee; but whilst the fight for peace and justice continues, tourism is very much on the increase.

Approaching the border, I felt my usual mixed bag of emotions. Sadness for leaving one country mixed with excitement for the next. I often felt guilty of rushing things because of that 'excitement for the next'. Like many who live on the road, I guess I suffered from what has often been described as 'white line fever'. An irresistible urge to keep moving and find out what lay down the road ahead. I couldn't help it. I still had that niggling fear in the back of my mind that the best times had passed, and that everything was in danger of becoming ordinary if I didn't keep moving. I guess I was yet to truly master the art of dwelling deeply in the present without some extreme feelings or events to anchor me there; excitement, joy, fear; even pain would do it. Colombia brought a vague feeling of nervousness; those butterflies I yearned for. It was stimulating. Real. Call me 'kamikaze', but I liked it!

My brief trip into the city of Cali for some new tyres and a fork seal was made far more interesting by the company of a local biker named Juan who had contacted me via Twitter at 11pm one night. A city is just a city without local knowledge and good company. Juan offered both, and within an hour I was hanging out in back-street salsa clubs and touring the city's monuments in between, including Cristo Rey (the 85 foot statue of 'Christ the King') and El Gato del Río ('The River Cat') – a sculpture by the Colombian artist Hernado Tejada. They were beautifully lit up at night and devoid of crowds.

The next morning, while treating my hangover by administering a crushed ice drink with lulo – or naranjilla (a citrus fruit with a particularly strong flavour), Juan invited me to ride on his Colombian horse named *Ronda*! I'd only seen the Andalusian horses on television; the way they move is incredible. Often referred to as 'dancing horses', their leg movement is double the speed of any other breed, whilst moving at the same pace. It's a controlled power, as if you're holding back a steam train with a feather while the wheels still spin at full speed. They have beauty, fierceness, athletic ability, and endurance and are still used today in bullfight rings in Spain.

I love a horse with spirit, and as I rode around this palatial stable complex (complete with horsey swimming pool), I was transported back to my childhood and a Welsh cob named Bonnie. Even as a child, I was a handful, so many a summer I got packed off to my grandparents' farm where I might stay out of trouble. My grandmother had many horses, but my favourite was Bonnie; with a strong spirit (and a 'cob trot' not dissimilar to the Andalusians) that could never be fully tamed, she often threw people off and would only jump fences for me, though even that depended on the day and her mood. I considered us both rebels; kindred spirits. I'll never forget those times we spent galloping along the beach with my dog (a lurcher named Kit) racing alongside us; the three of us connected in those moments; through energy, not thoughts.

Bonnie and I didn't always see eye to eye though, and during one particular show jumping competition, having committed to the take-off, she stopped dead, dropped her shoulder, and threw me into the jump. I landed badly on the wing; winded and with pain in my back and hip. My grandmother gave me a shot of whisky from her hip flask for the pain, then told me to get back on the horse and finish the round. You didn't argue with my grandmother, so I did just that. We went on to win the Champion of Champions medal that day – though not for that round.

Driving back to the farm that evening in the old green Daihatsu Fourtrak and smoking one of her sixty-a-day Berkeley Reds, my grandmother turned to me with a wry smile and said,

"You know, the reason you and that horse get on so well is because there is a bit of the Devil in both of you". I smiled now at her memory. She was a wise woman who didn't mince her words. She could also reverse park a humongous flatbed trailer better than anyone I knew. I missed her. She would have approved of my latest escapades, and she would have *loved* this horse!

From Cali, I rode the truck-infested winding roads from one beautiful Spanish colonial town to the next, camping in between, until I arrived at Bogotá, the capital of Colombia. A city I would discover over the next few days to be vibrant and colourful. The architecture, the night life, the people, and especially, the street art; the graffiti would have made Banksy proud.

On my second morning, I woke with an almighty hangover having enjoyed a little nightlife with some locals. It was also the morning that the 'win-a-Honda-like-Rhonda' competition ended, and I got to announce the winners. Rather than just announce it in a boring old blog post, I decided to make a video by wandering around Bogotá and asking random people on the street to announce the winners for me. I overcame the language barrier by getting the hostel receptionist to translate a message for me and write it on a piece of paper. The message said, "*I wonder if you could help me? I am announcing the winners of a competition by making a video. Would you be kind enough to read the following statement into the camera: The winner is… (insert name here).*"

I hit the streets on foot and wandered around filming myself saying things like, "And for the two nights stay in Motorbreaks…", then I would film a little interlude with the street performers, and approach a random stranger to announce the winner as per the piece of paper. People's reactions varied, but most agreed. I got taxi drivers, street vendors, young, and old to take part. I even tried several policemen, but none of them went for it. They were all very serious and moved me on quite quickly. I didn't argue; they had muscly rottweilers with them, and I didn't fancy my chances – not with a hangover.

By the time I got back to my hostel, I had loads of material to work with, and quickly rustled up a short video. I posted it online and then called the winners as well. The winner of the motorbike was obviously ecstatic:

"Wow. That's amazing. Thank you so much," he said. "I've just come off a really shitty night shift, and you have just made my year. Thank you!" It was such a great feeling to be able to give him that news. I went to bed feeling on top of the world, having brought a little joy into someone's life.

I worked my way north through a great many more Spanish colonial towns. The roads weren't much to write home about a lot of the time; often congested with trucks, but the places they connected were exquisite. Frozen in time, these little towns with their cobbled streets and white buildings often covered in rambling red roses and topped with terracotta tiles, offered a welcome break amongst the friendly locals. The farmers' markets were full of fascinating faces and a great place to sit and watch the world go by. The Saturday morning breakfast in Villa de Layva market was a must. I went for eggs, sausage, the Colombian version of black pudding, and hot chocolate – with a lump of cheese floating in it; not something I had ever experienced before, but it wasn't offensive. It tasted all the better for the atmosphere, as I sat amongst the hanging sausages and Panama hats.

In between towns, it was easy to find camping spots under the cover of the lush green foliage in the northeast. I would often wake to rain, and the sweet, fresh, powerfully provocative smell it brought with it more than made up for the wet outer layer of my tent. The sound of the rain on a tent on a warm night has to be one of the most comforting feelings in the world. It somehow puts everything into perspective. As I lay under my canvas next to a softly-scented wood and surrounded by nature, it reminded me with a soothing rhythm that the really important things are so basic – like warmth and shelter; the simple pleasures. Camping was a wonderful reminder; it took me back to basics *and* it saved me money. As it turned out, living on a budget enriched my experience far more than fancy hotels ever could have done.

I continued to enjoy these simple pleasures as I worked my way up to the coffee region around Medellín. I made fresh coffee with my little stovetop percolator while overlooking the plantations those very beans had come from and enjoyed the company of hummingbirds buzzing around me like mini drones spying for sugar. Eventually I found my way to Medellin itself, where – after 21,000 kilometres – I would put Rhonda on a plane over the Darian Gap and say goodbye to South America.

****

Sending Rhonda to Panama via Air Cargo Pack of Colombia turned out to be no easy option. We agreed on a price while I was still ten days out from the city, but they increased it when I arrived. Then they added charges I had previously heard nothing about. Once we renegotiated, they asked me to come to the airport to drop her off. The following morning, as planned, and with a severely loose chain that already needed replacing again, I set off through the city to the airport. It was a busy sprawling mess of fast-moving cars and people trying to get to their destination as quickly as possible. As usual, I found method in the madness, and went with it. Only this time, I didn't add quite enough madness to the pot.

I hesitated for a split second when choosing one of three lanes for my exit. The traffic was unforgiving, and the truck behind me didn't hesitate. I realised my mistake in that split-second before he hit. In the first half of that split-second, I glanced over my shoulder. Not fully, but just enough to see his grill out of the corner of my eye. In the second half of that split-second I threw my weight to one side and aimed the bike to the lane on my left. Then he hit me.

The impact sent me surging forward with the force of an impatient rhino. Rhonda and I twisted to the left and were sent sprawling into the lane opposite. Thankfully, that split-second warning gave me the time to aim our fall out of the path of his enormous wheels. I slid to a stop as an oncoming

214

car swerved around me. To this day, I have no idea how I wasn't hit in that left-hand lane where I fell. Moments before, there were cars all around me with no time for sprawling. The truck continued on. I was of no consequence; no more than a fly on his windscreen.

I jumped up and onto the central kerb that had stopped our slide; my eyes darted around the scene. Pedestrians on the other side stared, but it seemed like an age before anyone reacted. It must have been seconds, but it felt like minutes and I couldn't figure out why no one was coming to help. Then, there was action. A car stopped behind us offering protection from the flow of traffic. Passers-by ran over to help me lift Rhonda, and before I knew it, we were both safely on the other side with people asking (I assume) if I was OK. I was. Or at least I thought I was; I couldn't be sure. I was definitely in one piece. My helmet was scuffed, but at that moment, I was more concerned about Rhonda.

Aside from some holes in the soft luggage and a broken clutch lever, Rhonda was none the worse for wear. It was all fixable, although I no longer had a spare clutch lever. I'd used the last one after dropping the bike somewhere back in Argentina and hadn't thought to replace it. Somehow, I managed to bodge the broken lever, so it was workable enough for me to ride out of there. I just wanted to get away from the scene and assess the damage without all eyes on me – concerned as they were.

I limped into the airport cargo area half an hour later. By now, the adrenaline had worn off and I had a pounding headache from the impact of my head against the concrete. I could feel bruises developing on my left thigh and hip where I had slid; my shoulder, too, was aching; and the bodged clutch lever was not holding out too well. Neither was the chain, which complained and threatened to split with every rotation. All I wanted was to dump Rhonda and get out as fast as I could, so I could get away, lick my wounds and chill. The adrenaline spike had left me feeling depleted, and the shock made me feel as if I needed the release of a good cry, in private.

After some initial confusion, they led me to the office and asked me to sit while they found a translator. A meek looking woman arrived a few minutes later, and after a quick update from her colleague, she announced that I should come back tomorrow.

"You have to be kidding me!" I said, with the patience of a wasp.

The translator looked surprised at my reaction and turned to her colleague, who confirmed they were definitely *not* joking. They needed to get a mechanic to take the bike apart so Customs could check every bit of her, both inside and out. He added that there would be an extra charge for the mechanic.

"Please tell him that I was told to come *today* and my bike is damaged *and* I have just had a crash *and* there is no way I am coming back tomorrow *and* there is no way I am paying any more charges. You tell him that we are doing this *today*, and if anyone is taking my bike apart, it's *me*!". My anger rose with every word, and I realised I was now shouting. The translator's shocked little face struck a chord, and I calmed myself before adding, "Look, I'm sorry. I'm not shouting at you, but please explain to *him* that this is not acceptable. No way!"

I try to avoid being angry, especially when it comes to officials. Shouting at people doesn't often help. Patience and a smile usually work far better, but on this occasion I felt justified, and a Justified and Angry Welsh Person is a dangerous proposition. I was not going to take this *shit* today!

"You must come back tomorrow, and you must have a mechanic" she said nervously after another word from the guy in charge. I completely lost my cool and turned to shout at the boss man this time rather than the poor translator.

"I am not riding this bike anywhere and I am not paying any more!"

The translator rapidly translated my words, and after some debate and a quick phone call, she addressed me again,

"The police need to inspect your bike and we cannot get them here until tomorrow. He has agreed that you do not have to pay for the mechanic, but he insists you do have to have one. You can leave the bike here, and we will arrange for a taxi back to your hotel. You can come back tomorrow at 2pm for the inspection."

I guess they knew better than to mess with A Welsh Woman On The Edge. I took a second to calm myself, then turned to the boss, smiled, and said,

"*Muchas gracias, Señor*". I *had* lost my cool, but I still knew when to stop and be gracious in the face of a good compromise. I wasn't proud of my behaviour.

The next day I spent an hour taking the bike apart with the mechanic they had insisted on. First, the plastics and any covers that might hide drugs. Then the battery came out, the headlight came off, and we even had to take the wheels off and remove the tyres so they could check inside. Just as we finished, the police arrived with a sniffer dog. *Seriously?* What was the point of taking it all apart if the dog could just sniff out the drugs? But I said nothing; not even a huff. Today, I was on my best behaviour and determined to show these Colombian officials my best side. They gave us the nod of approval and left us to put Rhonda back together again, package her into a crate, and send her off to Panama. South America was over for her.

"See you on the other side, girl," I whispered as the lid to her crate was nailed shut with one final bang.

Chapter 21
The Price of a Quickie

My arrival in Central America in 2015 was made much easier thanks to a chance meeting I'd had in Peru back in 2013. At a fuel stop with my off-road group, we had got chatting with a guy from Panama called Arturo who was riding a BMW R1200GS on the adventure of his dreams. He was heading South to Ushuaia at the time and I told him about my plans to ride around the world the following year. Of course, I told anyone who would listen back then as if the very act of saying it made it real.

Arturo was lovely,

"You must look me up when you get to Panama" he'd said, quickly writing his email down and handing it to me on a tiny scrap of paper.

"I will" I said, and thanked him. I didn't believe for a second that I'd see him again, or indeed that I would make it to Panama. But that tiny scrap of paper felt symbolic, as if proving that it was not just a dream. The fact that he had given it to me meant he believed me, and I'd smiled as I tucked it away in my jacket pocket; the waterproof one with the zip. I guessed I'd probably lose it before I got home but I'd made a silent vow to try not to and to actually see this through. It was a 'warm glow moment' and I had thought about it for the rest of the afternoon as we followed our guide along the dusty trails of the Peruvian Andes.

I didn't lose his email address and Arturo was good to his word. Two years on, I received a very warm welcome from him and his large and exceptionally friendly family. As Rhonda's chain and clutch lever were buggered Arturo's dad kindly came to the airport with his pick-up and we stuck her in the back. I was put up in their enormous pool house for a week, given copious amounts of food, and treated like one of the family. Their monster-size American pit-bull Tyson became my companion while I was there and then Arturo and his dad rode with me to their summer house in the coffee region of Boquete for a short visit before I waved a fond goodbye and set off for Costa Rica.

****

"Rancho Burica lies where the bad road ends and the good life begins!"

Bad roads *and* good life? With a 'dualsport' bike and a constant craving for The Promised Land, this sounded like the perfect place to head for. Bad roads *were* the good life to me, but this little jungle hideaway in Costa Rica, nestled between the rainforest and the Pacific Ocean, seemed to have it all – certainly if their website was anything to go by. It promised peace, quiet, wildlife and beautiful surroundings. It also offered the chance to help with their turtle project, which was on a mission to protect the turtle eggs from poachers (one nest could fetch up to $100 – usually $10 an egg). Wildlife, beauty, bad roads *and* a mission? I was sold, and I was on my way!

Rancho Burica was indeed Utopia. It was practically on the beach and yet sat under the canopies on the edge of the jungle. The trees were full of monkeys, sloths, birds and butterflies and I read that the area was home to tapir, deer and many types of snake. I really wanted to see a tapir. According to my research, the sea was also teaming with whales, dolphins and of course, turtles.

I joined in the regular nightly beach patrol while I was there. Torch and walkie-talkie in hand. Any nests we found were moved to a hatchery where they would be safe and later we would release them into the ocean. It was a wonderful experience and a welcome break, but I was feeling a little melancholy. Despite being well rested in the pool house, I still felt tired. Everything ached, and the more I slowed down to compensate, the more aches my body seemed to find. I felt like an old banger on a used car forecourt. Too many miles on the clock and long overdue for an overhaul.

The only thing worse than feeling down in a *dump*, is feeling down in the prettiest place in the world while you do something productive and enjoy some of your favourite things ever. How could I be miserable in Paradise? When you feel less than ecstatic in Utopia, then what hope was there? Where else was there to go from here? I couldn't see that I was just worn out and a little lonely and that it didn't matter where I was, although I guessed I would have felt a whole lot worse if it hadn't been for the regular sightings of sloths, and the horse riding into the jungle, and the baby turtles, and the bar with the cold beer at the end of the day.

On my fifth day, I took some time off from turtle watch, borrowed a bicycle and took a potter down the trail. I thought I'd take a ride to the village a few miles down the road for some lunch. About a mile down the track a man jumped out from behind a large beach rock and onto the road in front of me. He was clearly a local and maybe in his late teens or early twenties. It was hard to tell. Everyone looked young and healthy here. He flagged me down and was trying to tell me something.

"*Delfin!*" he kept saying.

I leaned the bike against the rock and followed him, assuming my luck was in and he was going to show me a school of dolphins playing offshore. As we clambered around the boulders, I looked to the shoreline and there, in the breaking waves, was a stranded, bleeding dolphin, having clearly been thrown against the rocks a few times by the force of the waves.

"Shit!" I cried and stumbled over the last few rocks. But then I did something that I would find very hard to forgive myself for later. Before trying to help this beautiful creature and get him back in the water, I took my phone out and took a photograph. As if that were the most important thing. In that moment, I had prioritised sharing this moment on my blog over the safety of this animal who was clearly distressed and needed my help.

"What am I doing" I said out loud then threw my phone in the sand and ran over to the dolphin. "We must get him back in the water" I said, and together we gently dragged him deeper, fighting against the waves, in the hope that he would find the strength to swim. The waves were rough here though and although he tried, he soon tired and came reluctantly back to us. He was exhausted. We held him there for a while, giving him time. Stroking him and talking to him. Telling him everything would be OK. I looked into his intelligent and gentle eyes and I felt he knew we were trying to help. He seemed to be looking at me with a mixture of gratitude and sadness. He understood. What must he have thought of me taking a picture of him in his time of need? I felt utter shame and as I gently stroked him, I muttered, "I'm sorry"

His only hope was to get into calmer, deeper waters, although his eyes told me he had already given up. He was sick and there was a reason he had been beached here in the first place. We had to try though. I heard a pickup truck coming down the track and gestured for the young man to hold the dolphin while I flagged it down. Just as he had done to me, I jumped out from behind the boulder, then led the occupants to the dolphin. Making a series of hand gestures and pointing furiously to explain the plan, we set about lifting the dolphin into the truck and moving him a hundred yards down the beach to the calmer waters where we could hold him while he rested, and then, perhaps, with a little luck, he would find the energy to start swimming. If he still failed, then at least the sand would be softer than the rocks.

We waded out with him as far as we could go, guiding him into the safety of the deeper water. My heart jumped when, after ten minutes, he seemed to gain strength and began to swim out on his own. I cried out with joy as he jumped into a wave or two. The young man cried out too and we high-fived each other in triumph. He was going to make it! But seconds later he was spent again and came floating back to our open arms. He had looked so strong. This happened many more times, but despite his best efforts to survive, it now seemed futile.

After two hours, the dolphin gave up altogether and died. I was mortified. It couldn't have felt worse if it had been a human being that we had just lost. Nature could be cruel, and he was probably sick. I knew that, but it felt like we had failed him. I walked slowly back to where I had dumped my bike, dug out my phone and deleted the photo of shame. What had I almost become? A Facebook junkie, only in it for the 'likes'? The very idea was so abhorrent to me that it flowed down my body and rested in my gut.

I headed back to the ranch and lay silently in my tipi until dinner when I emerged with a fake smile. I said nothing about the dolphin. Afterwards, I half-heartedly played cards with the Dutch guys, then went back to my tipi overlooking the ocean, and fell asleep to the sound of the howler monkeys who played and chattered in the trees behind me as I drifted into sweet oblivion.

The rest of Costa Rica is a blur. I took few pictures and wrote little in my blog. The heat and humidity was getting to me, my energy levels were low, my joints ached and my shoulder was niggling. I felt ungrateful and my self-esteem had hit rock bottom. Heading for the border, I spent my helmet time dissecting my reasons for travel once more, and whether it was actually doing me any good at all, or whether, in fact, it was just ruining me, and my chances of future happiness.

****

Where Utopia had failed to lift my spirits, a little old man in the middle of Nicaragua succeeded – and in the unlikeliest of ways. I still ached and the shoulder I had fallen on in Colombia on the way to the airport had been getting gradually worse as I made my way through each country, but this brief chance meeting was just the tonic I needed to knock my perspective back into place – and in the unlikeliest of ways.

As I sat drinking coffee on a grubby plastic chair in the usual wooden shack on the side of the dusty road, I realised my chain was very loose again. It was brand new in Panama, so it wasn't really a surprise because chains often need adjusting once they have had time to 'bed in' and stretch a little. Rather than attempt to tighten it there and then, in the full glare of the midday sun, I decided to potter on slowly to see if I could find a garage with a jack to make life easier. I just didn't have the energy or the inclination to do this simple task that would only have taken ten minutes. Within a couple of miles, I found one. Sat in the shade on his own plastic chair, reading his paper, was an elderly man – let's say at least 75 to be kind, but I guessed more like early eighties. I pulled in and pointed to my chain. He saw it was loose and gestured for his... son? grandson? – to come and sort it while he insisted that I take his chair in the shade before ambling off to get me some water. He could see I was dripping with sweat, despite my recent stop, and cooking in my full riding gear – again!

While the son cracked on, under my watchful eye, I smiled at how much easier it was, rather than doing it myself on the side of the road. I then looked over and smiled at the kind old man who had so readily given me his seat and was now reading the list of countries on the front of my bike. "That reminds me", I thought, "I need to add Nicaragua". I always added it *after* entering each country; never before. I wasn't superstitious at all, but there was no point in 'poking the bear' of fate.

The boy finished quickly, getting the perfect tension, and then wacked a load of grease on to finish the job. I got up and thanked him very much. Then, I turned to the old man and asked,

"Cuanto?" (How much?). His gentle little wrinkled face turned and looked up at me (he was very short). Then quite openly, and without shame, he suggested that I keep my money and accompany him to the back room instead. I had understood some of the words but the gestures said it all. I smiled and said *"No entiendo"* (I don't understand) – but I did. He inserted his right forefinger into his left fist and moved it in and out, making it very clear what he had in mind, and took my elbow, trying to lead me into the back room. I pulled away gently and said, as best I could, "I'll pay with cash if it's all the same to you". The old man shrugged and seemed to take no offence at my refusal and stated his price – in cash this time. It was the equivalent of £2.50! That was what really made me laugh. Was that all I was worth? I like to think he would have given me some change, had I opted for the other, more physical, form of payment. I just loved the fact that there was no 'cloak and dagger' – just an offer. Hey, if you don't ask you don't get right? I respected him for that, and I loved his eternal optimism.

This moment really did brighten my mood. It had been another welcome reminder for a tired brain. *This* was why I travelled. There *was* no ultimate or constant high in life; no 'Promised Land', and I was OK with that. There was only a series of highs and lows and random moments like that, which, once stitched together, made a rich and beautiful tapestry weaved together by the road. The road. It brought so much – because you just can't make this stuff up.

"All part of the adventure", I laughed, as I rode out of the village, motivation restored and aches a little less obvious than before, at least for now.

A few miles down the road the next morning I found a puppy no more than five weeks old. Already riddled with mange and probably in pain but still wagging his tail at me. I left him some scraps and rode on. Before the border I found a horse with a broken leg. Snapped clean through with a gaping wound showing a jagged white bone and pink flesh. The leg swung loosely as the horse tried to move, then cried out in pain and gave up. He would probably stand there until he dropped. I gripped my handlebars, screamed in frustration and rode on to the border with Honduras. No amount of positive perspective could make these sights any easier to bear, but all I could do was ride on.

****

El Salvador is tiny. It's about the size of Wales and yet it has more than twice the population: 6.3 million. That year, 2015, it earned the ignoble status of having the highest murder rate in the world (not including current war zones like Syria). Compared to the UK as a whole (with a population of 64 million), you are 40% more likely to be murdered in El Salvador, three times more likely to have AIDS and the chances are that you would earn 80% less income. El Salvador's twelve-year civil war ended in 1992. Sadly, that was not the end of this little country's troubles. It is estimated that there are around 25,000 gang members at large with a further 9,000 in prison. Criminal youth gang members are estimated at a further 60,000. That doesn't bode well for the future. I decided to take up an offer to stay with a fellow biker in the capital of San Salvador for a few days and see if I could find some beauty within the reported 'beast'. I wasn't a gang member and I had no intentions of getting involved with any, so unless I was really unlucky, I was pretty sure I'd be fine – just like most of the people who went about their daily lives with no problems.

The roads in El Salvador weren't in bad nick; certainly, compared to those in Honduras, which had craters just waiting to swallow you up on every corner. You really had to look where you were going, but I had got across the small patch of Honduras between Nicaragua and El Salvador riding up the CA1 near the Pacific coast in just a few hours. Two slightly complicated borders in quick succession had got me craving for ice cream and I lasted just thirty miles into El Salvador before I stopped at a garage to try my luck.

It was in.

As I sat next to Rhonda, licking my cornetto and happily watching the world go by, five uniformed men came around the corner, each furnished with an AK47. They smiled and began asking me questions,

"Where are you from?" asked the first guy, as if practising his English.

"Wales" Blank faces "Next to England"

"Ah, England"

"Where are you going?"

"San Salvador"

"Where is your husband?"

"This is my husband" I said, pointing at Rhonda in the usual round of to-ing and fro-ing that often accompanied my stops. They guffawed at this, then spent a few minutes checking out my bike and chatting amongst themselves, before reverting back to a more official mean look, facing outward and standing guard over me while I happily ate my ice cream for the next ten minutes. I wasn't sure if this level of security was strictly necessary, but I was grateful for their concern and honked my horn as I left them in my dust, shouting, *"Adios. Gracias Amigos!"* as I rounded the corner.

Gabriel Escobar was a 20-year-old dreamer who lived in the heart of San Salvador, down another rambling and busy back street. He'd been following my blog and kindly invited me to stay with him and his family. His story was one of triumph over adversity. His father died when he was just 16 years old and, due to a long illness and some rather complicated circumstances before his demise, his family were left with nothing but the building that once housed the family business of grape import/export. They had no choice but to turn it into their home. The large industrial refrigerators in the heart of their home were rented out and Gabriel, his mother, his grandmother, and two brothers lived in what was once the office and storage space above. Gabriel's triumph came from the strength he had found within. Going from a private education and living a life of luxury, to a much more basic lifestyle that saw him selling mobile phone accessories on the streets at the age of fifteen, and living in a warehouse, had turned him into a very resourceful young man with a wise head on his shoulders. His optimism was uplifting, and he wore his hopes and dreams on his sleeves.

"This is your room" he said leading me to the old lock-up downstairs. It had not changed since its working days, only now there was a bed where once there were boxes. There was a desk in one corner and a chain that hung down next to it to open the big roller doors that led to the front street. Rhonda was parked just the other side of them in the main warehouse section, and I guessed the drone of the enormous fridges next door might actually help me sleep.

"This is great" I said, and I meant it.

"You will be safe here" he said and handed me a piece of paper with the Wi-Fi code. "Wi-Fi too? Perfect." I said. "Thanks Gabe."

The first evening we spent chatting and trying out some local food (pupusas – thick griddle cakes or flatbreads – pupusa has officially been declared the national dish of El Salvador, and has a specific day to celebrate it!). Gabriel told me about the troubles in his country and how he hoped to make a difference. He had considered leaving for a better life somewhere else, but after considering all the options, his heart told him to stay and try to make a difference. Listening to this charismatic and ambitious young man full of so many ideas forming, with his whole life ahead of him, I tried to imagine where he would be in ten years' time. With his attitude and dreams all focused on helping others, I had high hopes for him, and I looked forward to following his progress.

"It takes just one man to make a difference" I confirmed, feeling suddenly motherly toward him.

I was wrong about the fridges. They didn't help me sleep. I tossed and turned most of the night and then got up late. Gabe and his mum had already gone to work, but Naomi, the lady who helped to look after their 98-year-old grandmother, showed me around the kitchen so I could cook up some much-needed eggs and coffee. Over breakfast we chatted via Google Translate. Naomi told me how she hoped to join a convent one day, how she was learning the violin and how she used to have her own sewing business until the gangs began to pay her regular visits and take all her money. Gabe's mum took her in as part of the family. It was very funny chatting via the Google app. Sometimes it would translate incorrectly and on one occasion came out with a translation that was very rude – even by my standards! I tried not to laugh and explained that it did not translate correctly, but I didn't share the joke. I had a feeling she would have been mortified had she known. What I did understand though, was that Naomi felt pity for me because I had no religion and she was pretty shocked that my 22-year-old son had a girlfriend already. I explained that attitudes were very different in the UK and it really wasn't anything out of the ordinary over there. I didn't go as far as explaining that I was already a grandmother. I wasn't sure she was ready for that. Regardless of our differences, we spent a good hour, sharing lives and giggling over coffee. I knew instinctively that I would look back on that hour with fondness for a long time to come.

It rained all that day, so I spent most of it in my room, writing at the desk next to the roller doors. Gabe was going to take me to a nice spot by the crater lake that night to do some camping, but the rain continued and our motivation to sleep under the well-hidden stars waned. Instead, we went to a local bar. It was Friday night after all, and I was due a few beers. There, we came across some interesting characters. First, there was the lady dancing around with a beer bottle balanced on her head – much to the security guard's dismay! He tried to stop her several times but the fact that he had a Mossberg pump-action shotgun strapped around his neck did not put her off. She actually did extremely well considering her obviously inebriated condition. I was quietly impressed, although I made the mistake of making eye contact at one point and found her dancing in front of me demanding my full attention and approval. She was gently moved on by the barman, who was beginning to lose his patience.

We ended up drinking with a group of locals, some of whom spoke good English. The drinks flowed and the juke box was paid handsomely. It was clear we had the same taste in music. We were pretty much rockers of the same musical era despite the diverse ages around the table, and of course, the different continents. The conversation flowed easily although it soon became clear that we had many different opinions on things. For example, Salvadorians are very open about their intolerance of gays. They put this down to the culture they were brought up in. They are really quite vocal about it. However, I voiced my opposing opinions and then steered the conversation away to something a little safer. A few minutes later Queen was selected on the jukebox; clearly a popular choice as everyone started singing along. I laughed out loud and cried in jest and drunken bravado,

"Oh, so Freddy is OK because he can sing, right? Despite the fact that his group's called Queen, he has a massive moustache and wears spandex? You can't get much gayer than our Freddy!" I was about to cry 'Hypocrisy!' when Bohemian Rhapsody came on and...well what can you do but sing along?

The beer flowed as easily as the conversation and many glasses were clinked in harmony to cries of "Cheers!". We played all the old British and US '70s and '80s rock classics and sang along together, happy to have found this common ground. It was then that one of the guys turned to me and asked,

"What do the British think of Salvadorians". Without even thinking I replied with as straight a face as I could muster,

"They think you're all gay!"

The group fell silent. It was as if the music had stopped; paused, along with any frivolity. *Shit! Had I gone too far?* Then suddenly a roar of laughter and another clink of Pilsner bottles. *Phew! That was a close one!*

"You're all right Steph" they laughed, and the frivolity continued. The 'gay' issue was not raised again.

Gabriel and I rode to a crater lake and to the top of another volcano near Santa Ana before I headed for the border and waved farewell to El Salvador. I had fun, but I hadn't fallen in love with the place. I left for Guatemala wishing I had found more to shout about. El Salvador had a bad rep and I loved to back an underdog, but I just didn't find much to rave about. I really wanted to say "Hey guys, they've got it all wrong about El Salvador" but I couldn't. The truth was, I was glad to get out and as soon as I crossed the border, everything felt...well, nicer.

Chapter 22
The Ruin

I rode through Guatemala to a new playlist I'd compiled to suit my new scenery. It was light, summery and easy-going. It consisted of singers like The Cranberries, Simon and Garfunkel and Nina Simone. I also had a couple of classical tracks. Pavarotti's rendition of Puccini's Nessun Dorma always made me speed up, while Erik Satie's lilting piece for piano, Gymnopédie, (played by Finghin Collins), often made me cry and so was best reserved for the gentle roads through the hills. The power of a good playlist was still as strong as it ever had been and was essential for my 'helmet time'. It could alter my mood, add drama to an already dramatic backdrop, and give rhythm to an off-road section that required the body to work as one with the motorcycle.

Riding into the cobbled streets of Antigua I was instantly struck by its beauty. Its full name is Antigua Guatemala ('Old Guatemala') and in Spanish colonial times it was the capital city of Guatemala and most of Central America for over two centuries, until the capital was moved down to Guatemala City in 1777 after one too many earthquakes. Its 18th century population of 77,000 was about double what it is now.

Antigua was where all the tourists went and so prices were heavily inflated there, but that was my only criticism. Otherwise, it was the perfect rest place for a weary traveller. A picturesque, colonial town surrounded by volcanoes and beautiful countryside. Menus mostly had English translations, people were friendly, there was a great nightlife and I managed to find myself a quiet, fairly cheap hotel, a little out of the centre. The public parking in town was at a premium, costing a small fortune, so the hotel owner helped me get Rhonda up some stairs and into a spare room in the hotel. This was not unusual. Rhonda had been shoved into all sorts of places. In Bolivia she had been put in the fruit and veg shop next door while the customers reached over her to select their tomatoes. In Peru, customers had eaten their breakfast around her as she stood in the middle of the breakfast room, and on many occasions, she had slept in my room. After six nights I decided to ride to Belize via Rio Dulce and Tikal, spend a few days on the Caribbean coast and then ride back via Flores.

The problem with Tikal wasn't that it had a lot of extremely noisy and nocturnal wildlife that rummaged and screamed and rustled in the trees and surrounding areas of my flimsy canvas as I lay under it in the dead of night. Indeed, my months of training in this very department had paid off. The problem was that I was camping between a jungle and a Mayan ruin, and only three nights earlier I had inadvertently watched a horror film called The Ruin. It involved a load of teenagers going to check out – you guessed it – a forgotten Mayan ruin in the jungles of Mexico. Of course, they all died horribly, painfully and slowly by the growing roots of a killer alien vine. Not even one survived. There was no happy ending. Aside from my location being Guatemala and not Mexico, it was identical. It could have been filmed right where I was camping. I hadn't done this intentionally – I'd just happened to randomly pick *that* movie from a hundred or so that I happened to have on my hard drive.

Don't get me wrong. It was not a great film. In fact, it was almost B-movie cheesy, but lying alone in the dark, with the rather inconsiderate wildlife poking fun at me outside, logic had run away and hidden in the nearest inaccessible crevice. All the months of hard work to convince myself that there really was nothing to be afraid of, was gone and Little Miss Irrational had waltzed right on in to smugly hijack the empty stage. I was back to being a child again, frightened to look under the bed. Only this time, I was camping in the shadow of a Mayan temple surrounded by two million hectares of jungle.

To my surprise I survived the night and as I poked my head out of my tent, where once there were monsters, now grazed a mini capybara and the unmistakable screech of parrots as they went about their morning's business. I had read somewhere that there were three hundred species of bird in this area, several species of monkey and five species of big cat. It all seemed rather obvious and wonderful in the daylight. What a difference a few hours could make.

That morning I crossed the border into Belize, which was a real culture shock because everything changed. It could have been on a different continent from the rest of Central America. Formerly known as British Honduras, the country became a Crown Colony in 1862, changed its name to Belize in 1973 and was given full independence from the UK in 1981. It still carries the Queen's Head on its coins today. The mostly Afro-Caribbean-descended population, speak with a Jamaican-esque accent, and speak both English and Belize Kriol (or Creole), an English-based patois. It's a tiny country, with a big personality and a lot of history, which was clearly evident in its culture.

Having crossed the border with ease, I took to the Hummingbird Highway, a road that twists and turns through the lush rainforest that covers a large portion of the country. A hundred and fifty kilometres later, I had crossed the entire width of Belize, and had arrived at its famously beautiful Caribbean coastline. (In case you were wondering, they drive on the right, like the rest of the American mainland, and in contrast to most former British colonies, including the Cayman Islands and Jamaica, just across the water). I pointed Rhonda towards a small town called Hopkins. It seemed as good a place as any, and I was hopeful that I would find a fairly quiet community, with little tourism. I was not disappointed.

As I rode into town I was greeted with many calls of "Nice bike!" and big smiles. A few brief chats and enquiries later, I found myself at a small lodge called Kismet. I had been told I might find myself a cheap bungalow here and a safe place for Rhonda. It was so much easier than usual to get information because, of course, they spoke English and there was certainly no shortage of help on offer. I felt welcomed into this small town as soon as I laid tread on its dusty roads. I was greeted by a New Yorker called Trish, and a big pet pig, who I would quickly discover, was called Sergio. Sergio was eight months old and growing at a rate of knots! He had a bit of an attitude (like most teenagers) and could bite when in a bad mood. Not completely unaccustomed to pigs, I quickly found his soft spot, and a quick scratch on the back was all it took to calm him down.

Trish offered me a bungalow right on the waterfront for a good rate. As she was showing me the place, I noticed it had a sign above the door saying,

'Kit and Bob's Place'. I told Trish that this used to be the name of my cat and dog and added

"It's clearly meant to be". She laughed and said,

"Guess what Kismet means?", I looked puzzled, so she continued, "It means – 'it's meant to be'". That was it. Sold to the rather sweaty Welsh lady in the black pants and big boots.

Trish had been living in Belize for seventeen years, eventually marrying a local and building this little haven, set within the palm trees on its own tiny stretch of Caribbean coast. Later that day I met the two dogs – B.B. King and Taboo. Then I met Trish's husband, Elvis. Trish and Elvis turned out to be a right old pair. They reminded me of a black Ozzie Osbourne and Sharon. The banter between them, the permanent, slightly dazed-and-confused look on Elvis' face, the animals all over the place and the chaos that somehow – just –worked. Trish (Sharon) was clearly in charge and without her, the whole place would have fallen down long ago. Her attitude reminded me of a sticker my sister used to have on her bedroom door – *'This may look a mess, but you don't understand my system!'*.

Elvis, who liked his rum, as did so many of the locals, was drunk most of the time, from morning to night. However, he was actually a very good fisherman and came back most days with lobster or something equally tasty. Trish, having owned restaurants and written a cookbook in a past life, would take the offerings and turn them into something delicious. We would all enjoy a meal together around the big wooden table every night, along with a rum and coke to wash it down. We played cards and shared stories. Trish and Elvis were certainly a colourful couple. The only disagreeable thing about the place were the minute mosquitos that ate you alive. You couldn't even see them most of the time. You just felt them bite. After two days, I was completely covered in extremely itchy red lumps. It seemed there was a mosquito in *every* paradise.

Joining us around the table was a Norwegian guy called Frank and his son Chris. They had started an overland journey from the USA just eight weeks earlier. Frank in a 4x4 and Chris on a Honda XR650L. When Chris had announced he was going to take this journey, Frank, having done a fair bit of travelling before, had insisted he come along too. There was no way he was going to miss out on the fun. Sometimes they would lose each other, even for days at a time, but it was working, after a fashion. It didn't matter as long as they met up somewhere along the road, at some point. They planned to ride down to Argentina over the coming months so I shared any information I thought they might find useful and vice versa – as all good overlanders do.

Chris and I walked down the beach later that night and hit a bar. There was a band playing and we were quick to dive straight into the atmosphere with a couple of dark rums and a tequila each. As the night went on, the band ran out of ideas for songs and asked the audience for requests. Someone shouted *Hotel California*. The band started playing, but as they got to the bit about the 'mission bell' they realised they didn't know the words.

"I know them" I shouted. I ran up and took the proffered mic, but halfway through the second verse, I forgot them too. Someone else came up and joined me, convinced he knew the words, then another. As it turned out, none of us did. It's just one of those songs you would swear you know, until you try to recite it unprompted!

We all left Kismet and its crazy mix of inhabitants the next day. I left later than the boys, with only a vague idea of where I was planning to go. I changed my mind at the first junction and turned right instead of left.

A few miles down the road I bumped into Chris and Frank again, so we rode to the border of Guatemala together. Once again, we said goodbye but the next day the same thing happened, this time in Flores, a tiny town built all over a little island on Lake Petén Itzá, connected to the mainland by a causeway. There was nothing else for it – we would just have to have another beer together and figure out just who was following whom. We agreed not to fight it any longer and headed out together the following morning. A day that would turn into a long ride, and a great adventure – just the way we liked it.

Leaving the delicate and pretty little island town of Flores behind, Chris, Frank, and I mingled with all the tuk-tuk traffic and then headed off for the green mountains of Guatemala. We were going to ride to Semuc Champey, a place known for its beauty, but also for its inaccessible location. We had dirt bikes and a 4x4 though. *Bring it on!*

The first two thirds of the journey was paved. After crossing the river by moving platform-style ferry, we stopped for a breakfast of chicken and chips and pottered happily along, taking pictures often and occasionally filming as we wound our way along the warm tarmac that cut through the green farmlands. Stopping for directions in a small village, a pig truck we had just overtaken caught up with us. Two guys were sitting in the back above the pigs, on the metal framework and were in the perfect pointing position. We followed their finger and carried on, not quite sure whether Frank was now ahead of us or behind us.

The pig truck became a regular feature for the next hour or so, as we worked our way steadily upwards: overtake pig truck; stop for a junction or a picture or a drink; point or wave from the pig men as they pass; overtake, and so on. It was a little game we all seemed to be enjoying and on the last stop, we shared a cigarette together before our final turn-off. This was where the tarmac ran out and our pig men's directions pointed us up a rocky trail. I always love the beginning of a trail. It's a mixture of excitement and trepidation. Fear of the unknown. How will it go once we round the corner, and the next? After a brief assessment and conflab, we went for it.

Chris flew off on his XR650L, leaving me for dust, clearly excited to be getting off the black stuff. I followed his trail and smiled to myself. I too was quite pleased to see the end of the smooth road. Rhonda needed a challenge, and so did I. As I rounded the first corner, I caught up with Chris. It wasn't hard as he was on his side surrounded by a dust cloud of his own making.

I had to laugh when I saw him, and quickly fumbled with my headcam to turn it on. Chris saw me and quickly tried to pick up the bike in a race to save his dignity. He laughed with me

"What an idiot!" he said. "I guess I was going too fast". At that point he looked to have got away with some gravel rash. It was only later his foot started to hurt. Of couse, it would have hurt a lot less if he'd been wearing motocross boots rather than trainers.

The track ahead looked rockier and steeper, so we let some air out of our tyres. That helped a lot and soon, grateful to be on dirt bikes, we had found our rhythm. Chris stayed at a steady pace behind me as we headed onwards and upwards into the beautiful mountains of Guatemala. We had run out of water, so when we came to a little stall selling drinks and snacks, we were pretty pleased. The family who ran it were lovely, and confirmed we were indeed heading in the right direction for Lanquin, and that it would take us roughly two more hours. *Two hours? Wow! This was going to be a long day.* We had already been riding for hours and had thought we were nearly there. The Hondas had no trouble pulling us up and over though, and it turned out to be a wonderful ride. The scenery was breathtaking at times. We had to keep stopping to take it all in and I found myself repeating "I love this place" over and over again.

It turned out Frank and his 4x4 had somehow got ahead of us, and we met up with him at a roadblock in a small village on the trail. The villagers had blocked the road and would not let us pass without payment.

"What, for *this* road?" we laughed "You want payment to use *this* road?". Frank did the haggling and eventually we agreed on half the original asking price.

By the time we got off the bikes in Lanquin, Chris could not put weight on his foot. Thankfully, some friendly locals were there to help us with our luggage, a room and most importantly – a bar stool. By the morning, we had acquired some ancient crutches and we were ready for the next challenge – getting up to the limestone rock pools of Semuc Champey.

"I think this has to be in the top ten of beautiful places I have seen so far" I said to Chris as I helped him over a tricky bit. It was hard work for Chris as it was a fair hop on stony ground to get there from our drop-off point, and it was hot. His efforts on the crutches were rewarded with cool, turquoise rock pools waiting at the end – complete with natural slides and fish that nibbled off your dead skin. The best bit was that there was hardly anyone around. Just the local kids, who were a lot of fun. They advised me where was safe to jump and said "be careful lady" as I edged closer. Poor Chris could only swim and watch the rest of us jumping. He seemed happy to watch and shout encouragement or "Chicken!" at the appropriate times though.

After a couple of days in the hills, I helped Chris onto his bike before setting off again. We would part company when we got to Guatemala City. The bumpy roads were over and Chris just needed to ride another hour on to Antigua, where he could get checked out. I would have to reluctantly brave the capital city's traffic in the hope of locating a parcel with a camera battery I had ordered a few weeks earlier when mine had given up the ghost. I hoped it would take just a few hours.

Four days later I was still in Guatemala City waiting for my parcel. It had actually arrived in the country about three weeks earlier but the local jobsworths were not willing to release it without a fight and a *lot* of paperwork. I was sure I would have to give up. On the fifth day I finally got it, packed up quickly and headed for the Mexican border in torrential rain, on roads that were now rivers.

Chapter 23
Return of The Kiwi

The rain beat down mercilessly as I headed for the border of Mexico. I had two days to ride a thousand kilometres plus the border to get through, if I was to make it to Zipolite on time. The Guatemalan postal system had robbed me of the luxury of waiting for the weather to ease. It was time to zip up, turn up (the music), and crack on. My luggage was no longer waterproof since the crash in Colombia, so I optimistically wrapped my clothes in bin bags before packing them up and hoped that this time, I could avoid the lump of smelly, damp clothing that I had uncovered upon arrival in Guatemala City a few days earlier.

I had a date with Shane in Zipolite and I hate being late for anything. It's the British in me, although I don't always join the back of the queue these days. He was flying over from Japan, having made a detour there to go snowboarding en route from Australia. I had arranged to meet him in this little town on the Pacific coast – we had picked Zipolite at random off the map because it was in the far south of Mexico and seemed as good a place as any.

The rain stopped on the second day, but the wind picked up to a monumental strength as I hit the well-placed wind farms. Miles of enormous white blades, towering above me in perfect order. They looked like well-trained soldiers, standing to attention and ready for battle. I was not so ready. As soon as we hit this stretch, Rhonda and I began our own, now familiar, battle with Pacha Mama. If only I'd met her back in Peru, I might have had a word! We were being thrown all over the road. I had to rest under each bridge and then build up the strength to move on again,

"Ok you can see the next bridge", I'd tell myself, "You just have to get to that bridge. That's all". The phrase 'How do you eat an elephant?' popped into my head as I rode towards my next target. I kept repeating like a mantra "One bite at a time. One bite at a time"!

Luckily this only lasted an hour or so and just as quickly as it had started, the wind vanished. It seemed I had ridden out of the rainy season and into the hot and humid South Mexican winter.

Zipolite tuned out to be a quiet and curious little town. It was home to many ageing 'gringos' who I imagined came here travelling during the 1970s, got stoned and never quite found the motivation to leave. They still sat here amongst the colourful graffiti, sand and surf, smoking weed, listening to Bob Marley and occasionally going for naked strolls on the beach come the cool evenings, wearing nothing but a pair of shades and an all-over tan. It seemed this stretch of coastline had gone and got a name for itself as a nudist beach, and so perhaps this would explain the serious lack of tourists. They all seemed to be residing in Mazunte, just a couple of miles away.

I sat on the main street outside one of the restaurants and ordered a cheeky afternoon glass of wine. Shane arrived a few minutes later, half-drunk beer already in hand. He'd got a flight from Mexico City to an airport nearby, then a coach had dropped him off at the end of the town and he had grabbed a beer and walked along the beach, rucksack flung over one shoulder. He looked scruffy-as-ever, and sweaty, as if he hadn't washed for days or brushed his hair. I felt vaguely annoyed that he clearly hadn't made even the slightest effort to look good for me when I had just spent at least an hour preening myself before he arrived, (complete with new knickers!). Then I considered the fact that he had made the effort to get here at all and swallowed it.

Shane and I spent the next two weeks in Zipolite, reluctant to move. It was hot but there was shade, and in the evenings we would jump around in the powerful waves and regain the energy from them that had been sapped by the mid-afternoon sun. We developed a rhythm and fell in line with the other hippies – only we mostly kept our clothes on in public – albeit minimal. A dip in the sea then a beer on the beach at night. We wandered the lagoons together, named and fed all the stray dogs, lazed with the lizards who basked on every available rock, and *finally* took the time to learn *all* the words to

Hotel California. It was such a tranquil place, with an easy pace and a lop-sided smile. No one judged and no one asked questions.

Shane eventually found the motivation to take a bus the three hundred and fifty kilometres to the regional capital of Oaxaca in search of a motorcycle. I reluctantly followed on Rhonda. My shoulder had been getting worse and on my first morning in Zipolite I had struggled to move it at all. It had seized up and any attempt to raise my arm often ended in a yelp. I could now only sleep on one side. My wrist and elbows too were painful, and I was getting shooting pains into my fingers. Once I was in the city I could find a doctor and maybe some painkillers while Shane went bike-shopping. I had to ride there first though, and that would take about six hours by bike along Highway 175.

About halfway there, as I crossed the least shaded part through the hills, I felt an ominous front wheel wobble.

"Oh no" I said out loud, "Not now. I'm too hot".

There was never a good time for a puncture, but now? *Really?* No shade, plus my old friend the midday sun. *Perfect!* I pulled over and the less-than-perfect circle of my tyre confirmed my suspicions.

"Well it's only flat on the bottom" I laughed then whipped off my gloves, helmet and jacket before doing what I always do first in these situations. I dug deep into my tank bag, pulled out my tobacco, found a suitable rock and made a roll-up. Normally I'd find a bit of shade too, but there was no such luxury here.

After a few minutes letting the situation and the nicotine soak in, I whipped out my tool roll and rummaged through it for the 17mm spanner. Then I rummaged in my pannier and dug out the piece of old handlebar that I put on the end of the spanner for leverage. My dad had cut this down to the perfect size for me before I left home and, after several tests, we had concluded that this size of 'extension bar' was the minimum length required to create just enough leverage for my strength. It worked a treat, *normally.* I took my top-box off and placed it on the ground, then lifted the back of the bike, shimmied it over, and placed it centrally on top of the sturdy alloy container. Rhonda did not have a centre-stand, and this was my way of getting either wheel off the ground, as and when required. Using all my weight, sweat pouring down my reddening face, I tried over and over again to loosen the front axle nut. But nothing. It would not budge.

"Damn it", I cried, but I wasn't angry. I laughed out loud and looked around me, as if expecting the solution to have magically appeared in front of me by now. "My kingdom for a man with big shoulders" I laughed. I didn't doubt for one minute that there would be a solution along soon enough. The universe would provide somehow. I had no idea what form that would take but I was absolutely confident that something would come along. It always did. I'd earned that confidence; it had come from twenty months of positive reinforcement. Like Pavlov's dogs, I was already excited at the prospect that something positive was about to happen.

A car went by and whipped up a 'dust devil' that stung my optimistic eyes. I pulled out my water and languidly sipped the warm water from the nearly empty bottle. Ten minutes slipped by and the sweat pooled around my waistband in the small of my back. Something was going to happen. I knew it. *Any minute now!* Another five minutes and three more cars just pottered on by. I tested my strength once more against the might of the seized nut, but nothing. With my damaged shoulder and an increasing weakness in both arms, I just couldn't do it. *Damn, it was hot!*

I threw the spanner on the floor in defeat, just as a pick-up truck came flying around the corner, pulled in and came to an abrupt stop in a cloud of dust.

"Whoah there boy" I laughed, imagining the hero arriving on horseback in the nick of time. Then from the truck emerged the biggest Mexican I have ever seen, with bulging shoulders protruding from a checked red and white shirt that had the sleeves cut off.

Am I tripping? I laughed again and looked to the sky, "Ha. Never doubted you, Universe!"

The man came over. He didn't smile and he didn't speak English, but he was clearly offering the services of those big arms, so I picked up the spanner, put it on the axle nut and pointed at it, saying, "Push". He pushed. It didn't budge. I felt momentarily justified in my failed attempts. He pushed again. Nothing. He looked up at me as if to say, "WTF?", then tried again. Still nothing. Now the sweat was building up on his forehead and he looked almost embarrassed. I offered him some tepid water. He shook his head, and with one final attempt, he put all his weight onto that spanner. He didn't look as if he was used to being beaten by such things and clearly wasn't going to start now. Not in front of a woman – albeit a rather scruffy-looking street hobo girl with a dusty red face and reddish-beige riding gear that blended with the terrain.

My Mexican muscleman pulled off the spanner and held it up. We both looked at it incredulously. It had snapped.

"Oh!" I said and laughed. He looked mortified so I released him from his hero duties with compassion. "It's OK. Thank you. It's OK". I took the spanner gently from him, shook his hand and gestured for him to carry on with his day. This was *my* problem.

"*Lo siento*"(Sorry) he said, and then repeated maybe three or four times as, with head bowed, he reluctantly went back to his truck and sped off as quickly as he had arrived.

Alone again, I looked at the broken spanner in my hand and asked,

"What do we do now then?". The spanner remained shamefully silent.

I decided to use my hand pump to just pump the tyre up as best I could, then continue until I found shade. I would just have to stop and pump some more whenever the slowly deflating tyre needed some more air to be rideable. There didn't seem to be a tear so it might just hold for a few minutes at a time and I could take it really slowly until I found a spot to wait for help to come along. Now I needed a big strong man *and* a 17mm spanner! Things were *really* getting interesting. Shane would be in Oaxaca city by now, but I had no signal to call him and there was nothing he could do from there anyway. He wasn't expecting me for a while yet so he wouldn't be worried. I'd call when I could.

As I reached the brow of the first hill, I looked down to the bottom and saw a crossroads with a small wooden shack selling the usual roadside bottles of warm water, cigarettes and snacks that didn't melt. Next to that was a tree creating enough shade for a weary traveller and her motorcycle.

The crossroads led off to dirt tracks either side, probably leading to some little villages. This was perfect. Now I had the chance of more traffic, food, more water and shade. I could wait all night if I had to. Thankfully, within ten minutes a tuk-tuk came bumping along one of the tracks. A young man was driving it, looking quite smart in his jeans and surprisingly clean T-shirt. He was skinny and he didn't seem to have a hair out of place. Despite the heat and dust there was not a sweat mark to be seen.

"Are you OK?" He asked in perfect English. He wasn't the hero-looking type, but he'd do for me.

"Hi" I replied, "I have a puncture, but I can't get the wheel off and my spanner is broken. Any ideas?"

"Where are you from?" he asked

"Wales" I replied, "Next to England"

"*Ah, Gales*" he replied *"Ryan Giggs, no? The footballer."*

"Ah, so you've heard of us!" I laughed, "Yes Ryan Giggs. Tom Jones. Shirley Bassey. All the best people come from *Gales*"

"I am Juan" he said, offering a limp, yet seemingly well-moisturised hand.

"Nice to meet you Juan. I'm Steph. Do you have any idea where there might be a garage?"

"Follow me" he said. And without further ado we were off down one of the dirt tracks. About half a mile along it a lone shed came into view, covered in graffiti and surrounded by tyres of all shapes and sizes. There was nothing else around but bushes, rocks and the odd tree. I couldn't quite believe my eyes. The universe was really pulling out all the stops today. We pulled up and before I knew it, there were three guys changing my inner tube while I chowed down on some *Laughing Cow* and crackers. It did actually take all three to get the wheel off in the first place, but they had done it, we were done, and Rhonda was back in business. I can only assume it was all that Bolivian salt that made the nut so hard to turn.

Shane laughed when he saw me arrive at the cheap little motel he had found in Oaxaca. When I told him what had happened he said,

"All part of..."

"Yes, I know I know. All part of the adventure" I laughed, "Now shut the fuck up and give me a kiss". Shane was not the sort of guy to worry about sweat and dirt, so he let me pull him in for a big snog and a long hug without argument.

"Now where's my dinner?" I demanded playfully.

"How about a beer for now" he replied, "and we'll order pizza later to celebrate. How's that? Don't tell me I don't know how to spoil a girl" he laughed.

"Sounds perfect" I said, and meant it.

I wouldn't say our two and a half months together were perfect. Shane would often annoy the hell out of me, buzzing around Rhonda like a mosquito on his brand-new Honda XR150 and popping wheelies at every conceivable opportunity. He could out-ride me even on his smaller bike and often took off on the twistiest roads. I didn't mind that bit. I was happy for him to be ahead of me where he didn't bug me. When the wind picked up or we were on a long steep hill, he would tuck into my slipstream and ride an inch from my back wheel with his head down. That annoyed me too. Having company of any kind took some getting used to after so much time alone on the road, with no one else to worry about. We were both in the same boat on that score, but it did mean we had someone to share a beer and a view with. It did mean we could afford a slightly nicer hotel room as a treat every now and again, and it also meant the occasional breakfast in bed cooked on a one-pot stove and served with fresh coffee by a semi-naked easy-going freak with a personality not dissimilar from my own.

We'd left Oaxaca straight through the middle of two hundred riot police with shields and batons on one side, and protesters on the other. It's not uncommon to find protesters on the road in the Americas. Sometimes you have to gently negotiate your bike through the gaps with people lying in the road, rocks strewn on the tarmac to stop the cars, and fires blazing amongst the banners. Sometimes it could be a little intimidating and sometimes they took some persuading to let you through. My worst case had been in Peru, where I'd actually had things thrown at me, but I'd made it through without injury to me or the bike, and generally they saw it was not your quarrel and just let you through. This one, although apparently peaceful, did have a ridiculous amount of riot police who seemed prepared for the worst. A whole army of them. We cheekily stopped to pose with them before we left. It felt safer being one of a pair, and some of the police actually ended up posing with us.

Back on Highway 175, riding north east through the Sierra Madre de Oaxaca across the spine of Mexico saw us well and truly marinated in cold Mexican mountain rainwater, which penetrated every bit of our clothing and luggage. By the time we got down to Tuxtepec in the warmer climate, it had got to my important paperwork and well and truly turned it into *papier-mâché*. I managed to unpeel it, Sellotape the pieces back together, then hang it up to dry in our hotel room with our clothes, while Shane warmed up some spaghetti carbonara from a packet on my stove.

Onwards then, clothes and helmet still wet, to a place we had spotted on the map and liked the sound of: Playa Chachalacas. We knew nothing of the place except that it was on the Caribbean coast near Veracruz and it had a really cool name. As it turned out, it had much more than a cool name. We had inadvertently struck off-road heaven *and* it had stopped raining! The next morning, with our clothes all dry and a replenished sense of adventure that can only be found after a good night's sleep, we headed out on our unencumbered bikes to ride down the beach and into the dunes. In theory, Shane was still running in his new Honda, while, at the other extreme, Rhonda had carried me and a pile of luggage for 73,000 kilometres without complaint. We should have shown some respect, but sand has a habit of turning even the gentlest of riders into a hooligan before too long. With faith in the big H's engineering, we pushed the bikes hard and took respite from the long road miles. It was playtime – whether the bikes liked it or not.

Mexico proved to be as diverse in the experiences it offered as it was in its weather, as we crossed from the Pacific coast to the Caribbean and back again. We deliberately avoided ancient ruins as we had both had our fill of those lately, but we were lucky enough to come across *El Sótano de las Golondrinas*, The Cave of the Swallows, an open-air pit cave in the municipality of Aquismón instead. The elliptical mouth, on a limestone slope, is 49 by 62 metres wide and is undercut around its whole perimeter, widening to a room approximately 303 by 135 metres wide. The floor of the cave is a 370-metre drop from the highest side, making it the largest known cave shaft in the world and, understandably, an attraction for base jumpers from all over the world – but this was not *our* reason for the visit.

Long before people started jumping into the abyss, millions of swallows found a home here. Every night at dusk they fly home from the coast and circle above the 160-foot entrance, before choosing their moment to drop out of the sky in a synchronised free fall to the bottom where they sleep.

That night we set up camp amongst the trees near the entrance to the cave and sitting on the edge, we waited for the swallows to come home. It sounded like rain and looked like millions of black arrows being shot from the clouds. We sat for over an hour, mesmerised by the sheer volume and precision in this ritual that occurs every single night of the year. This night we had our own personal show as we sat observing the performance in silent awe! Later, armed with head torches, we crept back to our tent and slept well. We woke at 5.30am, keen to watch them leaving again for the coast. This time they spiralled upwards in a spectacular death-defying air display. We crept up to the edge and looked over to see them swirling below and shooting off as they reached the top.

Then to San Luis Potosi, where we learned that auto-hotels were not quite what we had expected. Arriving late, we had spotted a big sign offering rooms for a surprisingly cheap price of 250 pesos (about £9 or $11). It had a garage below the room and Wi-Fi. Perfect. The one-way mirror between us and the receptionist did not set off any alarm bells as we parked our bikes, and the electric garage door shut behind us. As we climbed the stairs, we found a very plush room full of mirrors, a very large bed, a menu for burgers and a menu for sex toys. All of which could be delivered through a discreet hatch next to the door. It turned out our 250 pesos had bought us four hours and not a full night. It was a moment that we knew would be funny later, but at that moment we were too tired for funny. This meant we would be thrown out at 2am. We were also too tired to use the room as it was intended, so we ordered a burger through the hatch, then left in search of a hotel that would let us stay longer. Later though, we would use similar places after first negotiating on the price for a whole night. Once, for a whole week! They were clean, often had a Jacuzzi and always had a secure garage for the bikes.

Then to Guanajuato, with its colonial architecture and silver mining history, where we rode the underground roads that ran like rabbit warrens under this beautiful world heritage city, followed by three days trying to recover all my photos after accidentally deleting my entire library, while Shane stayed out of the way and found more places to practice his wheelies.

Then to Guadalajara where we met fellow motorcycle travellers Chris, Francesca and Leonardo. Chris was a mild-mannered Swiss guy who started off on his solo adventure two and half years earlier, riding an old Africa Twin. In Goa, India, he met Francesca, an Italian. They fell for each other and within a couple of days of meeting, Chris invited the back-packing Francesca to join him on the back of his bike. He was heading for Myanmar and beyond. By the time they made it to Australia, they were expecting a baby (Leonardo). One had become first two and then three and the journey continued. They had attached a sidecar to the Africa Twin and a *lot* of luggage. Chris admitted that his journey had got a little more complicated than before, but he wouldn't change a thing. They had no plans to stop.

Onward and westward we rode, through miles of agave fields to the small town of Tequila, where we prayed to the Tequila Goddess statue and tasted her many agave-based nectars before praying to the white porcelain god a few hours later. I did anyway. Shane filmed me on my knees in my knickers while I threw up and swore at him to bugger off and close the door. He probably still has the video!

Further north, in Sinaloa state, we took the famous Devil's Backbone road en route to Durango. We were now heading into Cartel Central; Sinaloa is notorious in Mexico for some of the worst drug violence in the country. It is the heartland of some of the country's most powerful and dangerous drug cartels, and we planned to keep a low profile. We had especially been advised not to take pictures of anyone with a gun. *Not a problem!* A few days before we had arrived, the charred remains of two Australian surfers were found in their burnt-out camper van on our planned route. This played heavily on our minds and I think we were both grateful more than ever then, that we had each other.

The Devil's Backbone (*El Espinazo del Diablo*) runs through the centre of the mountain range in Sinaloa state and the marijuana and opium fields either side were a clear reminder of who lived there. The original road had since been by-passed by a toll road that is considered an amazing feat of engineering, with 115 bridges and 61 tunnels. We chose the old road and expected neglected tarmac, some dirt and a lot of speed bumps, or *'topes'* as they are called here, and are the blight of Mexico. Thankfully, our expectations were far from the reality. The road was empty and smooth as a racetrack. It was three hundred kilometres of perfect 'twisties', through layer upon layer of magnificent scenery. It was the sort of road that bikers dream of. As we passed the Tropic of Cancer sign, we stopped to make a coffee and waved back at every friendly V8-driving Stetson-wearing local who passed by.

It really was cowboy country – spurred boots and all. There were actually *lots* of V8-driving, white Stetson-wearing locals as we drove towards Durango. They looked as if they were about to wrestle a bull, and the whole area made me think of a cross between Texas and Alaska – though I had been to neither. Durango itself has a history of Hollywood Wild West filmmaking, and I could see why. More than a hundred films have been shot in the area, mostly after John Wayne 'discovered' the place in the early 1950s. Clark Gable, Kirk Douglas, Robert Mitchum, Rock Hudson, Anthony Quinn, and Chuck Connors all starred in Westerns made there.

And there weren't just cowboys. We also came across Tarahumara Indians as we continued towards Chihuahua. Originally these Native American people occupied most of Chihuahua but retreated to the High Sierra and places like Copper Canyon upon the arrival of the Spanish during the 16th century. It is estimated that there are some 60,000 Tarahumara still in the area, most of them following their traditional lifestyle. This includes traditional dress, farming livestock and living in natural houses such as caves or cliff overhangs as well as cabins and stone huts. They are well known for their long-distance running ability (as made famous by William McDougall's multi-million-selling book *Born to Run*). Much as we tried to get a smile or a wave as we passed, most were reluctant to communicate with us in any way. For this reason, we did not attempt to take any photographs, and respected their privacy – tempting as it was to pry.

As we neared the famous Copper Canyon (or *Barrancas del Cobre*), we stopped for a night in the town of Guachochi. It had an eeriness about it that I can't explain. Not scary exactly. Just different. It felt like a ghost town. The few people we saw there were mostly indigenous and still spoke their traditional language. We found a room for the night and spent the evening trying to decide which route to take to Batopilas and the bottom of the canyon the next day. The usual route would be all tarmac but half of it would be down a road that we had read was 'not for the faint-hearted'. It sounded like fun. However, there was another option. A mountain trail we had spotted on the map that promised a little more adventure than the paved route. The problem was, we had no idea what was in store down that route. We decided we would take it and see what happened, then ride north out of the canyon to Creel. This way we could experience both.

Batopilas had only just started receiving visitors. It had a history of kidnappings and violence by the cartel that ran the area, not to mention the fact that there was no easy way to get to it. It is said that the whole town (population 1,220) was involved in some way in the drug trade. We had no idea what to expect and no idea if any of what we had read were true. First, we had to get there.

The next morning, a little anxious at first, we pushed on down the dirt track that we had found and rode past the initial loggers' traffic of bulldozers and pick-ups. Soon the traffic died, and the trail became ours for the taking. We raced through the pine trees and stopped for the views which got better and better with each explorative mile. Deep in the forest we came across small settlements and families on horseback, driving their cattle to who-knows-where. We commented on the hellish rush hour traffic as we waited for the path to be expertly cleared of livestock by the cowboys.

The forest gave way to an extremely steep and exposed mountain track that would eventually take us down to the bottom of the canyon. I'm not good with downhill, slippery, hairpin bends that have no barriers and drops down to certain death. I can cope with one or the other – possibly even a combination of two. All three however, is a little too much for my rational mind, which tends to run and hide at the worst possible moment. I carefully negotiated the first few bends with relative ease. I'd been on a riding roll and felt good but suddenly, out of the blue, my fear of exposure kicked in and I stopped; frozen, with a very tight grip on my bars. I looked back for Shane who was still taking pictures a few corners behind me. I could see him as a small dot up on the trail above. Slowly, I convinced myself to edge forward, but as I was still holding my front brake on, the locked front wheel began to slide towards the drop on the loose surface. Clearly, I was going to die. OK, so the drop wasn't as close as all that, a few metres perhaps, but in my head, at that moment, I was inches away from certain death. Shane arrived and thought I'd just stopped for the view.

"I can't move!" I said, as he parked up and jumped off his bike without a care in the world.

"What do you mean you can't move?"

"I'm going to slide off the edge"

"Just ride around the corner. What's up with you?" he laughed

"I can't just ride around the fucking corner can I?" I snapped.

But we both knew that I had to ride out of it, because trying to get off the bike in that position, even with Shane's help, would have been difficult at best. It was too steep and slippery. Eventually, I ruled out the first options that ran through my head – which involved a helicopter rescue – and started pulling myself together. My fear was quickly replaced by embarrassment as I made it safely round the corner. I felt foolish for allowing my fears to take such a hold. I'd ridden many similar corners already that day but this one had, for some reason, just got to me. Embarrassment was then quickly followed by the usual biker piss-taking,

"You numpty!" laughed Shane "Oh my God I'm going to die!" he mimicked in a silly high-pitched voice.

"Fuck off" I muttered.

"Too soon?" he asked, with his stupidly boyish grin. I managed a smile before we set off again down the rest of the trail, through legions of the biggest cacti I have ever seen, and down to the floor of this magnificent canyon that had almost got the better of me.

Batopalis turned out to be a sleepy little village that hugged the wide river as it snaked its way through the canyon. As we rode into town, the cartel quickly made themselves known. They drove past us in pick-ups wielding guns and looking the part – fit and well-trained. I mean they didn't look like local thugs, more like soldiers without the uniform. However, they waved in unison as we entered town. An authoritative, non-smiling wave, as if they were under orders to do so by some unseen commanding officer. Tourism was good for the local families and happy tourists kept the military at bay. We were one of four visitors in the area that day, as it turned out, and it seemed our presence and money were very much welcomed.

Our ride out of the canyon two nights later was actually quite disappointing after the ride in. The road had been written about as if it were terrifying, with views to die for. It *was* a very pleasant ride. It did offer a fair few rock falls and a little off-road, I'd give it that, but it wasn't a patch on our ride in. As we made our way back out of the canyon, I silently hoped that 'our' road was never paved and remained a little-found gem in the beautiful mountains of the *Sierra Madre Occidental*.

****

The sun fought against the icy wind for temperature control of the day as Shane and I battled with the long stretches of solid ice lurking on every shadowy corner. We were 9,000 feet up and the wind was winning. It was epically cold. We stopped and looked at each other with concern.

"How much worse is this going to get?" I asked. Shane shrugged.

"Who knows" he replied, "All we can do is keep going"

"Just *don't* say it" I said.

"Say what?"

"It's all part of the *fucking* adventure!"

It was going to be a long day, with many long stretches done at walking pace, with our feet out. There were no gritters up here and there were certainly no prizes for being a hero – only a nice big mug of hot chocolate for those who made it out in one piece. We were on the long road out of the mountains of the High *Sierra Madre Occidental*. Our mission was to get out before the weather got worse and the snow reigned supreme. We wanted to be in the warmer climes of Baja California in time for Christmas and the thought of the warm beaches ahead kept us going. The thought of being snowed in up here was disagreeable at best.

Eventually, we made it out and over to the ferry port in Guaymas. After a thorough sniffer-dog search, a ten-hour crossing, and a hearty breakfast of eggs and tortillas, we made it to Santa Rosalia, Baja California (Baja just means 'Lower' and refers to the long peninsular in Mexico. Higher or 'Alta' California was originally all part of Spanish Colonial America but is now the richest state in the USA, and the 'Alta' has long since been dropped). After another lengthy dog search, where the dog actually bit into our bags on a couple of occasions, we headed for the nearest coffee shop to see if we could work out a plan. It was a couple of days before Christmas and we had no idea where we were going. We'd only got as far as "Let's have Christmas in Baja", when we'd discussed it previously. It was here in the café, drinking coffee and poring over our maps that we discovered how expensive Baja could be. The price of a breakfast was twice that of mainland Mexico. However, unperturbed, we picked a spot on the map and headed south, first to Mulegé and then to Conception Bay and our Christmas camping spot.

Baja didn't really feel like Mexico. It felt more like the USA and possibly had more Americans than Mexicans. The small towns were sparsely scattered along the coast and at that time of year, some of them felt like ghost towns. The best beaches were lined with motorhomes or recreational vehicles ('RVs') as the Yanks call them – a winter home to the affectionately named 'Snowbirds' (retired Americans) who head south when the cold weather hits up north. Unsociably, in true 'Baja Humbug' fashion, we turned down the kind offers of Christmas dinner with the Americans and found a little bay where the RVs couldn't reach. It meant we had to risk the tide coming right up to our tent porch, but the invasion of water was far preferable to people. We didn't need comfort and full size kitchens. All we needed was a pan, a stove, a pack of cards and some tin cups for the Jack Daniels and Polish 'Wodka' we had managed to acquire en route.

I spent Christmas Day cooking a stew in said pan, fighting off the seagulls and photographing the pelicans – nicknamed 'The Mexican Air Force'. Shane went riding with our neighbour who was camped in his one-man tent a little way down the beach. Pete rode a KTM and was road tripping from the States. They returned several hours later like conquering heroes with stories of fast tracks and near-death collisions,

"We were flying along the road..." laughed Shane with child-like excitement, "When this vulture flew up and hit Pete right in the head! I don't know how he managed to stay on the bike. The bird didn't make it I'm afraid", but he didn't look at all sorry. Pete, I now realised, had feathers protruding from part of his jacket and his helmet. There was blood and guts all over his front fairing and I guessed they weren't his. He reached into his pocket and pulled out a couple more larger feathers,

"I saved these" he said proudly, "It was brilliant".

"You should have brought the whole thing back", I laughed, "I could have made us more stew!"

That evening, the three of us sat around the campfire and played blackjack by moonlight while getting very, very drunk. It was the perfect Christmas Day.

Two days later, Shane and I continued north through the sand and hundreds of cacti that made it look like we were riding through a scene from *The Day of the Triffids*. Shane, of course, had to ride off the road and into the 'triffids', only to find himself, and his tyres, pierced by long, sharp and exceptionally thick needles; Thankfully not through to the inner tube.

"I did warn you" I said, like a mother to her child, as I pulled the last one out of his calf and mopped up the blood.

After a wonderful whale-watching experience in Guerrero Negro, and a very relaxed New Year's Eve in the run-down town of Bahia de los Angeles with some fellow-travellers, we headed for the place on the map that marked the home of the legendary Coco. It really is marked on the map as 'Coco's Corner'.

At 78, Coco was a rough and ready character with a heart of gold who had been looking after weary travellers passing through for the last twenty-seven years. He built a small wooden shack on a long dusty desert road, put a fence around to keep out the coyotes and started selling beer. He also provided water and shelter. We were even honoured with a cooked breakfast of eggs and tortillas in the morning, but the only charge was for the beer with the option, Coco explained, of leaving a piece of underwear for decoration. The roof of his hut was covered in knickers and underpants, and I think there might have even been a nappy up there. Messages and stickers were strewn all over his walls from people who had stopped in before us and enjoyed the company of the legend that is Coco. It seemed appropriate to donate a pair of my well-worn knickers to the cause.

Coco lost a leg many years ago following an accident with his truck. He lost the other leg some time later due to a diabetes-related problem. He still just got on with it and refused to move anywhere closer to 'civilisation' where life might have been easier for him. Instead, he just shuffled around on his bum and got messages to the local town whenever he needed to, via the travellers who passed through.

That afternoon we were joined by a KTM-riding traveller from Canada and that night we all stayed with Coco. Snuggled up in our sleeping bags we were treated to the most spectacular display of stars – a view of the sky that I only ever see in the desert – and serenaded by a choir of howling coyotes, as the tin cans on the fence rattled in the breeze to keep them at bay. It was a truly special night that gave way to the start of what I hoped to be another truly exceptional year.

Shane and I said our final goodbyes in Ensenada, just a day's ride from the U.S. border. It was time for us to go our separate ways again. Shane would go back to Australia to earn some more money to get back to his bike in Bulgaria, and I would continue ever northward into the United States and beyond. Shane made it easier for me by showing his jealous side just a couple of days before our parting. He flew into a strop at the mere mention of another man's name. He then suggested we find a way for him to continue with me across the border, but I turned him down.

"Let's stick to what we agreed" I said flatly. But it stung as I said it. After all we had been through together, I hoped I could find a way to change my mind, but as he pointed out, I preferred him better when he wasn't there! He was right in a way, and I hated myself for it.

Celebrating my 40th birthday on Ruta 40, Argentina.

One of the many dark tunnels I had to enter in Peru with no working headlight!

Rock formation in Atacama, Northern Argentina.

Playing with Mathieu at El Tatio Geyser Fields 4320 metres up, Argentina.

Claire, Matt and me smiling in front of the grumbling Villarica volcano.

Some of the damage caused by the flash flooding in Chanaral, Atacama.

Alone again in the Atacama desert.

Riding the twisties towards Tilcara, Argentina.

More beautiful Atacama. A place I long to return to.

Camping alone in Colombia.

One of the many military men seen on the roadside in Colombia.

I call this 'hotel camping'!

Another broken screen Rhonda (broken in anger!).

The security guard in a bar in El Salvador.

Posing on the Uyuni Salt Flats, Bolivia.

Pit stop with Shane, somewhere in Mexico!

Riding through the agave fields near Tequila, Mexico.

The lovely Chris, Franchesca and baby Leonardo.

Risking a quick pic in front of the riot police in Oaxaca, Mexico.

Playing in the dunes at Playa Chachalacas.

Having my hair pulled by the ever cheeky Coco of Coco's Corner, Baja California.

Chapter 24
The Gun-slinging Pastor

I knew I was in Los Angeles because the dogs were better dressed and had better haircuts than me. Stopping at the traffic lights, a leather-waistcoated 75-year-old riding a Harley-Davidson pulled up next to me. His wife was on the back in an all-in-one leopard print catsuit with matching ears on her helmet, and a rather non-supportive bra. He turned to look at my scruffy, dusty old bike and said, "nice ride" before winking and racing off with a grin. As I pulled off, I beeped at the grey bearded man with the baseball cap and the sign saying '*Honk for Peace*' in rainbow colours. Then I smiled, stopped and handed a dollar to the beggar whose sign said, '*I bet you a dollar I can make you smile*'. I knew I was in L.A. because I had a sudden craving for a burger, which I put down to the fact that the entire main strip along the coast smelt of them. I began counting the burger joints as I rode through. There was Fat Burger, McDonalds (times four!), In-and-Out and many more. Then of course there were the Mexican fast-food joints – Pollo Loco, Taco Loco and so on. There were more taco joints than in Mexico, though I doubted their wares were half as good and probably cost twice the price. I resisted the temptation and stopped at a 7/11 for a fruit smoothie instead. *Go me and my self-control!* As I drank it outside, standing next to Rhonda, I watched a homeless guy fight with the shop owner over the bins. Fists were raised and threats were thrown before the owner took his rubbish back inside. All the while the unmistakable sound of Harley engines rang in my ears as one after another rode past. *Yup! This was L.A. all right.*

I'd been in Latin America for exactly a year when I crossed my 38th border into the United States, and now I felt nervous. I believed my chances of being robbed had increased, my cost of living had increased, and I'd just run out of money again. I believed people would be less inviting there, and the culture shock was the worst I'd felt so far. There would be no parking Rhonda safely in the local fruit shops now. Coming back into the western world really did feel like just that – a different world, and one that I no longer felt part of. Physically, I was pretty broken now. The wear and tear after more than eighty thousand kilometres on a 250cc single cylinder motorcycle with a standard seat was showing in all my joints. My injured left shoulder was much worse, and my right shoulder was beginning to show signs of potential damage, as were my neck, back, elbows and wrists. In fact, everything hurt but my knees. I guessed that I had another fifteen thousand miles still to go at that point and estimated it would take perhaps another nine months. Mentally, I still felt strong, even if my body felt eighty years old. I had no doubt I could finish what I had started. In fact, I had never considered stopping in all the time on the road. I had questioned my reasons and forgotten my motivation at times, but never had I considered actually giving up. Now, I knew that if I just used a little common sense and got some rest, I'd find a way to keep going. But where the hell was I going to pitch my tent in these parts?

Rhonda the Honda was doing far better than I was. She *was* showing signs of wear though, so running costs could potentially become a much bigger factor now. The exhaust was already in desperate need of replacement, having rusted through. I guessed some of that could have been down to all the sea air in Antarctica, not to mention the waves in the Drake Passage and the Bolivian salt flats. The header pipe was cracked too, and could possibly break at any minute. Otherwise she was as strong as ever, as far as I could tell. The sub-frame repair we had done back in Java was still holding strong and the other parts she'd had replaced were just perishables that had predictably worn out. She'd had two new fork seals, seven sets of tyres, one new spark plug, one new battery, six chains and sprockets, and around ten oil changes. I could not have asked for a better bike on the reliability front.

It was February 2016. I was now nearly two years into what was supposed to have been an eighteen-month journey, nowhere near home and entering an expensive country. Of course I was nervous. I hadn't exactly come up with my next plan yet. Thankfully, the Californian welcome turned out to be as warm as the weather and my fears were quickly laid to rest. Before long I had invitations rolling in from biker after biker who had been following my blog and wanted to offer me a place to stay as I passed through.

The Dean family in Newport Beach were the first. The kids had made a sign saying *'Welcome Steph and Rhonda'* which they held up as I rolled into their driveway. Then there was Mitch in the San Fernando Valley, who showed me that musical films really can be entertaining. Tim and Anna, the British couple in Pasadena, and Eric over in Orange County who introduced me to a new sponsor called Sena, who provided me with a new helmet-cam with integrated communications and speaker system for my helmet.

These people were all so different. Mitch worked as a driver in the film industry, Tim studied as a motorcycle historian, and Anna, an ex-pat Brit had won awards as a director of feminist porn films. She also once ran for the Liberal Democrats before moving to L.A. Once again, my low budget was almost forcing me to take up these random offers of kindness, and in doing so enriched my experience far more than any hotel room ever could have done. I like to think we all gained something from my stays. We shared life stories and experiences. We became friends. I even cooked for them – occasionally. It was easier without a rolling galley and the threat of icebergs! Of course, we all started with something in common. We all loved motorcycles and that was what had brought us together in the first place.

While still in the L.A. area I got Rhonda's broken screen replaced courtesy of a company called Bajaworx. I'd broken the last one during a roadside strop in Mexico. The wind was blowing hard and my shoulder was killing me. When we stopped, I punched the screen with my good arm in anger. It comically snapped in half and fell to the floor. For a moment, I just stared at it in disbelief, and for a moment Shane just stared at me wondering which way I was going to go, before risking the line, "You have to admit that was funny". We had both burst out laughing and continued until we were doubled over in tearful hysterics. Gerald from Bajaworx had promised me this one would be stronger, and I could punch it as much as I liked.

I also started to receive and accept offers from people all the way up the Pacific coast, inviting me to present talks at various clubs and businesses along the way. I was amazed at how many people had been following my journey and wished I'd brushed my hair before shooting those roadside video diaries. These offers dictated my route and actually created a tight schedule.

One of the invitations involved leaving Rhonda in Orange County and flying to Baltimore on the east coast for the Timonium Motorcycle Show. An offer I accepted and an experience which turned out to be an education in the hard-core American chopper scene. A welcoming crowd none-the-less. There would not be much time to rest after all. However, I had to make hay while the sun shone. My body would have to wait. Money was the priority or there would be no way forward. Indeed, there would be no way home.

Despite the pain, I kept planning as if everything was going to be OK in the end. With a few adjustments to my route, and a little rest, I'd be fine. Africa though, was a bit of a concern as it required a lot of advance paperwork and route planning, not to mention finding an affordable way to get there in the first place. Whatever happened, I knew I had my hands full for the next few months and I wondered how long it would *really* take to finish this challenge. I did still see it as a challenge and in part that's what kept me going. That had not changed in me. I'd learned some time ago that without goals I had nothing. I needed them. I think we all do. Whether it be a working through our 'to do' list for the day, or conquering Everest. I certainly thrived on planning, doing and the sense of achievement that followed measured success. Besides, it kept my brain focusing on something other than the aches and pains.

I continued through Simi Valley to Morro Bay, about half way between LA and San Francisco, and home to another interesting character. Drew lived in a modest bungalow in a to-die-for location. He had coffee waiting and showed me around. Amongst his collection of relics, Drew had a signed photo of himself as a kid on set with Paul Newman. His mother had been an actress. He also had many old racing pictures from back in the day. The *pièce de résistance* though, was hidden under a tarp in the back yard. We were soon cruising down the road in a '53 Chevy and heading for the nearest bar to continue our conversation over a beer. The more people I met, the more I realised everyone had a story to tell. It was nice to have the time to listen.

The next morning, we headed north together along the two hundred kilometre stretch of Pacific Coast Highway 1 – a road famous for its cliff-hugging 'twisties' and spectacular views of the ocean. I'd been there before in the height of the summer when it was full of RVs. Now Rhonda and the

Chevy purred along the empty road, stopping only for photographs, and coffee.

My next couple of weeks was a whirlwind of forests, fundraisers and firearms. I presented to motorcycle dealerships, businesses and biker cafés. Each one brought me another fistful of dollars. Sometimes it was a case of passing the helmet around at the end, sometimes they paid a price on the door and other times it was a set fee paid by the establishment. I pushed on daily through the rain and seemingly endless stretches of road to the next destination, and the next performance, as I watched the bank balance rising. I smiled and told each audience how wonderful life on the road was. In part, the friendly, creative and dynamic people I met, who put so much energy into life, carried me forward. But riding, presenting, smiling, meeting new people, and dealing with a pain that felt like a big black dog gnawing at my shoulder joint, was in truth, exhausting.

My helmet time was more important to me now than ever. No matter how lovely everyone was, I was desperate for that time when I could be alone to recharge. I stopped often under the redwood trees and the Spanish moss that decorated them; the silvery grey strands dangling from each branch, creating a wonderfully spooky appeal. Rhonda looked like a Dinky toy as I parked her next to them and lay beneath, sometimes in the rain, drawing energy from their giant regal presence. I felt at peace here. I didn't feel obliged to smile and be engaging; something I was finding increasingly difficult to pull off.

I knew nothing about Frank except that he rode a Kawasaki KLR 650 and had offered me a place to stay and an opportunity to present to the residents of Lakeport. After a ride over the Golden Gate Bridge, along the dramatic coastline north of San Francisco and then a bit of dirt riding through the trees, we found ourselves in this sleepy little town on the west shore of Clear Lake. We were still in California, but it was in total contrast to the south end of the state. California had such diversity in terms of its weather and terrain.

Frank was a big man by stature and, as it turned out, a big man in his hometown too. He was in fact their pastor. A fact I wished he'd told me sooner; I might have watched my language. Still, he didn't seem to mind me swearing, and after meeting his family – his wife and two boys – he suggested we go and have some fun,

"How do you like shooting?" he asked.

"Um. Shooting what?"

"Oh, just bottles and cans. I can take you out in the truck if you like and teach you"

"My shoulder is a little damaged. I'm not sure I will be able to…"

"Nonsense!" he interrupted, "we can stick to the smaller guns. What do you say?"

How could I turn him down? The thought of going shooting with the pastor just felt so wonderfully American to me.

Frank was so enthusiastic, I couldn't help but get swept along, and before I knew it, we were hitting the off-road trails in his American-sized pick-up to a quiet little clearing in the woods.

Frank explained the guns to me, how to stand and how to hold them correctly. Although it was my left shoulder that was really in trouble, my right wasn't great either and I was a little nervous of the kickback, but he assured me this particular rifle would be relatively gentle. The first shot I took was a hit, and the bottle flew up in the air spraying water as it went. My face was a picture. Frank caught it on film with my GoPro as the bang went off. I looked shocked, then burst out laughing. I hadn't expected it to be so loud, especially wearing earplugs.

Our afternoon together was great fun and my whole time there was an education in the minds behind those who exercise their second amendment citizen's right to bear arms. Everyone in the town had at least one gun each. You'd be strange if you didn't. This act was very much the topic of conversation around the world at the time, as was the soon-to-be-elected Donald Trump. Should this right be revoked? It seemed the popular consensus of the majority in this region was a resounding NO! Donald Trump stood with them, and I could see why they voted for him. I had never understood why it shouldn't be revoked. More so with each new mass murder, but I came to see how important it was for many people to keep this right, and how it formed part of their culture in these backwater towns. Taking this right away would be like telling a Brit that they could no longer drink tea, though I knew of no tea that would kill if put in the wrong hands. Earl Grey is pretty bad, but not deadly.

The following day Frank had rounded up the whole town for my presentation. It was a great turnout of young and old. A couple of old dears sat in the front transfixed by my story and images. When I finished, Frank picked up my helmet and said,

"Right, time to pass the helmet" then as he began walking towards the applauding audience, he turned back and whispered, "I'm the pastor. I'm good at collecting money!" then winked and addressed the crowd, who did indeed give generously. The two elderly ladies who had been sat at the front came over to me and said,

"That was so wonderful to listen to. You are an inspiration" almost in unison.

"I'm glad you enjoyed it" I replied, slightly embarrassed by the compliment.

"We were just saying we wish we had travelled more"

"Well it's never too late"

They laughed and agreed that they did indeed have some life left in them yet.

"I look forward to reading your blog!" I laughed.

A journey of a thousand goodbyes pulled no punches as I tore myself away and headed further north once again – richer for my experiences, in more ways than one. The need to earn money en route had added a whole new dimension to the journey and led me down roads and into towns I might never have found otherwise. It allowed me to share my story with people from all walks of life and introduced me to communities that taught me something new. I found friends and experiences I might otherwise have missed.

By the time I made it to Corvallis in Oregon I was done. My shoulder had been steadily worsening and refused to be ignored. I could no longer raise my left arm beyond a few inches. I couldn't move it to the side either, and if I accidentally moved it backwards or in 'the wrong way' at all, I would find myself on the floor, screaming in agony. Sleep was difficult and broken. Riding was becoming more and more uncomfortable and the rest of my body complained as it tightened and twisted to protect and compensate. Slowing down or stopping was not an option as I had to make my presentations, but I couldn't take much more. The vibrations from the bike accentuated the pain and the movement in my left arm was so little, I could barely reach the handlebar with it.

Corvallis was home to Randy, a logger and BMW rider whom I had originally met in Baja. Randy found me on Facebook and stepped in when he heard mention of my shoulder problems. He offered not only a place to rest, but some treatment for my shoulder with a little help from his friends; one of whom just happened to be a bit of a shoulder expert having worked as a sports therapist with high league baseball teams, amongst others. I was desperate to get there and get the opinion of an expert, worried that I was causing permanent damage by continuing, and worn out by the pain. I had been riding on 'empty' for a while now.

Randy was an ex-military, ex-alcoholic, ex-Harley-riding, Trump-loving, born again Christian with a handlebar moustache to match. I wasn't sure what to make of him after my first night at his place. He'd got into his pyjamas, sat back in his chair and said,

"If you feel nervous at any time, don't worry, I've got this" and pulled out a pistol from under the chair in which he now lounged. I hadn't, but I did after that, and for the rest of that first evening.

Randy and I got off on the wrong foot. Perhaps it was a cross-culture clash accentuated by my weariness, but he certainly turned out to be man of his word, as were his friends. Within a few days of arriving I was being seen to by a masseur to loosen the muscles and work on the knots. From there I hopped straight onto the chiropractor's couch; he realigned my neck and straightened my lower back. The next morning, I was over to see Guido – the shoulder expert who now worked at Oregon University – who examined me and checked my range of movement before announcing that I had a frozen shoulder, probably brought on by a rotator cuff injury back in Colombia, and the vibrations of the bike.

"This is the worst case I have ever seen" he announced, "You are going to need an operation to release this. Otherwise it could take years". This was a concern, especially as he added that riding was potentially going to make it worse – although I knew that already. I had to have treatment and

I had to rest. I could afford neither, but I accepted Randy's offer of staying for at least a little longer.

I stayed two weeks in the end and grew to understand Randy a little better. He was a kind man. He just had a different way from me, but where it mattered, he was there. I continued to get treatment for my shoulder. Massage, physio and a cortisone injection at the local hospital; driven to each appointment by Randy.

Rhonda too was being sorted out. Harold Olaf of *Giant Loop* (a company that makes motorcycle luggage) lived in Bend, Oregon, 160 kilometres away, and not only offered me a chance to speak for a group he had organised, but also offered me a whole new luggage system. My Kriega saddlebags were holed by the crash in Colombia, so were no longer waterproof. The new system also had the potential to make Rhonda as lightweight as possible and would distribute the remaining weight in a more balanced way around the bike. If we could make her less top-heavy, with less weight on the tail, the manoeuvring and handling would be smoother. It would also give me a (very) slim chance of being able to pick her up with just one good arm.

For me, changing luggage was like moving to a new house. I was anxious. I knew where I was with the old system. It may not have been perfect and yes it was now leaking, but I knew how to pack and unpack; I knew where everything was, and it was comfortable. It had taken a while to get it that way. Even after all these miles and all the changes in my life, I still found it difficult to let go of what I knew. I was, as someone once described me, "a minefield of contradictions".

Harold had to practically prise the old kit out of my hands and coaxed me gently into the new system. I was not convinced at first.

"Where will I put my coffee so I can get to it easily?" I complained. Harold was patient and just kept on building up the separate compartments onto Rhonda's front and rear. It wasn't until I rode her that I was fully convinced. The difference was wonderful. It felt as if there was nothing on the bike. I knew it would take time to settle in and get things how I wanted them but soon it was coming together, and I was smiling again, relaxed in my new 'home'. The only problem I could see now was that I didn't have a centre stand if I got a puncture (I had used the top-box as my stand in the past). Luckily, my short visit to fellow bikers Kris and Nathan over in Hillsboro provided me with the answer in the form of a parting gift: a neat little gadget, simple yet effective. It was a little adjustable bar that fitted under the footpeg and provided the lift required to get either wheel off the ground; easily storable and simple. Perfect.

My final visit was to Don Webber of Mr Ed's Moto in nearby Albany. Don had been customising seats for people from all over the world for forty years. We had a real giggle in his workshop as he worked his magic to produce a perfectly fitting seat. His wife plied me with painkillers which also helped. Not only would the new seat help with my posture and support my hips, but it would also look good. It was brown faux alligator skin with red on the side. I loved it and my bum was now in safe hands. Don also replaced my handlebar risers with a set that went backwards as well as up. This reduced my reach to the 'bars. I left feeling a little stronger, but in all honesty, little was helping with the pain. If I could just make it to Canada, (where I had citizenship, because I was born there), I believed I could get treatment and stay as long as I needed to rest. I had friends there too, and no visa worries. *If I could just make it to Canada.* It was only five hundred kilometres away now, and yet it seemed so far.

Chapter 25
The Fall of British Columbia

I trusted Pete Bog. I always had. Instinctively. A big-hearted family man with so many talents. Music, comedy, and being there at the right time. We'd been through a lot together and I hadn't seen him since he had escorted me out of the country two years earlier. So, when he looked at me over a single malt whisky that night at his brother's house in the Okanagan area of British Columbia, as the tears rolled down my face, and said,

"Steph. It's time to go home." I knew he was right. I guess I just needed someone else to say it. Then the sobs came. Deep, soulful sobs. At that moment, I let go of the months of pain and heartache. It was time to face facts. It was time to stop beating myself up and listen to my body. It was time to go home. I had failed.

I had been in Canada for a few weeks now and crossed the border to find Jeremy waiting for me with a place to stay in Yaletown, Vancouver. Jeremy and I had stayed in touch since our get together with Shane in Nimbin, the 'hippy town' in Australia, and it just so happened he was home for a while as I passed through his part of the world. I had my first full-blown medical crisis in his apartment, and he had taken me to the hospital Emergency Room in a taxi. I had another one in the hospital and, after an examination, the doctor suggested I might have an impingement of the lateral nerve, "a bit like carpel tunnel syndrome but in the legs". She believed the damage to the shoulder and this new problem was a result of a combination of the crash back in Colombia plus prolonged exposure to vibrations.

After a visit to a neurologist and an MRI scan, they concluded I had a narrowing of the spinal canal, causing pressure on various nerves. This, added to the constant vibrations, had also caused sensitisation of the entire nervous system, which had resulted in 'deep somatic pain'.

"Deep somatic pain occurs when stimuli activate pain receptors deeper in the body, including tendons, joints, bones, and muscles" the doctor explained. Her description certainly matched how I felt.

An ultrasound scan of my shoulder then confirmed there were many bone 'spurs' holding the joint in place "a bit like Velcro". I was also suffering from 'pain anxiety' after several months of chronic red-hot-poker-in-my-shoulder syndrome followed by what felt like electric shocks and episodes of spasticity in my legs. When the electric shocks came, it started a domino effect of spasticity, tightening of the chest and dizziness. It would last anything between twenty minutes and an hour and those 'episodes' were pretty scary. I worried about having them when I was on my own and was so grateful to have my friend Jeremy watching over me.

Jeremy was in his late-fifties and a real Vancouverite; always well-presented and on-trend. He'd founded the global outdoor clothing company *Arcteryx* back in 1989 and sold it for a tidy profit several years later, before hitting the road on his motorcycle in search of the simple life in 2012.

I spent my forty-first birthday, March 27th, in hospital, but when I got out, Jeremy had a ridiculously large chocolate cake waiting for me. Cookies and Maltesers on top; the lot. It was way too big for both of us, but he didn't care. He was out to cheer me up and spoil me rotten on my birthday. The next day he took me shopping for a birthday present and bought me a new Leatherman multi-tool to replace the one my dad had bought me, which I'd frustratingly lost in Australia. The perfect gift for a girl on the road. We then went to watch an ice hockey game; my first, and the only place you'll find rowdy Canadians. Then we pottered down Commercial Drive, a wonderfully 'hippy' part of Vancouver. It was there that I found myself joining in with a street performance karaoke. I took the mic and, holding my arm crookedly against my body for protection, sang Janis Joplin's *'Mercedes Benz'*. *Ha! Only Africa to go and I will have sung karaoke (kinda) on all seven continents!* It was good to know I still had it! Not necessarily in the singing department but, even in pain, I still had my dreams and more importantly, the desire to chase them.

"Right, let's get back to Yaletown and get you some medicinal cannabis from the weed shop" suggested Jeremey after a sidewalk coffee.

"I've got opioids from the doctor for that", I replied as I downed the last of my flat white.

"Yes, but it'll be fun" he said, "They've got all sorts in there". And they did. It was a very high-end affair with expert assistants hovering over clean glass counters offering everything from Acapulco Gold (for reducing fatigue), to Purple Kush (ideal for reducing pain and muscle spasms), to Maui Wowie (for energy). I was sent away with a prescription of Purple Kush for the spasms and Blue Dream for the inflammation. Annoyingly, they both just made me feel even more anxious, though it did help us finish off that chocolate cake. Even a cigarette made me dizzy, so I stuck to the opioid tablets after that, and only when I was at my worst.

After three weeks of rest, Jeremy escorted me on his BMW HP2 1200 Enduro through the Canadian Rockies and up to my friends, Mark and Jackie's house in the region of British Columbia known as 'The Okanagan' after the lake and valley of the same name. It took us two days, at a very steady pace, with only two minor events. The first was when my legs turned to jelly and I threatened to turn back for fear of another full-blown 'episode'. The second was when a member of the Royal Canadian Mounted Police in an unmarked car gave us a ticket for overtaking him on double lines, having followed a log truck several miles up a steep hill! My first ticket, and a proud moment for both of us, when a photograph of the Mountie giving me the ticket (taken by Jeremy) was later used on the local news announcing my arrival in Canada!

Mark was Pete Bog's brother and we had been on a couple of adventures together in the past, in South Africa and Morocco. Mark and his wife Jackie were so kind and so positive and offered me a place in their home for as long as it took for me to recover. So much so that I felt bad for feeling bad. I wanted to withdraw most of the time, even when they gave me permission to be miserable,

"You can cry and be miserable as much as you want", Mark reassured me gently, "You don't have to hide". But most of the time I found myself confiding in their dogs, as I always did at times like this. I hated myself for being so down. I just couldn't find my mojo. Eventually I started going for walks. I'd use a stick to support myself at first because even the slightest incline made my legs go weak. I wondered if I was ever going to be right again. Perhaps I had *really* done it this time, but I was desperate to find a way to rebuild my strength as quickly as possible.

I took weekly sessions of stem cell treatment and prolotherapy; both relatively new procedures, altogether involving several injections in my spine, groin, and shoulder joint.

"You are going to have twenty injections per session" the doctor explained, regarding the prolotherapy procedure.

"They won't hurt that much, though right?" I said looking at the ridiculously long needle.

"Oh yes. I'm afraid they are going to hurt like hell" he replied.

And they did. They really did. Each one was agony and only compounded my pain anxiety. It was all so aggressive, when what I needed was the opposite, but my desire to repair quickly and get going again, far outweighed my fear of those needles, or indeed my rational mind.

I spent several months in Canada like this. I did manage one presentation for a charity event, and I did get a little strength back. I would take very short rides on the bike, but only down into town for coffee, or over for more physio or injections. About a week after my presentation I received an email from a guy called Ryan. He had brought his mum, Lorraine, along to my presentation with the intention of introducing her to the positive side of motorcycles. Lorraine had lost her other son eight months previously, to a motorcycle crash. He had been home on leave from service in Afghanistan when it happened. Killed by a man in his eighties who just pulled out of a junction without warning. At first I wasn't keen to meet up, and explained that I wasn't really myself right now. I was hardly a positive representation of the motorcycle world, but in our first couple of emails, he seemed to know just what I was going through, and I agreed to meet him for coffee. I ended up spending a lot of time with Ryan, and indeed with his Mum. We became firm friends and I like to think we helped each other through some tough times.

When Pete 'Bog' arrived from the UK for a family wedding, I was still a mess. He could see I was really not myself. Withdrawn and with my emotions always so close to the surface, I felt as if I was ready for a full-on breakdown. I had now spent all the money I had earned in the US on treatment and I just needed to stop. A short 'tester' ride out to the Kootenay region of southeast B.C. with Mark, Jackie and Pete saw me on my knees, legs like jelly, on the side of the road gasping for breath through my tightening chest. It was shortly after that, over a whisky with my old mate Pete, that I agreed, after two years on the road, to go home. I needed to see the mountains of Wales again. I needed to be with my family, and I needed a hug from my faithful dog Chui, who was waiting patiently for me with my parents. Most of all, I needed time. It was July 2016.

****

Back home in Wales I had physio and I walked daily. I stretched and took up Qi-gong/Chi-kung (related to Tai Chi, but different). I ate tons of spinach and chicken and I even gave up smoking *again*. I was the perfect patient and decided that when I could climb a mountain, I would be ready to go back to Rhonda in Canada. That would be the sign that I was strong enough. Winter was coming though, and friends and family suggested I wait and rest through the winter then go back in the spring. They were worried that if I went back too soon my recovery might prove superficial under pressure and the strength I had built up might quickly disappear.

But after just eight weeks in the UK, I could wait no more. I had to get back to Rhonda. I'd worry about the winter when I got there, and work something out. I had promised myself so many days, so many long nights when I was at my worst and contemplating the futility of it all, that I would respect my body from now on. I had come too close to disaster, and I would listen when it told me to stop. I would take things slowly. But like a child who had got what she wanted, I now forgot those promises and at the first opportunity I got a flight back to Rhonda.

The last few months had not been easy, but those unfortunate circumstances had led to the rather wonderful accident of being in British Columbia for autumn or 'the fall' as North Americans call it. Rather than worry about the snow and try to race across Canada before winter, I made the decision to do a big loop around B.C. and check out this exceptional part of the world at a steady pace, before settling in and waiting for winter to pass. Africa could wait. This would be a great opportunity to test my body and do some filming at the same time. Having been born in Toronto, I felt a connection to Canada, and I wanted to get to know it better. My parents had moved back to the UK when I was five, having moved there from Wales ten years earlier. They had done a 'big loop' with my two sisters and me in a converted school bus (they called 'The Blunder Bus') before they left the country in 1980, so this felt like 'Round Two' for me (although I remembered nothing of the first time around).

It was October 2016 and I was finally back on the road after what had seemed like an eternity of injury and recovery time. I had returned with a small income from *Ride* magazine who had given me a monthly column. I also had a series of articles to write for *Canada Moto Guide* as I travelled across the country. I had been contacted by a journalist called Rob Harris and after a few Skype calls, we had agreed on a price, a theme and set dates for each piece. Rob was a Brit living in Canada and we got on immediately. We agreed to meet when I got to 'the other side'. He offered to put me up and show me around the east coast when I finally arrived.

The combined income from both jobs didn't amount to much money, but along with the odd freelance article for various other magazines, it was enough, if I lived frugally. Canada was an expensive country though, with B.C. often said to stand for 'Bring Cash'. The budget was low but the feeling of adventure had returned and was coursing through my veins once more. I was back on the road and it felt as if I'd had a total transformation of perspective on my entire trip. I was no longer running away from or to, anything. I was just running. I was completely submerged in the 'now', and once again, it felt liberating.

Armed with bear spray, backroad maps, tent, and GoPro, I headed out into the wilds of the Canadian Rockies in search of dirt roads and grizzly bears. I was determined to find me a bear and capture it on film. I was determined to have fun!

****

As I sat cradling my coffee and watching the puddles form around my tent next to the lake in the one-horse town of Clearwater, I debated which way to go. I had worked out a rough route before I left but I wasn't sure if I would have time now. It was getting colder already. Still, I had been told by many Canadians that the Icefield Parkway between Jasper and Banff was a 'must ride', and so I had to go and see for myself.

The weather was unpredictable as I rode higher into Alberta, along the forest trails and into the Rockies. By the time I reached Jasper the night-time temperatures were reaching minus five Centigrade. This was the coldest I had ever camped at, that I could recall. It wasn't too unpleasant with a decent North Face sleeping bag and my 'Vernadski' socks from Antarctica, but the morning was bitter. I snuggled back into my sleeping bag after making my morning coffee on the stove outside and decided any further attempts to get up would have to be postponed until the sun arrived over the surrounding trees. The bear hunting could wait.

Returning to Canada had been a big decision for me, but the last two weeks had been a blast. I had been feeling stronger by the day. All the hard work and spinach back home had paid off and once I got the wheels turning and the blood pumping, I found the remaining aches and pains became much less noticeable.

The Parkway (Highway 93) climbs to an altitude of about two thousand metres and is lined with ancient glaciers, waterfalls and rock spires. There is no wonder it's one of Canada's national treasures. My hands froze but the sun soon warmed me as I stopped for photographs at every corner where a fresh view hijacked my senses and forced me into an admiring gaze once more. There was no denying the beauty of this area. The roads were good too and bears were often to be seen here, apparently, but not by me; only mountain goats and sheep. Still, the threat of bears and cougars while camping all added to the excitement. Every evening I carefully placed my kitchen bag in a tree and huddled into my tent, half expecting to find it torn to shreds in the morning. Nothing. I was almost disappointed.

There were signs everywhere saying, 'BE BEAR AWARE'. I giggled as my imagination went wild, and one night, while camping in the woods, I decided to strip naked and take a shot of myself – with the help of a tripod and a timer – from behind. Arms stretched out and baring all for the falling sun through the opening of the trees. Once back in signal I posted it on Facebook with the quote, '*I thought the sign said, 'Be BARE aware'*. It got more 'likes' than any of my previous photos! It was lovely to have time to be so playful and allow the imagination to run free. To me, this was the biggest difference between transitory life and life on the road. It's easy to forget to make time for play when you are sitting still and dealing with the onslaught of everyday life. Movement stimulates the whimsical gene and restores our inner child.

From Canmore in Alberta, I took the Kananaskis Trail which is two thirds well-graded gravel and occupied only by hunter camps and the odd unsuspecting deer. It was a road well worth riding and the colours of the changing leaves added to the magic as I wound my way south back into B.C. and the little town of Fernie. As I parked my dusty motorcycle and wearily sat down in the coffee shop for a much-needed Americano, I was approached by a man in his late fifties wearing a baseball cap and looking like he was trying too hard to be hip in a scruffy 'aloof' kind of way.

"Where are you from?", he asked in a voice that boomed across the café. For some reason I answered in decibels that appeared to match his – which surprised me just as much as the guy on the next table.

"I'm from North Wales"

"Oh RESPECT!", he replied and proceeded to grab my hand and reach in for what I initially thought was going to be a kiss! Thankfully he stopped before I could react with a swift move and stern voice (the one I usually reserve for naughty dogs, drunk Turks, Iranian taxi drivers and semi-conscious British military men). He lined up his nose with mine and looked me expectantly in the eyes, just inches away, holding that pose for what seemed like an eternity. I clearly looked confused. He pulled away with a disappointed look and said,

"You don't do the nose touching? Which island are you from - North or South?"

"Um I'm from North Wales not New Zealand", I said apologetically. "I'm afraid we don't do the nose touching thing there. We're more the quick doff of the cap type".

I stayed in Fernie for two nights in the almost empty Raging Elk Hostel. Once a busy coal mining community encircled by the Rockies and nestled in the beautiful Elk Valley, it survived a disastrous fire in 1904 which levelled most of the town. In the 21st century Fernie had turned to adventure tourism to save itself from becoming another of British Columbia's ghost towns. In contrast, the neighbouring village of Coal Creek did not survive. It too had had a disaster a century earlier, in 1902, when an explosion in one of the shafts had left 128 dead in one of the worst mining disasters in Canadian history. During the 1950s most of the residents left when the mine closed. Some parts of the town remain in the form of ruins, but most have been overgrown by forest. I rode up the trails to see if I could find the remains of Coal Creek but found very little other than a few piles of rocks where buildings once stood.

From Fernie I rode the now familiar-looking trails, just a short ride to Marysville. I camped out once more and spent a lovely evening in the woods. Just me and Rhonda and our campfire. It was the perfect evening and it was warm at last. The hostel had been fine, but *this* was the life, and I was looking forward to my ride tomorrow. I had been told by a biker from the Okanagan area that there was a pass over the mountains between Marysville and Gray Creek on the edge of Kootenay Lake. He had said it was a dirt road that climbed over a 2080 metre summit and was only passable for a few months of the year due to the conditions. (That's nearly 7,000 feet, and I later discovered that it's one of the highest roads in Canada). It sounded like a challenge and I was really up for challenges at this point. I packed up early and hit the road once more in search of my daily dose of adventure.

As I began the ride, the sign warned me that the road was not maintained, and only high clearance vehicles may pass. I looked down at Rhonda, smiled and said out loud,

"Check! We are good to go!".

That sixty-kilometre stretch represented what biking is all about, for me. Any cares or worries melted into oblivion as my mind was given the equivalent of a 'factory reset' and the pure, uncomplicated emotions that biking can bring were restored. The nervous anticipation gripped me as I climbed higher into the bear-infested mountains, not knowing what terrain I would find. The feeling of solitude washed over me as I was swallowed up by miles of colourful trees going through their seasonal change. My confidence was building and my smile growing as I picked up speed with a devil-may-care attitude and headed joyously into the unknown. This was my kind of riding, and I lost all sense of reality and danger and rode faster and faster along the trails. I couldn't stop smiling and shouting "Yeeeehaaaaah" as we rode like hooligans through our own patch of wilderness. Sometimes I had to stop and kill the engine, just so I could listen to the silence. It really was golden.

"Not bad, eh Rhonda?", I said as I soaked in the atmosphere and surveyed the miles of colourful landscape around us, before setting off again. I was disappointed when the main road appeared ahead and I knew it was over.

The disappointment soon faded. The paved roads around Kootenay Lake are perfect biking roads. They are quiet, twisty and smooth, in contrast to the undulating dirt I had just ridden. I could see I was going to have some fun in this area too. I was heading for Kaslo, a place I had visited with Pete, Mark and Jackie before going home to Wales.

Kaslo is a small, old silver mining town with a population of one thousand, and a place I had fallen in love with last time I was there. That night I decided to stealth camp by the lake so I could use the money I saved to buy myself a pizza and a glass of wine from the rather special little pizzeria I had appreciated on my previous visit. Pleased with myself for the idea, I set up camp once it started getting dark and raced over to claim my reward. The pizzeria was closed...

The next day I headed over to Toad Rock, a bikers' campsite I had heard about on my travels. Mary, her overgrown pig named Happy, and her five dogs, were there to greet me. It was the end of the season so there were no other customers.

"I've had a good summer", she informed me when I asked her how much it would be to pitch my tent. "Take a cabin", she continued, "and just pay me whatever you can afford".

The cabin was basic, with no electricity or heating, but it was a step up from the tent, with a large and extremely comfortable bed, an abundance of blankets, pillows, and even a hot water bottle. The nights were pretty bitter now and so I was grateful to be off the ground. Mary had run this campground for forty years. It had an old-fashioned biker feel to it. The cabin area was also dotted with old buses and vehicles of all shapes and sizes that had been roughly converted into sleepers and set in their final resting places, scattered amongst the trees and the fire pits.

It was my dream home, and I spent a happy two nights there before moving on to a couch-surfing host in the town itself. When I left, Mary refused any money at all,

"Let's just swap stickers" she laughed (it's a biker thing), "I've enjoyed your company, and so have Happy and the dogs". I'd loved their company too.

My couch-surfing host, Luke, was such a dude. So laid back, in a typically Canadian way. We got on immediately. Being there felt homely, not least because I kept stealing his slippers; the ones his mum had made him. I missed my slippers, and now Luke missed his too! It was as if we had known each other for years, and we enjoyed many beers together in the evenings as we sat on his balcony that overlooked the town and the lake. That's when I finally saw a bear! Walking casually up the street below the balcony, it turned into the neighbour's garden and looked into their window before helping itself to some apples off the tree before wandering off again. Of course, I was too slow with the GoPro!

Filming was more fun on this leg of the journey than it had ever been before. I enjoyed playing in front of the camera. Capturing moments and coming up with silly things to talk about. I'd done a lot of video diaries all the way through and had put some serious effort into some of the epic bits, like Antarctica, but anything more was time-consuming and often got pushed to the side in favour of just riding. Now, it was all part of the fun. The Sena system I'd been given in California allowed me to narrate as I was riding as it had a built-in mic and speakers. I also had my main Sony camera and a tripod that I could set up in the woods to film some camping scenes, or on the side of the road to capture a piece of riding. A five-second clip of riding could easily take half an hour to shoot, and quite often at a great risk to my camera. When you're on your own, it's a laborious procedure: Stop the bike, gloves and helmet off; set up tripod and camera angle; ride away, ride back past the camera; turn around; come back; pack up, and so on. If you want a shot of the bike going away from the camera as well, that's an extra whole procedure to turn the camera around. The clip would then be saved to be edited into a compilation later, which in itself could take hours.

For a few days I edited, wrote, chilled out, split logs and took long walks with Luke, then went riding when he was at work. Everywhere was just so beautiful. It was like Wales on steroids. Bigger mountains, bigger lakes, fewer people and more wildlife. My dream place. The riding was outstanding, with world-class paved roads and dirt tracks in abundance. I had found my Utopia for sure this time; but soon the passes would be laden with snow, and so with a heavy heart I gave Luke one last hug and said goodbye, hoping that one day I could return.

Chapter 26
Trucks to Tuktoyaktuk and Yetis in the Snow

I'd had no concrete plan for the winter, but almost as soon as I got back to the Okanagan after my 'test loop', I landed myself a house-sitting job overlooking the lake, with a log fire and a cat to stroke and feed on a daily basis. I packed my tent away, ready for a four-month winter stay in a town called Summerland, (ironically); it was half way between Mark and Jackie Jennings-Bates' place, and Ryan's. I had a big kitchen, a massive fridge and a coyote that visited me in the garden most mornings as the temperatures dropped and the snow piled up outside.

I spent a lot of time writing while I was there, including an online-only book on how to ride around the world on a budget, which I called 'Embrace the Cow'. (Remember? Some days you get milk, some days you get shit!). I'm pleased to say that sales were good enough to have a significant positive effect on my finances while I was still in North America. I visited Ryan and his mother Lorraine a lot too. Lorraine insisted on plying me with all kinds of sweet goodies every time I saw her. She was always baking and seemed determined to fatten me up to keep me warm. She became known to me as 'my Canadian mom', while Ryan and I spent a lot of time in his workshop, building things and coming up with ideas to reduce the vibrations during my ride; my body was still super-sensitive to them. We fitted rubber pads on the footpegs, fatter grips on the 'bars and a new top yoke (triple clamp) with built-in neoprene dampers in the handlebar clamps that my friend back home, 'Smiley Steve' had made for me. Amazingly, he created the whole yoke out of melted down old beer cans and a few old pistons, in a forge he made himself. He'd even engraved Rhonda's name on the top. We also fitted a new exhaust donated by a local guy I'd met named Dennis, who had upgraded his own and so had a spare! I really did have some amazingly talented and thoughtful friends.

Ryan also spent time on a rather spectacular custom bike he was building in memory of his brother while I made a holder for my tripod out of a plastic drainpipe, two end caps and an old zebra-patterned belt – and generally hovered about annoying him as best I could. I decided he was a lovable grump, and a surprisingly good listener! It was nice to spend time getting to know someone, rather than rushing off all the time.

By mid-November I was finding it increasingly difficult to ride Rhonda due to the snow, but I needed some kind of 'fix' to stop me getting 'cabin fever'. My only experience of skiing had ended in a desire never to repeat the experience, so what else could I do? When I was offered the opportunity to ride and write an article about a 'snow-bike' called the Yeti MX, I jumped at the chance. Packing an avalanche beacon, radio, and lunch into my rucksack, I headed out to Hunters Ridge in the Monashee mountains to meet with Kevin Forsythe (the founder of Yeti MX) and a photographer who was going to capture the day for my article. The Yetis were converted Husqvarna 450cc motocross motorcycles – the front wheel was replaced by a pair of small skis and instead of a rear wheel, drive came from a single caterpillar track. They were much lighter and more manoeuvrable than the traditional snowmobile or 'skidoo' – but also easier to fall off!

Blasting through this sensational scenery was like nothing I had ever experienced before – the nearest comparable experience was dune riding. A similar feel and just as exhausting, but somehow easier – the bike driving through anything with surprising stability.

We travelled up trails, off trails, across meadows, and were kept on our toes by the occasional snow hole or creek that had been invisible until a split second before we hit it! Face-planting was surprisingly forgiving *most* of the time but getting upright again was not so easy. The powder snow was soft and enveloping, but on such a beautiful 'bluebird day', who cared? The snow ghosts (trees that are layered in a type of ice called rime) gave the landscape a wonderfully mythical look and we played like children as we carved our way through this sparkling dream world.

Over Christmas and New Year, I spent a lot of time snow shoeing, taking wine-tasting tours at all the local vineyards and eating copious amounts of cheese, thanks to my new friends Vic and Linda. Vic, a biker himself, had contacted me having followed my blog for some time. It turned out they were just around the corner and I adored their company that winter. I even tried skiing again with another new friend who turned out to be a ski instructor. I enjoyed it but stuck to the appropriately named 'granny' runs. *This* granny was not up for any more injury and recovery time!

Being snowbound in Canada was not so bad after all, although I must admit I was finding it harder and harder to stay still. Logically and logistically it made total sense to stay as long as possible. The roads were treacherous, it was way too cold for a tent, and I had lots of paperwork to prepare for the next leg. However, Adventure Deficit Disorder (ADD) was an affliction I found myself having to deal with on a daily basis, and the only thing to do was to keep planning ahead while I waited it out and ate copious amounts of cookies.

I emailed Tim to let him know I had decided to skip Alaska as it meant I would have to hold out until June at least, and much as I loved being there, just the very act of moving had become addictive; the newness of relationships and the ever-changing scenery. Seeing the best and moving on, never settling long enough to get into a quarrel, see the dark side or, God forbid, think too much. The thought of actually finishing the journey now became another challenge to face, perhaps even more daunting than the start. I had made sacrifices. There was little left back home now, but what price would I really pay for this new addiction? How would I cope sitting still? Was I ruined? I guessed the full extent of any damage would not become apparent until I reached the end of the road. Something I knew I would have to do eventually. For now though, I had plenty more to go at, and Africa was waiting.

Tim replied saying he was concerned about my African leg and that when I got to Toronto in April he would come out and meet me. We could talk it through over a beer. By now Tim and I had become firm friends and had been vaguely discussing a few business ideas together. I wasn't worried about Africa at all. I wasn't worried about my body or what lay ahead. Of course, it would be sensible to keep it short, but there was as much chance of me being sensible at this stage as there was of scientologists finding The

279

Mother Ship. The future, like the past, was all about perspective. Africa was a whole new adventure just waiting to happen and I was well and truly 'in'. There would be no convincing me otherwise. Still, I wanted to see Tim and I had a lot of other things I wanted to discuss with him.

By early March, the owner of the house had returned. I still had some time to waste before I could ride across Canada, so I was pretty excited when I was given the opportunity to join Mark J-B and his auto-journalist friend Budd to the northernmost point of Canada. Our mission – to review and test two Nissan Titan trucks to their limits; one diesel and one petrol. Mark and Budd were both rally drivers and our plan was to work our way through the Yukon, into the Northern Territories, and up the Dempster Highway to the ice roads that would take us all the way to an Inuvialuit hamlet called Tuktoyaktuk. *A truck to Tuktoyaktuk!* It just rolled of the tongue. This was going to be fun.

First, we drove south to Vancouver for a promo shot in the city. The Titan Arctic Challenge would officially begin there. Driving north again, much of it seemed the same initially; trees, snow and long straight stretches. Of course, we had some beautiful stops, like Whistler, and the hidden gem of Terrace – the road to which took our breath away. A stunning display of snowy mountain ranges, and a friendly community nestling in the middle, built by the gold rush pioneers in the late 1880s. It's not a place you can get to quickly without flying in, but get there we did. That night we camped just outside the town, in the woods, under a beautiful moonlit sky. Our tents were on top of the trucks and each night the boys would set these up while I dug the stove out of the back and heated up a previously prepared meal. We had no worries of them perishing, as the air remained well below freezing the whole time.

Towns became even more remote as we entered the Yukon and the Northern Territories. The winter is unforgiving up there and just about as rugged as you can get. In my experience, the more remote the settlement and harsher the conditions, the warmer the community. Whitehorse and Dawson City were no exception. We picked a friend of Budd's up in Whitehorse. Brian was another journalist who worked on the same magazine. He was a jolly sort and a great addition to the team.

Dawson City was a highlight in itself. Situated on the Yukon River, it is a colourful old town with a rich and interesting history, having been at the heart of the famous Klondike Gold Rush of the late 1890s. Its population exploded from 500 in 1896 to 30,000 by 1898! On the day we were there, in March 2017, it was home to fewer than 1500 friendly and adventurous people. 30% of the population were native aboriginal, and the rest were made up of mostly French-Canadians and Europeans who had been drawn there by the rugged lifestyle, and relative shelter from the rest of the world.

The city is not just famous for its Gold Rush history. It also has some rather well known oddities, one of which just had to be experienced by our team – The Sourtoe Cocktail. A dead man's toe in a shot of whiskey! Established in 1973, the toe had been kissed by over 100,000 people who had come from all over the world to do so. As we were presented with our drink, our friendly barmaid reminded us of the golden rule, "You can drink it fast, you can drink it slow, but your lips have gotta touch the toe". Fortunately it only tasted of whiskey, not toe!

The legend of the first 'sourtoe' dates back to the 1920s when alcohol was prohibited throughout North America (but much less strictly in Canada than the USA, including adjacent Alaska) and features a feisty rum-runner named Louie Linken and his brother Otto. During one of their cross-border deliveries, they ran into an awful blizzard. In an effort to help direct his dog team, Louie stepped off the sled and into some icy overflow – soaking his foot thoroughly. Fearing that the police were on their trail, they continued on their journey. Unfortunately, the prolonged exposure to the cold caused Louie's big toe to be frozen solid. To prevent gangrene, the faithful Otto performed the amputation using a woodcutting axe (and some over-proof rum for anesthesia). To commemorate this moment, the brothers preserved the toe in a jar of alcohol.

Years later, while cleaning out an abandoned cabin, the toe was discovered by Captain Dick Stevenson. After conferring with friends, The Sourtoe Cocktail Club was established, and the rules developed. Since its inception, the club has acquired (by donation) more than ten more toes!

While we're on the subject of alcohol, I was amazed to discover that while Alaska officially repealed the Bone Dry Law on April 6, 1933, the act ended prohibition in Alaska for whites only. Strict anti-alcohol measures for Alaskan Natives, first put in place in 1867, remained in effect until 1953.

The North is a mysterious part of Canada that few will ever visit, and certainly not at that time of year. Often referred to as 'the frozen wasteland', none of us was prepared for the beauty hidden within this vast and unforgiving wintery beast. The Dempster Highway, between Dawson City and Inuvic, was breathtaking. The road followed the winding frozen river along the valley bed, occasionally popping up and over a crest to give us a different perspective. The hot springs melted the ice in places, creating spectacular steam displays that rose up invitingly into the cold air. The warmth could be felt from metres away. The mountains engulfed us in white and rugged beauty that left us speechless – in-your-face and untouched by man. On the first day, we managed less than 200-kilometres, as we soaked it all in to a regular chorus of *"Wow!"* and *"No way!"*. I don't think any of us was prepared for its beauty. Mark put both our thoughts into words perfectly when he said,

"I have fallen in love with the world all over again". Two years previously I had been in similar conditions in Antarctica and I had never expected I would go from the bottom of the world all the way to the top, as well as around it. This was truly beyond my wildest dreams.

I placed the triangular-shaped hood of my Canadian bivvy bag over my head, tied the cord around my neck, and that night, for the first time since I was a child – I prayed. This time it was not for the life of my sick dog, but for my bladder to stay strong! It was minus 30° Centigrade and, for reasons that seemed hard to fathom, I was camping on the Dempster Highway in conditions three times colder than your average freezer. I watched as my breath landed, then froze, on the material that was keeping me inches away from total congelation. That's when I tucked my head into the bag and vowed that, no matter how strong the urge became, the call of urination could wait.

News of our arrival spread as we arrived in Inuvik, and there was much interest in our trucks and what exactly 'The Arctic Challenge' was. This was just called 'living' to them. No big deal! We told them about our camping experiences in minus 30°C, and they did not hesitate to inform us that it was a relatively mild winter this year. I'm sure they were thinking, "All the gear, no idea!", but they were far too polite to say it out loud.

Driving onto the 170-kilometre stretch of ice road between Inuvik and Tuktoyaktuk felt wrong to start with – very wrong – but we soon found our stride! 'Tuk' could only be reached by air during the summer months, but in the winter the mighty McKenzie River and the Arctic Sea froze over to become a rather dramatic 'road' into this small community. This place was about as remote as it got. It was a beautiful, out-of-this-world experience – as if we were driving on the moon. I was sharing the driving with Mark, and stopping took a little practice and corners were interesting, but doughnuts were easy, and boy, was it fun! We actually had a phone signal up there too, so I called my Dad,

"Dad guess what?" I said excitedly,

"What?" he said, wondering what the hell was going to come out of my mouth this time.

"I'm on the Arctic Sea. I mean I'm driving on it. *Right now!*"

"As long as you're not *in* the Arctic Sea" he replied. Nothing surprised him anymore.

Tuktoyaktuk (population: 850), commonly known simply as 'Tuk', is the last Arctic village on the edge of mainland Canada's frozen northwest wilderness. The settlement has been used by the native Inuvialuit for centuries as a place to harvest caribou and beluga whales. The name itself means 'resembling a caribou'. According to legend, a woman looked on as some caribou, common at the site, waded into the water and turned to stone. As if divorced from its own country, the village sits segregated on a spit of land jutting out into the Arctic Ocean. There are no sugar maple or spruce – or any trees at all for that matter – only endless ocean, the cold, grey waves frozen and paused. The streets were empty from the cold. Only the husky kennels had life, and the sled-dogs, whose thick fur kept them well-insulated, were keen to announce our arrival.

Most of the houses were old and weathered with tin roofs heavily laden with snow. We drove around slowly for a while taking it all in and stopping to chat to some locals, who pointed us in the direction of an igloo that we could check out. It was a short and sweet visit to the town but one that I would never forget.

In a few weeks, the ice would melt, and the ice roads would never open again because a land road was just being completed (2017). That would change things forever, I guessed. Would it mean cultural erosion for this forgotten outpost? Would the locals welcome or reject this new world? Only time would tell.

I watched the steam rising as I peed unceremoniously in the middle of the Dempster Highway on the way back down to Dawson. Just a few metres away, my three teammates were busy tending to the fire and counting fingers to make sure they were all still there – for they could no longer rely on sensation. "Oh, for my Shewee right now" I thought as my bum began to freeze. Sadly, I had lost that civilised aid to female comfort somewhere in Asia. Now I wondered who had found it and how it had been recycled! *A funnel perhaps? An ear horn for an old dude? The Asians were good at recycling!*

I heard the distant rumble of a truck, and it occurred to me that perhaps it was not so distant. *Why does this always happen?* I could be in the middle of nowhere, not see a soul for hours, and just as I drop my pants and get beyond the point of no return, someone shows up. Hurriedly, I finished my business and pulled up my extensive layers, just as the big red truck came charging angrily around the corner, snorting steam into the cold night air. My dignity was spared this time as my near-frost-bitten bum was hidden from view once more. It wasn't always easy being a woman on the road – but it sure was fun.

The Titan Arctic Challenge was almost complete. In twelve days, we had covered nearly five thousand unrelenting kilometres. Back in Dawson City, in a nice warm hotel room, crammed full of four unkempt hairy humans – and their socks – I sipped my coffee and reflected on the highs and lows of a wonderfully icy, and suitably challenging experience, with a glow of satisfaction that only a successful in-the-bag expedition can bring. The small comforts in life, like heating, beds and hot showers were once again appreciated to the max.

We had landed in Dawson during a spring festival called ThawDiGras (obviously a twist on MardiGras)! The locals were busy axe-throwing, snow-carving, and serving up free food for all. We soon joined in the festivities until the early hours of the morning and were treated to a glimpse of the Northern Lights, *Aurora Borealis*, as we left our last bar. This really was a special part of the world. The journey up there had been worth every moment to see it in winter. I had thoroughly enjoyed being part of a team again too, and I was extremely grateful to Mark J-B and Budd for giving me the opportunity to make it to the Arctic Circle, and the top of the world, having spent the last two years travelling up from the very bottom.

Now, it was time to head back to Rhonda and start working my way east across Canada.

****

Before leaving B.C. I had to go back to Kaslo. Who knew when I would see it again, and now I would get to see it in a third season: spring. I went straight to Luke's place, put on his slippers, opened a beer, sat on the balcony and took in the stunning view once more. Yup, it was beautiful, whatever the season. I really did love that place and I enjoyed Luke's company immensely.

One evening, maybe three nights in, I was making dinner. Luke's friend had popped in and they were chatting at the table. I put the meatloaf in the oven and set the timer for ninety minutes. Grabbing a beer out of the fridge, I went to my laptop and opened it up to check my emails. Great. There was one from Tim. But as I began to read, I gasped in horror at the words. Luke and his friend looked over at me surprised.

"Are you OK?" asked Luke with concern.

I didn't reply at first. I just read it again, trying to make sense of the words. Surely this was a mistake? But it wasn't.

"It's my friend" I managed, "He's dead". With that I got up and headed for the balcony doors, "Sorry. Excuse me. Just give me a minute" I added as I went outside and closed the door.

The email was from Tim's wife, Karen. I had emailed him a few days earlier telling him I was now back on the road and heading for Toronto. I expected to be there in a couple of weeks, and I was looking forward to that coffee with him. Karen had replied,

Hi Steph,

This is Tim's wife Karen. I'm really sorry to tell you like this, but Tim was killed in a bike crash on Saturday 25th March. I know you are a very long way away but just for your information the funeral is on 19th April 2017 at 1.45pm at the Worcester crematorium.

Safe journey

Karen x

And with that he was gone. A few lines in an email and he was whipped off the planet; gone forever. *Oh God! His poor wife. His poor children. Oh no! Why him? Why now? Why not me?* I was the one that everyone was worried about. Tim was always worrying about *me*. That was the whole point of his visit after all. *It should be me. This is just not fair.* All the usual questions went through my mind as I sat there in shock staring at the beautiful spring view before me. Then I cried. I cried for Tim. I cried for his wife and for his children. Lastly, I cried for myself. I had failed him. I had promised I would see him at the end in France, at his bar. I would buy him a drink and we would celebrate together. I had so looked forward to that day. I wanted to repay him for his kindness and belief in me. He really had stood up when others stood down. It was all in that moment. That beer. So many times, I had dreamed of that moment where we would celebrate together. Now, because I had taken *so fucking long* about it, I would never get to do that. I would never get to show him, to tell him that he was right to believe in me.

Luke came out a while later and silently handed me a beer before sitting next to me in his usual spot. Together, we sat quietly staring at the view. Then we talked. Long into the night, beer after beer, we sat there questioning the meaning and apparent futility of life, trying to make sense of it all, as the meatloaf, long forgotten, burnt to a crisp.

****

The Rockies gave way to the rolling hills of Alberta and then suddenly there they were – the Prairies. They say you can see your dog running away for days there. I didn't have a dog – only Rhonda – and a new drone I had acquired over the winter, ready for some filming in Africa. If it would just stop blowing, it would be the perfect place to practise my flying skills.

Some days it felt as if I was just floating along in a dream-like state, with nothing much externally to catch my attention. I actually liked these days. Just me, the road and my thoughts. On other days the weather forced me into reality with ever-changing winds and gusts that played with me, occasionally throwing me across the centre line of the two-lane highway, and trying my patience. On the good days, I would smile to myself and blast the '70s disco music in my helmet. If you can't beat it, get in the groove, right?

Go with the flow. On the bad days I would curse my aching body, tense up and go head-to-head with Pacha Mama once more. I really should have 'purged' more for her in Peru. Several hundred miles of this was exhausting, with not even a tree to hide behind.

The weather was unpredictable to the point of being surreal. I would go to sleep in my tent one warm evening, listening to the prairie dogs (which are really ground squirrels) shuffling around me, and wake to snow blizzards with sub-zero temperatures chewing at my fingertips, as I frantically tried to pack my gear back on the bike. Of course, it never fitted properly in a cold rush.

It was not for the faint of heart and it was at times like these I wished for a bigger, heavier bike that could cruise faster than Rhonda's maximum *comfortable* speed of about 60mph. After all, there was nothing technical to worry about. No dirt or tight spots where small bikes rule – just miles and miles of straight, flat, open road. The beauty and warmth of the prairies though, lies within the people and *they* were worth suffering the beating to meet. The man who provided me with fuel when I ran out (rookie mistake); the Cairns family, who invited me to stay on their ranch to ride dirt bikes and horses (and film it all with my new drone) after their 13-year-old son Casey, found inspiration in my blog posts; the man who saw me crawl out of my tent on the side of the road and ran over shoving $20 for breakfast before I could explain that I wasn't really homeless – this was by choice! I really must have looked bad.

Thanks to a chance meeting with a guy at a gas station, who filmed me and put it on Facebook, I was found by Rick Bradshaw of Shrader's Honda in Yorkton. Rick, as it turned out, had Welsh heritage and was the great-great-great-grandson of Catherine Roberts-Davies, a middle-aged woman amongst the first group of 163 Welsh settlers that set sail on May 28, 1865, from Liverpool to Argentinian Patagonia. She was the first to die after arriving. I had learned of Catherine while passing through Trelew, in Patagonia (back in Chapter 14).

Catherine, her husband Robert, and their three sons, left Llandrillo in North Wales in search of a better life. However, the youngest of the children, eleven-month-old John, died on the voyage to Patagonia and was buried at sea. Less than a month after landing in New Bay, Catherine also died of black fever. Robert was to follow in 1868, and the eldest son William died at the age of fifteen in 1872. The only surviving member of the family, Henry, who was only seven years old when he travelled to Patagonia, eventually emigrated to Canada. I had been intrigued by this woman who had risked it all for a dream and saddened by the brevity of her taste of her Promised Land. Now, standing before me, was her only surviving son Henry's great-great-grandson, and it was a real pleasure to meet him.

Rick also happened to be friends with André Laurin. André ran a shipping company (OTSFF), as well as a motocross team (Rockstar OTSFF). He just happened to have crates of the right size already made up for Rhonda and so offered her free passage to South Africa. This was amazing news. The cost of shipping, on top of the new carnet, agents, health insurance, visas, servicing, and flights had been a weight on my mind. I had been working hard to raise the money through organised talks, and writing as I went, but the costs were mounting faster than I could talk or type. I must admit I had been finding it hard to write of late. I had so many mixed emotions at this stage that it had been hard to clarify to myself, let alone write it down, just how I was feeling.

The busy mind of someone preparing to ride across Africa; the money worries; the excitement of starting a new chapter; the sadness of finishing the current one – especially without Tim – and the fact that I could see the lines on my map getting closer to meeting up and creating the circle, as had always been the mission. Despite lots of company, loneliness played a part at times, too. Making new friends and constantly leaving them behind took its toll after a while. I liked riding solo but more often than not, lately, I was craving company.

Still, I was ready for Africa. I would make it through. I would ride to Tim's bar and I would bloody well buy him a beer as planned. *Nothing* was going to stop me from doing that. Tim might not be there to see it, but it somehow seemed more important than ever. I needed to prove that he was not wrong about me. I'm sure he had known that already, (even though I'd been unsure myself), but I still felt strongly that this single act had to be done. I had to make friends with misfortune; make adversity my ally and use it as a strength. If I gave up, I would just remain angry at the universe.

Despite these distractions, I emailed my first article to Rob Harris and Canada Moto Guide on time. He replied saying he was just off for a weekend's spin on his motorcycle. He said he'd take a look when he got back and added that he was looking forward to seeing me soon – but I heard nothing more. I chased it a week later worried that it hadn't been good enough. Sadly, a hauntingly similar email to the one I'd received from Tim's wife came back; this time from Rob's colleague. Rob had never made it back from his weekend ride. He too, had been killed in a motorcycle accident.

From Winnipeg and on to the Great Lakes, I found a network of bikers who reached out to each other to provide a tight network of pit stops and warm welcomes for any bikers passing through. I was given the full Canadian experience with rustic lakeside cabins, where I could rest and wait out the next short-lived snow blizzard, and great company. I was overwhelmed with the kindness of my fellow bikers once again, always being sent on my way with a full belly, a big smile, and in many cases, much more.

****

Before I had left Luke, he had given me the address of his identical twin brother Dan, in Ottawa. When he answered the door, I was a little taken aback as to just how identical he was. That night we went to a great little bar that was decorated with old things, like record players, old phones, gumball machines and so on. Everything was for sale too. It was a really quirky place and we were the only people in there for some time. Dan pointed out to the barman that there was a spelling mistake on the menu. He was just being helpful rather than pedantic. It was a new place and he thought they would appreciate someone pointing it out – which they did. Luke had done exactly the same thing at a café we had gone to together in Kaslo. Dan smiled the same smile, hugged the same hug, laughed the same laugh and talked the same talk. It was comforting to be with someone so familiar despite the fact that we had never actually met before! A caring, albeit fleeting friendship that distracted me from my sadness and helped me focus on the future. Before I left, Dan handed me a parcel. It was from his mum, Lina, whom I had met with Luke back in Kaslo. It was a beautiful pair of ruby red slippers with a note saying, *"I heard you missed your slippers. These should see you home"*. She had made them for me especially and sent them on to Dan ready for my arrival.

In Toronto I stayed with the brother of a friend of mine from the UK. Pat was in the Navy and a couple of days after my arrival, he had to leave me to it. He had no problem just letting me stay in his home, while he went off on some diving mission or other. I loved him for it.

I sorted my personal items into two piles: 'Can't live without for six weeks', and 'Can live without for six weeks'. I said goodbye to my trusty steed Rhonda, as she went into the box with the second pile of items. The first pile went into an old Canadian Navy issue rucksack, donated by Pat. From there we would have to make our own separate ways to Africa. First, Rhonda would board a train to Montreal, and then a ship to Durban, where we would eventually reunite. I would go by train to New York, and then fly to our final continent. Until then, I was just another tourist.

I cleaned out the hard-earned grime and grease from under my fingernails and tried desperately to get rid of my 'helmet hair'. The flip-flops came out, and I wasn't afraid to use them. If I was to survive, I would have to become a tourist. The McDonalds-munching, queue-forming, mindless zombies that will eat your brain as soon as look at you. They meander without purpose or direction. They bump into you without apology. They live life through a lens, and never stop clicking long enough to actually enjoy the moment. We've all said it, 'Bloody tourists!'

I braced myself, as I put on my disguise and entered the madding crowd and its ignoble existence. As Rhonda set sail, destined for a dark ride on the big water, I waved goodbye to my individuality and headed for the nearest attraction – to get in line.

Drew with his '53 Chevy on the Pacific Coast Highway 1.

Having a beer with the Nimbin Collective crew after one of my many presentations in California. I stayed with Rex (second from the right) and family. Rex is a biker and a true gent.

With Jeremy en route to the Okanagan in British Columbia, Canada after a three-week rest in Vancouver.

Getting my first ticket of the journey from a Canadian Mountie en route to the Okanagan, BC.

Back on the dirt and testing out my body after a few weeks R&R.

Camping at Clearwater and heading north towards the Canadian Rockies.

Looking out over Trout Lake during one of my ride-outs from Kaslo, BC. Such a beautiful place.

Taking in the view somewhere in Alberta, Canada.

Camping in the ever-changing conditions as we head into a Canadian winter. Bear spray, waterproofs, and firelighters at the ready!

Testing out the amazing Yeti MX snow bikes with the designer, Kevin Forsythe, and crew in BC, Canada.

On the Dempster Highway in the Northern Territories and heading for Tuktoyaktuk.

On the ice road (Mackenzie River) en route to the Arctic Sea and Tuktoyaktuk.

Camping on the Dempster Highway at -30ºC with Mark Jennings-Bates (left) and Bryan Irons (middle). Photo was taken by Bud.

Mark taking a dislike to the Sour Toe Cocktail in Dawson City.

The current toe used in the famous Sour Toe Cocktail.

Finally, we made it to Tuktoyaktuk in the North West Territories, Canada.

Chapter 27
This Is Africa

As I sat quietly updating my blog in a little Airbnb on the outskirts of Durban, I heard the door tentatively creak open. I looked up from my screen to observe a monkey with an injured arm walk in. Surprised by his cheek, I just said,

"Hey!" (in the voice I would normally reserve for naughty dogs, drunk Turks, Iranian taxi drivers, barely conscious British military boys, and *almost* one nose-kissing Canadian dude in a dodgy baseball cap), assuming he would run. Instead, he just looked at me nonchalantly as if to say, "What's your problem?" and, without hesitation or deviation, walked straight past me, hopped onto the dresser, grabbed a banana out of the fruit bowl (no more, no less), and walked back out again. Half expecting him to close the door behind him, I laughed and said out loud, "Oh! T.I.A indeed. **This Is Africa**, all right!".

'T.I.A' is a commonly used phrase in South Africa. Usually said right after warnings of my impending doom, as they learned of my mission to cross the continent alone by motorcycle. When I explained I had made it this far alone across six continents, people would often shake their head and say things like,

"You ain't seen nothing yet!", to which I would reply,

"And have you been out of South Africa yourself?". The answer, more often than not, was "No". One guy looked at me dead in the eyes and said in a strong Afrikaans accent,

"You will be pulled from your bike, raped and murdered". I'm sure he believed it too. It wasn't a threat, merely a matter of fact, (in his mind), and I did love that accent, so I let him off with a smile.

The topic of corruption and violence over a *braai* (barbecue) in South Africa was on par with the subject of the weather over a brew back home.

Now was not the time to be fazed by the naysayers, though. After 100,000 kilometres through forty countries and on every continent on earth, I had made it to Africa – the last long leg of my journey. My worries about camping with bears in Canada seemed trivial now when faced with the abundance of 'tooth and claw' that awaited me across Africa, but as I kept repeating to the naysayers, "I choose to believe I will be OK!". The words of Nelson Mandela's favourite poem appropriately lodged in my head like a mantra, *'I am the master of my destiny! I am the captain of my fate'*.

Having already run into a glass door and broken my nose in Johannesburg (to which I'd flown from New York), it felt as if that sentiment was being painfully tested by some force with a sick sense of humour, *but* my ambition to enjoy and savour every moment of the ride for what it was, had never been stronger. In the last few months, something had crept up on me and I hadn't realised it until now. I was much more relaxed. Losing Tim, more than anything, had made me realise that life was so short, so precious, that it shouldn't be wasted on internal struggles and self-doubt. After all, those things made no difference to anything, or shouldn't. I knew I would always have my 'down in the dumps' moments like everyone else. And trying to be positive all the time, and not letting things get to you, is easier said than done. But something definitely felt different about my outlook. The big stick I had been constantly beating myself with had been reduced to a twig.

With a healthy respect for the road ahead, my sense of humour, adventure, and optimism were all still intact – unlike my nose! I still believed my last leg was going to be truly magical with, no doubt, the odd scrape along the way. That was 'all part of the adventure', and I expected nothing less.

My bike was still lost at sea somewhere on the Atlantic Ocean. It would be three months before I saw her again, but while my 90-day visa ticked away, I kept busy riding around S.A. and Swaziland on a borrowed example of the newly-launched Honda CRF250L Rally, courtesy of Honda South Africa's Head Office in Jo'burg. Swaziland (or Eswatini as it's now called), is a beautiful country, where I spent a couple of nights camping around a fire that was still burning forty years after it had been lit, with a load of almost tame warthogs who were snuggled so close to it they smelt of bacon by the end of the night.

I also had an enjoyable visit from Dani, (the unfortunate friend who, through no fault of her own, crashed her hired bike when riding with me in Nepal, suffering a really nasty leg injury). We stayed in a rather 'dodgy' neighbourhood in Durban. In fact, it was described as 'the roughest neighbourhood in the city' and I was offered money by one of my previous couch-surfing hosts *not* to stay there. An offer I declined, but not before asking 'How much?'! In truth, the more people protested, the more I wanted to stay there. They saw a mousetrap; I saw free cheese and a challenge. Dani and I ended up having a great time in 'the hood'! I mean, it *was* as rough as hell and, as two white women, we stood out a mile, but we kept our heads held high and smiled a lot! I think we gained respect for having the balls to stay there in the first place, and people seemed to be watching out for us in the end; telling us things like which roads to avoid and even which side of the pavement, at times.

I finally picked Rhonda up from the shippers in Durban and rode northwest, through the valley of a thousand hills to the Drakensberg, then skirting around the eastern edge of land-locked Lesotho (a country I had previously visited), I rode on through the Golden Gate Highlands National Park to Clarens and Bethlehem in the Free State Province (formerly the Orange Free State). From there, I rode 1,250 kilometres south and west right across the country to Cape Town to meet up with a 'cheeky chappy' Scouser by the name of Billy Ward whom I'd first met at the London Motorcycle Show a month before I'd left. I was being interviewed on stage about my planned trip and Billy had been in the audience. When the interviewer asked if there were any questions, Billy put his hand up and asked,

"Will you marry me?"

"Bit busy right now" I'd replied with a smile, "maybe when I get back".

Billy worked with Charley Boorman, who had become a bit of a household name as a motorcycle adventure rider after the success of his TV series with Ewan McGregor The Long Way Round, and the follow-up series, Long Way Down. Charley would be arriving in a couple of weeks and the pair of them were guiding a motorcycle tour from Cape Town to Victoria

Falls. Billy had borrowed an apartment in Kommetjie, a small town on the coast just outside Cape Town and had offered to share it with me.

"You'd make a rubbish bloody wife" he said when I arrived, "You're never here".

"Good to see you too mate" I replied and moved in for a big hug.

A few weeks earlier I had been contacted by a popular local TV programme called *Carte Blanche* who wanted to do a feature on me, there in Cape Town. I told them that Charley would be in town too and perhaps they might like to include him. They thought it was a great idea, so when Charley arrived we met up with them and did some filming together on Table Mountain. They created a lovely little eight minute film from the day and aired it the following week.

Billy and Charley suggested I ride with them for a while when their group arrived. However, my standard 90-day visa was due to expire in a couple of days and I had to get out of the country before they were due to leave. We agreed to meet across the border at the famous and spectacular Fish River Canyon in Namibia a few days later.

****

There is one tarmac road that cuts through the middle of Namibia – the rest are dirt – and for now I was on it, chasing the newly-painted white lines and allowing myself to believe in the mirage that lay before me: any minute now I would be splashing through the magical river that had engulfed the empty road ahead. Once again, while away from it, I had forgotten the power of the desert in playing with our minds and reigniting our imagination. The space and lack of people always gave me a sense of freedom; the beautiful simplicity in this unprejudiced, levelling landscape that still brought with it a certain comfort.

I turned off the black stuff and headed east towards a campsite near Fish River Canyon, which is an awesome and humungous 'groove' in the earth's surface, like Southern Africa's answer to the USA's Grand Canyon. It wasn't long before I suffered a puncture en route, thanks to the sharp rocks that protruded from the sand, and spent the following hour fixing it, with the desert sun burning down on me.

As I was wrestling with the final, and most annoying stage of getting the back tyre back on, a local guy turned up and placed his truck in a position where it threw its shadow over me. Then he gave me an orange and an extra pair of hands to hold the weight of the tyre, while I ensured that the spacers, brake pads and chain were all aligned. I've never understood why we haven't come up with a simpler solution to roadside tyre changes – preferably one that didn't ideally need six sets of hands!

Anyway, it was done, and I was happily on my way again, although I had now used up my spare inner tube (the punctured one was not repairable) and would need to source a new one before I could safely have another incident! Not an easy task in the middle of the Namib Desert. I decided to call Billy as soon as I got to camp and asked him to bring one over the border for me. *Sorted!* It was nice to have a back-up buddy.

Waking up in my tent the next morning, I jumped out of bed, made the coffee, and shared an energy bar with the exotically yellow birds who had come to prey on my good mood. I sat watching the sun move across the adjacent hill, slowly working its way towards me. My departure time was all down to how quickly its rays could reach me and dry the dew from my tent. I smiled and savoured the moment, completely submissive to the control nature had over my day.

The pace of life slowed down on the road. I didn't really feel part of society any more. Moving all the time brought with it a certain detachment from society's restraints and going back to basics felt good. The journey had evolved over time into something far more beautiful than the challenge. So slowly, that I hadn't even realised until the transformation was almost complete. Far from feeling as if I didn't belong anywhere, I felt as if I belonged *every*where. The world was now just one big neighbourhood; *my* neighbourhood, and I was free to roam it as I pleased. I had not taken the journey. The journey had taken me.

I grabbed my toothbrush and stood idly in my own little world, enjoying the peace, when suddenly an angry-looking baboon ran into the camp and tore me from my daydreaming. He was *big* and I was not going to challenge him, instead choosing to shuffle behind Rhonda, using her as my shield as I continued to brush my teeth and just watch as he frantically ripped through my neighbouring camper's belongings. I guessed he was looking for food. I also briefly mused that I should try to do something. Instead, I watched and I brushed, and I considered myself quite fortunate to have finished my energy bar in good time; his big yellow fangs convincing me that toothpaste wasn't his thing. As I watched the oversized primate race away from the scene of the crime, complete with what looked like a bag of cheese and onion crisps, I suddenly remembered that I was due to meet Charley and Billy today. It had clean gone out of my mind. The universe had prompted me with a reminder in the form of a big hairy ape!

I got to the hotel earlier than the gang and decided to chance my luck on a cheap room at this establishment in the heart of the Namib Desert. Dumping Rhonda at the front step, I dusted myself off and walked in through the grand entrance, immediately sensing myself out of place.

"Can I help you?" asked the smartly dressed lady at reception.

"Ah yes hello" I replied in my best Queen's English, "I am riding ahead of the Charley Boorman group. I'm one of their guides. I believe you have special rooms for guides. I forgot to book ahead. Do you have any free?" (Hotels in Africa often have basic rooms at a fraction of the cost for guides).

"I'm afraid we don't Madam, but we do have a couple of standard rooms empty. Let me speak with the manager and see if we can get you into one of those at the same price. You look like you could do with a shower." She wasn't wrong! That night I had a proper shower, followed by dinner with the gang, and of course, the obligatory after-ride beer, before falling asleep in a four-poster bed on a beautifully soft mattress. It was terrible behaviour on my part, but I had zero regrets in the morning.

Having fifteen riding buddies was a total contrast from the norm, but one I thoroughly enjoyed. Rhonda and I managed to keep up by maintaining our usual 'Keep Calm and Potter On' approach. I envied the smooth speed of the big bikes (mostly Triumph and BMW 1200s) on the long stretches, but I got the feeling the tables were turned on the softer, more technical bits. Billy managed to get me into a couple more hotels under a similar 'blag' after that first night and so we made good use of the bar together in the evenings.

One afternoon when we weren't riding, Charley asked me to do a talk on my journey to his group (who were all Brits). Of course, I was happy to, having had so much fun with them all. I had already done a couple of talks in South Africa while waiting for Rhonda. One had gone well and the other, not so well, with the Nelspruit (SA) group not seeming to appreciate a *woman* spouting tales from the road on a silly little motorcycle that had no manly power. The other, in Cape Town, had been very well received and had restored my confidence enough to give it another go.

When I finished, each rider came to me and shook my hand with a discreet deposit of cash in each one and encouraging words to continue my journey. It was so touching and completely unexpected.

That night, Billy and I 'entertained' the group further with a few songs and a guitar at the bar. When there were just three of us left, the barman offered us a go on his 'snuff machine'. A contraption not easily explained, but best described as a desktop torture device consisting of two nails, a hammer, and a seesaw! Imagine a block of wood with a small seesaw on top. At one end were two nails sticking sharp end up, with one end of the seesaw between the two nails and a pile of snuff on the top of the seesaw. You had to position your nose over the two nails then breathe in quickly as someone hit the other end of the seesaw with a wooden mallet. The seesaw flew up to just below your nose, the snuff flew up your nose while your brain told you the nails had followed. Of course, they hadn't, but the sudden bang of the hammer, the very presence of the nails, and the feeling of something shooting up your nostrils, was horrific! After the initial shock, it felt as if you had just snorted a hundred extra-strong mints, which sent an explosive *zing* straight to the brain, quickly followed by the rest of your body. It was excruciating at first, and then quite refreshing! I wouldn't recommend it!

I was with the guys long enough to miss them when we parted company. That fork in the road came too soon for me, but it was time to head back to my solo life of roughing it – just me and Rhonda – cooking dinner for one on my little gas stove.

Still, I was recharged on company and I smiled as I set up my tent behind a gas station in Sesriem. I was desperate to get to the petrified trees in Sossusvlei, but I wasn't prepared to pay the inflated prices at the nearby campsite. I didn't need the extras. A patch of sandy ground would do me just fine. The Namibian workers from the garage disagreed, and that night they ran an extension out to my tent so I had power. They even brought me a table and chair to make me more comfortable as I caught up on my blog. My space was now so inviting that I was later joined by a couple of Aussies on a 4x4 adventure of their own.

"Can we join ya?" asked blondy with a smile, "We bring fresh fish for a barbi and a bottle of red".

"Then of course you can" I laughed.

Rowen and Andrew were on a three-month adventure and were clearly having a lot of fun. These guys were good at fun and I enjoyed their company that night over dinner. "Not bad for a camping spot behind a garage in the middle of a desert", I thought.

I could not take the bike into Sossusvlei, so one of the petrol attendants came with me to see if we could find a way in. We came across a couple from London, Wagner and Tatiana, who were planning on driving there in the morning. I tentatively asked if I could join them. They happily agreed so we met again at 6.45am the next morning for the sixty kilometre drive over to Dune 45, Big Daddy, and of course – the petrified forest.

Estimated to be 260 million years old, the trees stand like an army of ghostly shadows on the cracked surface of the salt and clay pan; as if waiting for the order to stand down, from time itself. Framed beautifully by the red, pink and orange of the dunes around them, they were a magnificent and eerie sight.

The following day, Wagner, Tatiana and I took the same long stretch of sand and corrugation out of the area and so kept meeting up whenever one of us would stop to take in the view or take respite from the vibrations. Originally from Brazil, Wagner had a great sense of humour, and Tatiana was so relaxed and it was clear that very little fazed her. It was a real pleasure to share the road with them, and I must admit, their offerings of sugary treats at each stop were probably what kept me going when my patience wore thin from the relentless jarring and front wheel wobbles. The scenery was breathtaking at times and all the better for earning it. I restocked in Walvis Bay and travelled up the Skeleton Coast – the name comes from the whale and seal bones that once littered the shore from the whaling industry, although in modern times it is also littered with the skeletal remains of wrecked ships caught by offshore rocks and fog. Then I made my way inland to Windhoek, (the capital of Namibia). over the course of a couple of days.

****

I hadn't intended to go to Angola, but it started calling to me as I got closer to its border. The advice on the UK government website said quite clearly, 'DON'T GO', but in my experience, those websites were often out-dated or over-cautious, so I had decided to chance my luck at the Angolan embassy in Windhoek.

It was hard to see the woman clearly behind the thick Perspex barrier. My own reflection was in the way, my face merging with hers like a double exposure photograph. By a careful process of elimination, I realised that it was *her* eyes I could see rolling in irritation, and not my own. I was nervous as I knew that few people were granted visas, and certainly not when applying from outside their home country. Her voice came through the little circular holes in the pane with great depth and authority,

"Why do you want to go to Angola?". I replied with a long-shot and as much confidence as I could muster,

"I am on a record-breaking mission to ride a motorcycle on all seven continents," It really wasn't relevant, but it sounded good. "I emailed you in advance and I was told I could have a few days to record my journey. It won't take long." The woman stared at me so hard I was convinced I would turn to stone just like those petrified trees back in Sossusvlei. Then she rolled her eyes again, huffed, and slid a form under the Perspex that I had assumed was to protect *her*, but now I wondered if it might be to protect her 'customers'…

"Fill this in and come back tomorrow with the documents listed," hissed the pen-pushing Medusa, and wafted her hand at me, clearly indicating that I was dismissed. After a couple more days of paperwork, eye-rolling, and open irritation at my very presence, I was granted seven days. It wasn't much, but I took it and ran.

Arriving at the Angolan border, the usual crowd of touts greeted me, offering to guide me through for a fee. They were friendly but quite pushy, until one of them recognised me. To be more accurate, one of them recognised my bike and exclaimed,

"Rhonda the Honda! You were on television! I saw you!"

"Yes, I was" I confirmed, "I'm Steph by the way". I was just the sidekick, whose name was so easily forgotten, but I was getting used to it by now as Rhonda had been recognised everywhere we went following the broadcast of our *Carte Blanche* TV item.

Soon, I was on my way with lots of advice and the usual warnings of my impending doom as a woman alone in Angola. The general consensus was that I was crazy. The border guards were of the same opinion, but I was used to this too. At my final gate to the open road, I was stopped again and asked for the confirmation stamp to show everything was in order. Then the guard saw the skull I had found on a deserted sand trail a few days earlier – appropriately near the small town of Springbok – just before the Namibian border. I'd strapped it on the front mudguard in one of my 'Mad Max' moments, before recording a live video on Facebook, telling everyone I had called him Skully and that 'Skully' was coming with me all the way back to Wales.

"What is this?"

"Springbok" I replied proudly. I'd originally thought it was a sheep but had been corrected by some helpful locals.

"I should arrest you for this" he replied.

I laughed assuming he was joking. He gave a cheeky grin and let me go.

I weaved my way through the usual throng of moneychangers and SIM-card sellers shouting, "No thank you!" at each one lurching forward with his/her offerings. Finally, I was out and onto the open road and feeling that familiar buzz of anticipation at what lay ahead.

After the border came nasty foot-deep potholes, anxiously waiting with open mouths for me to ride into them. Any, even momentary, loss of concentration and one of them would eat me whole. The sun was slowly cooking me alive in my riding gear – like a boil-in-the-bag meal – whilst promising the potholes to make me easier to swallow, and more digestible.

After the first few kilometres though, the roads became surprisingly smooth. The traffic was practically non-existent, and the only buildings visible were made of sticks and mud. And of course, with Angola being a former Portuguese colony, I was now having to ride on the right hand side of the road again, having been back on the 'British' left throughout Southern Africa.

Angola has a rustic beauty that is uncorrupted by tourism. Venturing deeper into its arid and struggling landscape, I suddenly became aware of one, then two, then three familiar outlines in the distance. As I got closer, the picture came into focus and my hopeful suspicions were confirmed. The barren soil suddenly gave way to a forest of magnificent trees towering over the other pathetic shrub life. To the locals they were merely a welcome shade for their cattle and goats but to me they were so much more.

I stopped my bike and feasted my eyes on the host of baobabs that stood regally before me. I was here. They were here. This time it was no dream. There was no razor wire in my way; no indents in my face from cold steel bars. *Oh, please tell me this is not a dream!*

I got off the bike, removed my helmet, and allowed the overwhelming sense of pride and happiness to wash over me. Tears rolled down my grubby face as I laughed out loud in a state of triumphant baobab-induced euphoria. *This* was the reason Angola had called me. I was indeed the master of my destiny. The captain of my fate. I'd made it!

Baobabs became my friends as I travelled on. Whenever I needed rest, I took refuge from the sun underneath their welcoming branches, resting my back against the enormous trunk, and sipping from my hydration pack. I was loving being in their presence. They were more than symbolic of my 'ladder', the image that had helped me find my freedom so many years ago. There was something magical, almost spiritual, about them too. I was in total awe and almost felt as if we had some kind of connection; as if they were old friends watching over me. In a way, I guess, they always had been.

The one thing that the baobabs couldn't help me with were the flies – highly aggressive flies that you will find only in arid conditions like this. The flies that go for the moisture in your eyes and mouth with a steely determination, emphasising the constant battle for survival that comes with that kind of uncompromising environment.

Whenever my hydration pack ran dry, I would stop and top-up at one of the community wells dotted along the roadside at fairly regular intervals. This was where life happened, and each was a hive of activity; watering cows, washing clothes and kids to-ing and fro-ing with their buckets of water back to their homes. There were always chores to be done. The young boys seemed to be the predominant cattle and goat herders, while the girls did most of the water-carrying. Occasionally when I stopped, some of the younger kids would run, screaming for cover, not sure what to make of the strange sight! I must have looked like an alien to them.

I didn't come across a petrol station for a while, so I stopped at the next village and shouted, "*Gasolina?*" at a group of young men who were busy propping themselves up against a wooden hut. They pointed over my shoulder at an old man sitting outside another hut, surrounded by green wine bottles. The old man looked leathery. He had a kind face that told of a life of labour under a baking sun. A semi-toothless grin spread across his face as I approached, and he popped a cork in anticipation of my steed's requirements. We fed Rhonda four bottles of his finest petrol before the small crowd that had gathered to check out the new girl in town waved us goodbye. People there live off the land, but not the *fat* of the land by any means. It's a tough life in a harsh environment, and I respected them immensely for getting on with it against all the odds. They knew no other way. In truth, they had no choice.

Angola's twenty-seven-year civil war officially ended in 2002. It continued for so long in large part due to ethnic tension in the country after its 1975 independence from Portugal. Power struggles raged on for years and wiped out most of the infrastructure as the dominant liberation movements refused to share power in a multi-ethnic society. Old tanks still lay abandoned on the roadsides as a harsh reminder of a very recent history that had brought the country to its knees – and kept it there.

I entered Lubango, (Angola's second city after the capital, Luanda), where I found shocking contrasts between rich and poor. Barefoot kids with ragged clothes and vacant eyes competing with mangy dogs while scavenging for food amongst the rubbish that piled up on every dusty street corner. They looked resigned and emotionless, as if life had already beaten them into frightened submission, while the rich were the only ones who could afford even the simplest of commodities. The price of groceries was way above that in the UK, and a simple one-course meal in a very basic restaurant set me back the equivalent of £20.

My hotel room was in the scruffiest part of the city and cost nineteen thousand Angolan kwanza (£25) for the night for a very basic room with shared bathroom – although there were no other guests. There was no water until 7pm. In fact, there was no electricity when I arrived either. I hoped it would come on soon as it was getting dark and my torch was still on the bike. Perhaps it wouldn't come on at all. I had no idea and couldn't ask as no one spoke English. Other than the rural tribes – who have their own languages – people speak Portuguese there. Some of my Spanish phrases were working, so that helped, but I had nothing in the word-bank for "When will the electricity come on?"

My room was small with mint green curtains down to the floor and a big, old-fashioned wooden cupboard in the corner with the smallest TV ever, which I doubted would work even *with* electricity. It must have been older than me. I didn't stay in Lubango long. It was beyond my budget, and with credit cards being of no use, I was in real danger of running out of what little cash I had on me.

Heading back onto the country roads, I pointed my wheels west, on course for some serious switchbacks in the beautiful *Serra da Leba* mountain range. A steep 1.7-kilometre climb with seven switchbacks; this road was surely built with bikers in mind. Perhaps, due in part to its beauty, it's the site of many fatal accidents, earning it the nickname 'The Beautiful Precipice'. It was a short, wonderful ride that begged to be ridden both ways, so I did – twice!

My ride back to the Namibian border the following day was long and hot. I stopped regularly to drink some water or make sweet coffee under the baobabs, as was now my ritual. After disturbing a snake at one tree, I developed a ritual of stomping around loudly and shouting,

"Go away snakes. I'm here now!" before settling down to enjoy my beverage. While stomping and shouting on one such occasion, I suddenly became aware of a host of eyes watching me from behind a tree on the other side of the road. It was a group of young girls, unmistakably of the indigenous Himba tribe. The tribe is famous for the red mud-like mixture of animal fat and ochre they wear in their hair, and often on their skin, possibly to protect them from the sun.

After a little persuasion, they came over and joined me. Everywhere else I had stopped until then, I was greeted by inquisitive eyes and big friendly smiles. Not these girls! It seemed at first that they had made a pact *not* to smile before coming over. This all changed once the drone came out though. That *did* raise some smiles, and a great deal of excitement. At first shocked by my flying machine, they soon giggled and chattered away as they crowded around to see themselves on the screen in my hand. Sadly, I later realised, I hadn't pressed 'record'. It was an amazing moment destined to be kept between the Himba girls, my friends the baobabs, and me. To adjust the old adage for the location: 'What happens in Angola, stays in Angola!'

With that thought, I put my gear away, jumped on Rhonda, and headed back across the artificial line to the comparatively easy life of Namibia.

Chapter 28
Elephant Hide and Seek

I woke up next to the Okavango River to the unmistakable sound of grunting hippos and shrilling cicadas. I packed my tent, said goodbye to Namibia and followed the flow of the river south across 'the Caprivi Strip' to the Mohembo border crossing into Botswana. This is the fourth-longest river system in Southern Africa, and like me, it was on a mission to get to the Okavango Delta – a swamp brimming with wildlife slap bang in the heart of the Kalahari Desert. (The Okavango is very unusual in that it doesn't run into the sea, it just flows into a 'tectonic trough' in the Kalahari and evaporates). Botswana was a country I'd been looking forward to immensely, purely for the wildlife and abundance of elephants that were no longer hunted there, and could roam free across the land.

It was a lovely dirt road ride in, through a small section of the Caprivi Game Reserve, with signs warning of lions and elephants, and an extra big sign saying, 'Disclaimer – Enter at Own Risk'. It reminded me of a sign I had on my bedroom door as a kid and I smiled as I imagined lots of teenage wildlife leaving clothes all over the floor and mouldy cups under their beds. I saw neither elephants nor lions however, but I did see lots of antelope-type 'things' and made a mental note to myself to 'brush up on antelopes'.

Just after the border, I stopped in the small town of Shakawe to refuel and to see if I could get a sim card – a usual first job in most countries. I ended up changing $100 there at a reasonable rate and got the sim card from the shop next door. While I sat there on the appropriately bench-shaped tree inserting the card, a man came over and sat next to me, so close our arms pressed against each other's.

"Hello" he said, and may have given his name but I forget what it was as I was more concerned with our close proximity and the unmistakable aroma of unwashed man in the midday sun.

"Pleased to meet you", I replied as I made a point of moving away slightly. As I did, he reached over and brushed the hair out of my face. I stood up coolly but said in no uncertain terms (in a voice I normally reserve for naughty dogs, drunk...you know the list by now!) that touch was definitely *not* cool,

"It's OK because I like you"

"It is *not* OK because I don't want you to". I clearly had a special case here.

"I like white women"

"Good for you". It was the best I could come up with at short notice.

"I love you"

"You're crazy"

"Yes I *am* crazy!". I thought it best to leave whilst we were in agreement and reached for my jacket and helmet. He watched me for a while as if pondering what to do, then stood up, grabbed my arm roughly and announced,

"You must stay with me". His smile had gone and his eyes looked desperate now. I pushed him so hard that he stumbled backwards and fell over. That surprised me as much as it did him. As he was getting up I got on my bike to leave but paused long enough to say,

"Good luck with those white women" and turned the key.

He ran at me and tried to drag me off the bike. At least I think that is what he was trying to do. He could have been trying to get *on* the bike. It was all a bit awkward. Rhonda was already off the stand, so I was trying to keep her upright while trying to get him off me. *If this guy makes me drop my bike, I'm going to punch him in the face. It's too hot for this shit.*

Just then a group of four guys came running over. I wasn't sure for a second which side they were on and my stomach lurched as I remembered those words of doom in South Africa, "You will be pulled from your bike, raped and murdered". Thankfully they were on my side and started untangling his arms from mine, then dragging him off me while I wrestled to keep Rhonda upright. I thanked them ardently, fired her up and promptly set off. I watched in the rear-view mirror as they let him go and he chased after me through the traffic like an impassioned jilted lover. His arms waving frantically, getting smaller and smaller as the distance grew between us and our beautiful yet pithy relationship lay shattered on the African-red soil. It hadn't been a great chat up line!

The road to Maun was fairly straightforward after that, unless you count getting fined for speeding after racing two ostriches. They reached 40mph (according to my speedo) before peeling off and disappearing into the bush, leaving a cloud of dust and narrowly avoiding a warthog and a giraffe as they went. Just after that there was a dip in the road, and in the dip was Mr Policeman hiding behind a tree with his speed gun. I was doing 70km/h (43mph) in a 60 zone, and it turns out ostrich racing is not a valid defence for not noticing the signs. The fine started at 800 pula (about £55).

"I don't have that much and wouldn't pay it even if I did. That's ridiculous". I exclaimed.

The policeman repeated in a jovial mimic,

"That's ridiculous", emphasising my scandalised high pitch.

We both laughed, and the ice was broken. We agreed on 200 pula. I left saying I had learnt my lesson and would pay more attention in future. I did in a way. Within five minutes I was already over the speed limit again but paying more attention to the dips and shady spots under the trees where police might be lurking. It was too hot to hang around.

The Old Bridge campsite just outside Maun offered shade and refreshments. I set up my tent after quickly calculating which tree would give shelter from the sun for the longest time, had a cold shower, and quickly followed it with a beer; a ritual I have repeated too many times to remember. There is no better feeling in the world than holding your first cold beer after setting up camp and removing sweaty riding gear after a hard day's ride.

This wasn't a bad place. Situated right on the river, it was a typical backpackers' stop, mostly full of the hired 4x4 brigade and twenty-somethings who often believe that they invented travel and say things like, "Oh, I don't do the tourist thing", as they sit in a tourist spot and book onto their next guided tour. It was the kind of place I could enjoy for a short time only, but enjoy it I did. I drank overpriced beer in the shade by the river, took a sunset boat ride amongst the abundant waterlilies, birdlife and elephants, and later played the card game 'Shithead' with some fellow travellers, including

Rowen and Andrew, my Aussie friends from behind the petrol station in Namibia, and Ben, a young American I had met back in Cape Town. I had recognised his beat up old Kawasaki KLR650 in the carpark as soon as I'd pulled in. It was great to see them all again.

I took a horse ride while I was there, and as I jumped off the horse at the end, my right shoulder screamed out in searing pain. I fell to my knees and stifled the urge to scream while Cash, the horse I had just been riding, nestled my hair and offered a comforting touch as I crouched next to his gigantic hooves and waited for the worst to pass. It was touching to see that he clearly felt my pain and whilst staring at his massive hooves just inches away from my face, I was glad we had bonded during the ride. I vowed then that if my shoulder got any worse, I would ship home – happy that I had made it to Africa and counting my blessings for getting to the seventh continent in one piece. Tim would understand. Still – I cried that night while having a beer with my young KLR-riding American friend Ben. The thought of not finishing was heart-breaking, no matter what I tried to tell myself. Ben had already decided this was where he got off. After three months, he'd had enough of the road. Two weeks later, he flew back to the 'States, leaving his faithful old KLR behind with the hope that one day he would return to finish his ride across Africa. To this day, he has not returned.

I refused to dwell on my own worries as I made my way further east towards Nata, then took a left, north, towards the appropriately named Elephant Sands Eco Camp. I knew the last five kilometres up to the camp was going to be tricky going on deep sand, but Billy Ward had told me this place was well worth a visit if, like me, you love elephants. Elephant Sands was once an old hunting borehole, now used by tourists to camp out and get close-up-and-personal with the thirty or so wild elephants who dropped in most days for a drink and a bath. The only shooting there now is with cameras, and the elephants have become relaxed enough to ignore the tents that come and go around them.

At dusk, while crouching silently behind a stick-thin tree, well away from the safety of the viewing point where others had gathered, I held my breath and looked on helplessly as the imperious herd approached. My plan for getting a photo of Rhonda and some elephants had been sketchy at best and now it was too late to change my mind. The elephants were everywhere and Rhonda looked like a Matchbox toy as they towered over her. Having spotted the well-worn path and elephant dung, *I had parked Rhonda right next to their regular route through to the watering hole, hoping for a dramatic photo opportunity.* This was a one-shot deal. If I had misjudged their behaviour, I was going to pay, and so was Rhonda, but I had decided it was worth the risk and promised myself I wouldn't cry if she got crushed. We'd had a good run after all. At least that's what I told myself. On reflection, I'm pretty sure I would have cried like a vegan in a butcher's shop.

I watched from my stake-out with a sense of overwhelming elation and respect as, one by one, they passed by us, just metres away. The youngsters stopped, with a look of shock, before shuffling off as fast as they dared without losing face. Others took a closer, inquisitive look at this strange two-wheeled animal, and some displayed a mild annoyance at me, shaking their heads and flapping their ears before continuing. Now *this* was a 'life hit'! Thankfully, none of them touched the bike and none of them charged. Of course, they could have crushed us both in a flash. It had been a crazy risk but sometimes those, 'sod it' moments came up and I was helpless to deny them. More reason to try it, than not. Our luck had been 'in' that day.

Botswana was beautiful and I spent my last nights on the river Chobe near Kasane taking in the abundance of wildlife it attracts; dozens of elephants, hippos, crocs and birdlife. I even got to see my first drinking giraffe. What a sight it was as it splayed its gangly legs apart and reached down for a sip, before springing effortlessly back up again. I was having quite a few 'first' experiences lately – a mother and baby elephant swimming, a pied kingfisher catching a fish, and who could forget, the best damned Irish coffee I had ever tasted from one of the nearby cafés. (The latter wasn't technically a 'first', but it was so good I had to give it a mention.)

****

I crossed the border by ferry at Kazangula, the 'quadripoint' on the Zambezi river where Namibia, Botswana, Zimbabwe and Zambia all meet (and there is now, since 2020, a bridge). I was spoilt rotten In Zambia – I mean *really* spoilt. Billy Ward had once again come up trumps and had put me in touch with a guy I would later call, 'The Crocodile Man of Zambia'. Ian, as he's otherwise known, owns a crocodile farm in Livingstone and Billy had suggested he might let me put my tent up on his land.

"No need for that" he announced when I asked where the best shade for my tent was, "you can have the castle!".

The 'castle' was actually a modern-ish square tower (complete with fire escape) set amongst the trees with a pool out front. It was 5-star luxury and totally unique. "Ewan McGregor stayed in that very bed whilst filming *The Long Way Down*" Ian said as he showed me around my new home. "It's all yours for as long as you like". *Great!* I could now boast that I had slept in the same bed as Ewan McGregor from *Star Wars* and *Trainspotting*! (the two events might have been several years apart, but let's not get caught up in the details!). I had the whole place to myself, including that lovely inviting pool out front *and* a maid who tidied up for me and tucked my labels in whenever they poked out the back of my vest. *Surely, I couldn't be this lucky?*

I ran up and down the spiral stairs, jumped on the four-poster beds, and ran up to the turrets where I had a great view and a roll-top bath. Life did not get any better than this. These days I was ecstatic if I had *any* bed and a flushing toilet. *This* was something else!

The next few days were spent in a whirlwind of sightseeing based mostly around Victoria Falls. First by walking across it, then by walking underneath it, and finally by flying over it in a helicopter, thanks to my host, Ian's good connections in the area. As Dr. Livingstone himself once said, it is 'a scene so lovely, it must have been gazed upon by angels in flight'!

Zambia has its problems. It is one of the poorest countries in the world. The deforestation due to its rapid population increase and charcoal usage, as well as its over-dependency on copper, is stripping the country of its natural resources and leaving its people, wildlife, and economy, exceptionally vulnerable. It's no wonder Zambia has so much illegal wildlife trafficking. The Chinese are always on hand to buy more than the odd pangolin or illegally poached elephant tusk, and who can blame the local man who steals them to feed his starving children? The problem, as ever, is global and it's going to take a lot of time and hard work to put it right.

Thankfully, there are people who care, and I was lucky enough to stay with such a couple in Lusaka, the capital, as I passed through. Dedicated conservationists and former police officers from Amsterdam, Marianne and Remco had given up their comfortable lives back home to head for the less convenient Zambia, in the hope of making a dent in the wildlife trafficking trade.

When I arrived, they got a call from one of their team, saying a pangolin and her new-born baby had been rescued from a man who was carrying them in a sack. Pangolins are quite rare, and very shy. We still know very little about them and few get to see one close up. I was invited to come with them to check it out. My excitement at meeting one, was tinged with the sadness of its situation. The adults often fared badly once they had been captured and the baby, we discovered, was barely a day old – probably born in the sack in which it was found. This poor shy anteater was just a tiny ball of scales, and mum was not planning on uncoiling herself to feed her baby. Clearly traumatised by their experience, we didn't hold out much hope for them. The baby later died, after every effort was made to save her. The mother survived and was flown out quickly to an area where she could be released again. These creatures are so sensitive that any prolonged captivity often kills them.

Unbeknownst to me, my hosts had more than one or two nights' B&B in mind for me during my stay. Before I knew it, Marianne was rushing me off for a couple of relaxing days at Mukambi Lodge, a luxury bush camp owned by their friends and situated right on the river in the heart of the Kafue National Park. This was the place that dreams were made of. Elephants, lions, hyenas and hippos roamed free and the sounds at night while we were safely tucked in our four-poster bed were magical. Marianne and I enjoyed a game drive (where the highlight was five lion cubs) and a sunset riverboat cruise (with rather large G&Ts), whilst we stayed in a luxury cabin with a bath on the decking outside, overlooking the river. On our last morning we packed our bags and walked towards the car, just a few metres away from our cabin. Breakfast was in the main building maybe five hundred metres away. It was not advised to walk alone due to the abundance of wildlife. The elephants were a particular threat there as they had a long memory and hadn't quite forgiven humans for some particularly harsh atrocities in their not-too-distant past. Only the year before, a man had been killed while walking to his cabin.

Half way across to the car, I looked up to see a big bull elephant standing just a few metres away from us. Marianne still had her head down and amazingly, neither of us had noticed it until that second.

"Marianne" I whispered. She looked up.

"Shit!" she replied and we both slowly and carefully tiptoed to shelter behind the car. Quietly, we opened the door and put our bags in. "We can't get past it in the car" said Marianne. We had already been warned not to even move around in vehicles if one was close. After ten minutes, with no signs of the elephant moving, we decided to try to make it on foot, using the round huts along the way as shelter. They were spaced about fifteen metres apart.

With our hearts racing, we made our way to the first one as slowly as we dared. No sudden movements – run and you're as good as dead. I dared a sneak look around the side,

"Shit! He's coming this way!". We tried the door of the hut but it was locked. Images of being chased around the hut by a wild bull elephant popped into my head and I let out a nervous giggle, "What do we do now?". We waited, then peeked again, to find his back was slightly turned, having spotted a nice bit of mango tree just inches from our hut.

"Let's go for it!" Marianne whispered. And so, we did. From one hut to the next, slowly and steadily all the way to our well-earned breakfast. Elephant-dodging was hungry work!

Two weeks later, I found myself sitting by my campfire overlooking the Luangwa river that separates Zambia from Mozambique, and reflecting on my last three weeks in the country formerly known as Northern Rhodesia. My life had been made so very easy in this country. This had led to the odd feeling of guilt. I should have been roughing it, surely? Not taking up these good people's time. The Zambian community, both bikers and ex-pats alike, would hear nothing of the sort though. With a dodgy shoulder, and Welsh blood boiling in the ever-increasing heat and humidity, I had been taken in and overwhelmed with kindness, good humour, soft pillows and copious amounts of whisky. I only had to put my tent up once across the entire length of the country. Rhonda too, was taken care of and I felt we were both coming out the other side, having had a good overhaul of our major components – ready to take on the rest of Africa, and whatever it might throw at us. Sadly, I'd had to leave Skully (my springbok skull) behind after my new friends had warned me it was not worth the risk of taking him through any more borders,

"We will keep him safe until you return one day" promised Remco as I tore myself away and waved my thousandth goodbye.

My last ride in Zambia, to the border, and out of the country, was a beautiful one. The roads were in great condition and the countryside stretched out for miles around me in the cool morning breeze. Dotted with rustic villages and the usual fruit stalls, I pottered on through, waving to the kids as I went. I felt relaxed and happy. My shoulder wasn't even hurting. All that luxury living had paid off. I stopped at the next lot of kids to see what they were selling to the truckers on the side of the road. I hoped for a nice bit of fruit for breakfast, but as I got closer, I realised they were cooked mice on a stick, like a kebab. They were whole; eyes, feet, tail, the lot. This was not something I had seen before, but they seemed popular with the truck drivers. A quick snack on the road to keep them going to the next stop. Tempting as it was to try, I decided against another 'sod it' moment, but gave them a few kwacha anyway and took a photograph for my collection before jumping back on Rhonda to the usual waves and cries of, "Mzungu! Mzungu!" (a term usually used for any white person in Africa, but literally translated means 'aimless wanderer', so I liked it), and heading into Malawi.

Once across the border, I aimed straight for Nkhata Bay in the north, after reports came in that there had been a spate of blood-sucking vampires on the south side of Lake Malawi. More accurately, there had been problems with the locals who *believed* that blood-sucking vampires were amongst them, and that it was probably the fault of the foreigners who must have brought them – or *were* them. This would have been funny, were it not for the fact that eight people had already been killed by vigilante groups in the area. The deadly mob violence that resulted had been so alarming that UN agencies had relocated their staff away from the southern districts. The US embassy also moved out its Peace Corps volunteers, and following attacks on ambulances believed to be carrying blood-suckers, hospitals also halted these services to the worst-affected areas. In response, President Peter

Mutharika visited Mulanje, the district with the highest number of incidents, and condemned the lynch mobs but also the suspected vampires. "If people are using witchcraft to suck people's blood, I will deal with them and ask them to stop doing that with immediate effect," he said. I imagined him then sitting down with a group of vampires and asking them nicely to stop.

All seemed peaceful at the northern end of Malawi though, so I spent a week there writing and trying to stay cool in a little hut overlooking the lake, before heading into the mountains of Tanzania.

After nearly twelve thousand African kilometres, seven countries, five speeding tickets, one puncture, one broken nose, and absolutely no blood-sucking vampire sightings, I sat in the bustling city of Mbeya listening to the Islamic call to prayer of the muezzin resonating through the cool evening air, signifying the beginning of my East African adventure. From the majestic towering Baobabs in Angola, to the regal elephants in Botswana, the unrivalled sunsets of Zambia and, of course, the long desert stretches in Namibia – Southern Africa had given me all the ingredients for a great adventure, and I suspected there were plenty more adventures to come, as I inched slowly but surely closer to home.

****

"Are you ready?", shouted the man over his shoulder. I braced myself, held onto the bars, and looked ahead. I was not ready at all and didn't think I would be any time soon.

"Yes, ready!", I lied. The rope tightened, and I felt myself being dragged off the dirt path and straight into the chaotic traffic of Dar es Salaam. Rhonda, for the first time in our 110,000 kilometre journey, had broken down, or to be more precise, had refused to re-start after a stop.

Four years earlier I had spent some time with my friend Tony in Wales practising for this very moment, but being towed now felt like a far cry from those green fields where the only obstacles were the slow-moving sheep and the fast-pursuing sheepdogs. I had no control and had to trust in a guy whose towing skills were yet to be determined. The rope had been tied to my bike and so there was no quick release. My horn was not working and my screams of, "Nooooo, I can't fit through there!", were left trailing behind me as we squeezed through the gaps between the moving cars with millimetres to spare. *Why was he filtering? Had he forgotten I was attached to him?* All I could do was hold on, keep the bike upright, and hope that he had not lied when he said we only had two kilometres to go.

I left Rhonda with Dennis the mechanic (and reckless tow-er of motorcycles) and caught the ferry over to the island of Zanzibar (aka Unguja) while he figured out what was wrong with her. It was still hot and humid in Stone Town, the old part of Zanzibar City but I managed in my flip-flops and sarong (the same purple one I'd used as a hijab in Iran). I did feel strange without my motorbike, and more importantly, my riding gear. It was what defined me, it was what I felt most 'me' in and in all honesty, it made me feel less 'tourist', more 'adventurer'! I had to stop myself telling anyone who'd listen that normally I'd be wearing big boots and body armour! Quite ridiculous really, but there was a genuine sense of something missing without Rhonda. In part, and quite ironically, it was nice not to have the burden and responsibility of a motorcycle with luggage, yet it was the very thing that set me free. But mostly I felt naked without my sidekick. It was as if someone had just surgically removed my Siamese twin.

Zanzibar was fun. I danced with the locals until 3.30am, I read a book on my Kindle, wrote my blog and recovered from my hangover overlooking a white beach and turquoise sea, and I got happily lost in the historic city, with its medina-esque narrow winding streets. If you are a single white woman (of any age) and you need an ego boost – go to Zanzibar. The men are young and beautiful with good skin and surprisingly white teeth. They are very keen to admire you and tell you how beautiful you are. They are also, no doubt, very keen to find an affluent wife who will provide for them – but don't let that put you off. Once they realise you are skint and you have no interest in taking home a souvenir anyway, they soon get down to the important stuff – dancing! And *boy* can they dance!

It turned out that Rhonda's ailment was merely a defunct starter relay. Dennis the mechanic found me a working second hand one and we were soon back on the road and heading for the cooler climes of the Usambara Moutains in Northern Tanzania. I was happy to leave behind the humid and grimy port of Dar es Salaam (now a city of over four million people), with its busy roads and aggressive mosquitoes that buzz in your ear at night with high-definition clarity. After a few hours of working my way north through the crazy traffic – narrowly avoiding the bus, and the motorcycle (or piki-piki in Swahili) with a wardrobe on the back, coming the wrong way up a dual carriageway – the climate and environment completely changed.

The cool air enveloped me as I climbed the series of switchbacks up the mountain into the little village of Lushoto, and the moist green foliage gave me an overwhelming sense of home. As I climbed and tried to take it all in, I saw a movement in the verge. Suddenly, an eagle flew up clutching a snake. I ducked and narrowly avoided both as they wrestled with each other in mid-air. As I composed myself and laughed with delight at this unexpected treat, a crowd of people (clearly members of a local tribe) with painted faces and brown robes carrying spears, came into view, marching as one, with a real sense of purpose. I was glad I was not their 'purpose'! "This place is like something from a Terry Pratchett novel", I thought. "All we need now is Rincewind and a giant turtle!". It was totally surreal and a wonderful breath of fresh air.

Whenever I thought I had seen it all, the world provided a whole new scene, and with it, a fresh perspective and the energy I needed to keep going. My shoulder even felt mildly better and it was the respite I so desperately needed before taking on arguably the toughest section of my African leg. I had missed out a couple of countries – Rwanda and Uganda, leaving them in the pile marked 'Come back when re-energised physically and financially' and I was pushing on towards home now. I wasn't in any great rush, I was enjoying Africa immensely, for the most part, nor was I up for any major detours either. My mission was to get home safely and happily while still enjoying the ride. Once again, the nay-sayers had been wrong.

CHAPTER 29
KENYA:
Paperwork, Police and Waterworks

Having crossed the border into Kenya, I found myself in the bustling and frankly unpleasant capital city of Nairobi, where I had to get on with some serious admin for the next leg. This would involve working on my route, sorting a visa for Ethiopia, writing my column for *Ride* magazine, and catching up on my blog. It would also involve my eleventh and *final* tyre change. That's eleven pairs – I always changed front and rear together to keep workshop time to a minimum. And in any case, the back tyre was often so worn that the front was catching up fast by the time I could get a change. By the time *these* treads had worn down, I would be home. This seemed a significant and sobering thought, that was perhaps as reassuring as it was unsettling.

Thankfully, I had found a place to gather my thoughts and lay down my tent at a campsite called Jungle Junction in Karen, south west of the city. Not a bad place. It had a workshop, a tiny kitchen to use, five dogs to play with, and some good company in the other overlanders who, like me, came to rest up and take stock before moving on in whatever direction their hearts desired.

That night I sat with a Canadian couple called Mike and Sue who were in their seventies and a 35-year-old Frenchman called David. They were all in 4x4 vehicles. Mike and Sue were in an old converted German ambulance, and David was in a younger (but still old) Mitsubishi L300 camper; a funny-looking thing he had named Mr. Fox for reasons I never discovered. We shared a beer and a nice stew around the campfire while comparing elephant encounter stories and discussing the route ahead. We were all going in the same direction – towards Ethiopia, Sudan and onwards, eventually, to Europe, but the more we compared notes, the more barriers we discovered.

It seemed I wouldn't be able to get into Israel to catch the ferry to Italy as I had originally planned. Going from Egypt across Sinai was currently forbidden for motorcycles and 4x4s due to several recent bombings. Going through Saudi Arabia to Jordan was forbidden for women riders without a husband, and the ferry from Sharm El-Sheikh to Jordan no longer took vehicles due to more recent security scares. The only option left was a vehicle-only roll-on-roll-off service for the bike from Alexandria in Egypt to Salerno in Italy – which was quite expensive at $800 – plus the flight for me at $300. My first hurdle though, would be dealing with the Ethiopian Embassy. A mission, I now knew, from our campfire conversations, that would require my utmost patience, and a very large slice of luck.

I spent a lot of time with David over the week that followed. He was fun and a good source of information, having already done a lot of the research on the road ahead. We got on well and he allowed me the pleasure of taking the Mickey out of his lovely French accent regularly, often retorting with some wisecrack about the Welsh – usually involving, rather predictably, sheep! Despite the restrictions, David was still hoping to go the Sinai route to Israel. His master plan to fool the authorities? Removing his 4x4 sticker and claiming the vehicle was only two-wheel drive! It was so simple, it might just work. I once got fifteen motorcycles through the Moroccan border with only photocopies of the original documents (because Honda owned the originals and didn't want to risk handing them over). When the officials at the border had looked at the file and said angrily in broken English,

"These copy *not* original", I simply replied,

"Ah yes but they are *laser* copies so it's OK". It's amazing what you can get away with, if you put enough confidence behind it.

David was now trying to convince me to take the Turkana route into Ethiopia with him. This route was all off-road and quite technical by what we could tell from the information we were getting. It was also a tribal area, and there was no border post, so you had to get stamped out of Kenya before you actually left Nairobi. It sounded like a challenge, but I had already chosen to miss out two countries (Rwanda and Uganda) because of my weak shoulder, and now I was considering a 100-kilometre cross-country endurance route. It was a dilemma. On the one hand, I had a back-up truck in David and we would get to meet many of the rural tribes. I was really into getting portrait photography by now – trying to capture the ever-changing faces that told a story of their ancestors and their environment – and knew this would be a hard one to beat. On the other hand, I wondered if it was worth the risk after already deciding to 'take it steady' from now on. I decided to sleep on it.

My tent was a bit of a soggy mess. The rain had come down hard over the last couple of days, and my faithful canvas, that had kept me dry all these years, was failing. Still, there was a dry corner and I tucked myself into it with a towel down one side acting as a dam, with my riding gear over the top. Podcast on…skin covered…and sleep.

One hour later I was woken by mosquito 'engines' approaching over the sound defences of Radio 4's World Service. A single rogue and ambitious kamikaze fighter had infiltrated my bubble and was out for blood. How could something so small be so aggressive, so annoying, so relentlessly loud?

It struck at the unprotected juicy skin of my forehead above the sheets and my attempt at a defence manoeuvre was far too slow. A rookie could have done better. Within seconds I was itching. Strike one to the mozzie. I tucked myself further into the sleeping bag and under my jacket. Silence descended on the swamp-tent; long enough that I dared to hope that it was spent. Just as my mind started to drift into beautiful oblivion – *it was back*! *Dzzzzt!* No warning. Another attack. This time, straight into my nostril. *No mercy. No mercy!* I blew out like a raging bull. I shook my head. I swore just a little too loud, then whispered,

"Please let me sleep", just as the neighbourhood hounds began their nightly routine.

I woke the next morning feeling rough. Rougher than even a mosquito attack, a damp tent, a midnight hound chorus and a 5.30am call to prayer could explain. Still, there were things to do. Everything had to be ticked off my list before I left Nairobi. First on the list was a trip to the dentist. Joy! It seemed my teeth were falling apart, along with the rest of me. I had visited dentists in Nepal and Thailand due to a recurring niggle with a particular tooth. I'd paid a small fortune for treatment in Canada on the same tooth, before having root canal treatment in South Africa, and now it was painful again. This single tooth had already cost me more than a flight home, although I was comforted by the fact that I could perhaps recoup some of my money by writing a book entitled *Dentists Around the World*.

The dentist's was a scruffy little place. From the outside, it looked derelict. The car park was a rubbish-strewn patch of dirt with enormous muddy puddles that looked worth avoiding. The less-than-artistic, graffiti-covered rendering was falling off, and as soon as I parked the bike I was surrounded by touts trying to sell me CDs and fruit. I told the CD guy I had no player on the bike and I told the fruit guy I'd buy some bananas when I came out. I shooed the stray cats off the metal fire escape steps and made my way up to the first floor.

Reassuringly, the door opened into an organised and clean environment, in total contrast to the outside world. The dentist was Asian, and very chatty. He had all the latest technology at his disposal, far more advanced than what I had seen back home, and he quickly deduced that the tooth where I'd had the root canal treatment should be removed and a crown made on his in-house 3D printer.

"Let's get this over with" I said.

"You won't feel a thing" he replied as he leaned in with a juicy needle. He was right. I didn't. Barely a thing, anyway. He was far gentler than any dentist in any other part of the world so far.

Back in the car park, I picked up my bananas as promised, and set off. At the entrance there was a crowd of motorcycle taxis with the drivers lounging idly around and chatting. I hadn't noticed the policeman amongst them until he stepped out in front of me with his hand up.

"Is there a problem officer?" I slurred through swollen lips.

"Insurance" he demanded in a deep, authoritative voice.

Shit. I had the wrong bag. The campsite was only a few miles away and I hadn't thought to bring any paperwork with me other than my passport. I sucked in a line of drool that was trying to escape before replying,

"Sorry officer. I don't have it with me".

"This is illegal" he said angrily. "I can arrest you".

He then turned his head and looked back over his shoulder at the taxi drivers, who smiled in response. One giggled. I just caught the tail end of the nasty little smirk he had shared with them before it disappeared, and he faced me once more with his 'stern' look. He was playing up to his crowd.

"Yes, I do have insurance but it's at the campsite. I'm sorry I forgot to bring it."

"Where is your high visibility vest?" he asked.

I had no idea this was a requirement of the law here. I had seen more without vests than with, including most of the taxi drivers behind me. Most didn't even have helmets.

"Passport" he demanded.

I handed my passport over. He flicked through it quickly and said,

"You must come with me to the police station".

"What? Wait. I can go and get the paperwork. You keep the passport. It's just around the corner. I will be back in ten minutes".

"Come with me now." he said, and turned back for another encouraging giggle from his crowd. There was little I could do. He was not going to back down with an audience to impress. So, with an aching jaw, a numb face, and a little bit of drool making another run for it down my chin, I reluctantly followed. The police station was just around the corner. It looked more like a stable, and outside on a bench sat three more officers, who looked amazed when I took my helmet off to reveal that I was in fact, a woman. They all confirmed that having no high-vis vest was indeed a serious offence.

"Why doesn't one of you jump on the back with me now?" I suggested, irritated at where this was going, "Let's take a little ride and we can count together the amount of people who actually do have vests. Seriously, hardly anyone is wearing one". This was bullshit. My mouth was hurting, I felt like shit and I was in no mood to be pushed around.

"You pay a fine now and you can go", he said, to no-one's surprise, least of all mine. "If I take you into the station we will have to put you in a cell and wait for a court date which could take days". I knew Kenya did not have on-the-spot fines, and so what he actually meant was, "Give me a bribe and you can go". Blatant corruption.

"This is very serious" said the other guy backing him up and sounding more and more like I was up on a murder charge. Their intimidation tactics didn't scare me though. I'd dealt with more uniformed bullies in my time than they'd had hot showers.

"I'm filming this by the way" I said pointing to my headcam. And I was. I'd switched it on as soon as the policeman had stepped out. Something I always did. I had never paid a bribe, and I didn't intend to start now. "I am not paying your fine outside. You want money? Take me into the station. Let's do this properly. I want paperwork."

The policeman tried to hide his disappointment, then puffed out his chest and led me inside.

Damn it! He was calling my bluff, but I too was holding my nerve; playing the game to see who buckled first.

The policeman led me down a dark corridor and into a maze of medieval-looking rooms and corridors which felt as if they were going on forever towards some deep and damp dungeon – but it probably lasted just a few seconds. I really thought I was getting put in a cell and, quite honestly, I have no idea how I managed to keep walking. I had not been in one for over twenty years, and I had promised myself I would never enter one again, but there was no way I was buckling now. The policeman stopped and turned to face me before saying,

"Last chance. You pay $20 now". I considered the money in my inside pocket. How easy it would be to pull it out now and be on my way.

"Not without an official charge." I replied, with as much confidence as I could muster.

Instead of a dungeon, we turned into a dark room with no windows, lit only by a bare bulb of no more than 40 watts. Immediately beneath it was a lady in uniform hovering over an ancient typewriter on a wooden desk. He said something to her in Swahili. First she looked at me, then back at him, then back at me again – all the while fingers poised mid-air over the well-worn keys.

"I have paperwork" I tried, "I can go and get it if you would like to keep my passport?"

Her eyes moved deliberately and silently from mine to the policeman before saying something in what sounded like irritated tones. He didn't look surprised by what she said; more as if he was expecting it. Resigned to the fact that he would get no bribe today, he handed my passport back and told me to leave. I did, and quickly, but triumphantly.

"How much did you pay?" asked one of the policemen on the bench outside as I passed them to get back to Rhonda.

"Not a thing." I replied, "This must be my lucky day". All three officers looked surprised, and now it was *my* turn to smirk!

I had decided to take the Turkana route with David, and the following day, we went in his truck to the Ethiopian Embassy, armed with letters from our embassies. Ethiopia required a letter stating our embassy would take full responsibility for us while we were in the country. Of course, no embassy would do this. Until now, the preposterous process had worked like this: the Ethiopian Embassy requested an impossible letter; the other embassies wrote a letter stating they would *not* provide the sort of letter required. The Ethiopians would then stamp *that* letter and process it anyway. A totally pointless exercise, but everyone was happy.

I was feeling terrible when we arrived. I had become feverish the night before, but I had to get this sorted to be in with a chance of leaving with David. I didn't have much patience, so when David's letter was accepted and mine wasn't, I lost my temper quite quickly. They had sent me from the back of the building to the front and then back again, in the Nairobi heat. In the end, I began shouting. This did not help, and just as I was debating a sit-in protest, sense prevailed, and we left; David with his three-month visa (when the norm was one month), and me with nothing. It didn't make sense.

David left the next day. He planned to head for Lake Turkana and then come back towards the main road for the border. Without back-up he was not going to risk the whole route alone and there was no point waiting for me. This could take some time to sort – if at all. I stayed in bed for most of the day and tried calling both the British and Ethiopian embassies, with no luck. It turned out they were all on some national holiday.

After a couple of days I felt a tiny bit better and so started calling around again. I couldn't avoid Ethiopia because South Sudan was closed, so if I couldn't get into it, I had a major problem. There would be no way to finish my trip. The British could not help. The Ethiopians said the same thing when I appealed to their better nature over the phone, but then they seemed to soften and told me to come back with the same letter. As quick as a flash I was on my bike and racing through the horrible Nairobi traffic. It still took me a whole hour to do nine miles. I filtered like a hooligan, climbed kerbs and had crazy *Mad Max*-style *matatu* buses racing up my arse trying to intimidate me off the road. With a notoriously bad reputation, (due to the frequent fatalities), the 14-seater *matatus* were designed to intimidate: aggressively loud horns, teeth drawn on the front and even spiky hubcaps to bully people out of the way. At one point, I rode for a mile behind an ambulance which forced its way through. That was a little hairy too, as the traffic flowed back in behind it pretty quickly, but I didn't care. I was on a mission and I had no time or patience for cautious driving.

At the embassy once more, the woman behind the counter took one look at my sweaty helmet hair and red face as I popped the letter under the screen, raised an eyebrow and, after a quick glance at the letter, turned me away again with the same response as last time. The letter wasn't good enough. They had completely wasted my time and I wondered if they were doing this deliberately to wind me up. I was so tired, hot and frustrated I wanted to cry. Instead, I tried to ask if it would help if I came back with a letter from the Canadian Embassy, but she shook her head. I tried to speak with the ambassador but no chance. I couldn't get past the woman everyone called 'the dragon lady'. I thought they had been joking when they said she had barbed wire tattoos around her neck. Actually, I'm not sure it was barbed wire, but it certainly gave the same effect. She was a pit-bull terrier bitch, and she took no prisoners. She clearly didn't like me and once again, I left empty-handed, with no idea what I would do next.

A few days later, I decided to try going back to the Ethiopian Embassy, but this time with my Canadian Passport, and a letter from the Canadians, despite what had previously been said.

The Canadian Ambassador was late to the office and I had to wait over an hour. When she arrived, she apologised profusely saying her cat had been ill. I laid on the empathy and told her all about every cat I'd ever owned. Boy, did I love cats that day! In the end she could only give me a very similar letter to the one the British High Commission had given me, but when I took it back to the Ethiopians, it was accepted immediately. No fuss, no argument. All the previous nonsense had cost me nearly two weeks' delay in the end. My journey could now continue as planned, and so, armed with all the paperwork I needed, I happily waved farewell to Nairobi.

****

Archers Post is a ramshackle, one-horse town, with a bank, a butcher's, and a small hotel called The Camel Inn. The locals were all dressed beautifully in their beads and tribal robes. Some carried guns, others spears, and some dragged unwilling goats behind them. People smiled at me as I arrived and gave me directions down the sandy track to my abode for the night. I had come to visit Eric and Paul from The Samburu Project, on my way north to Ethiopia. I had raised money through my blog for this charity and I was keen to see the area. Eric would come and collect me in the morning and take me out to some more rural areas where I could check out the wells we had helped raise money for and meet the people whose life depended on them.

I woke at 5am to the sound of beautiful singing. It was clearly tribal, and I could tell that something was going down in Archers Post. A wedding perhaps? Maybe a religious event? I lay in my rickety bed listening for half an hour before dragging myself into the cold trickle of brownish water they called a shower and washed off the weariness after a half-sleep night thanks to another mosquito torture session. The singing continued as I headed out to meet Eric for breakfast. Eric grew up there, living off the land in one of these peaceful Samburu tribal villages in Northern Kenya. He now helped to run the Samburu Project, a community-run charity set up to help the local tribes with basic human needs like sanitation, education and especially, fundraising for more wells. I was keen to see the work they did and the difference they made to the communities of this magical place.

As I took my place at the table and poured the coffee, I asked Eric what the singing was all about.

"Someone is being circumcised", he said as he tucked into his eggs, "They are singing a motivational song. When they stop singing, then you know it is being carried out". I winced at the thought. He laughed and,

325

speaking from personal experience, he added, "Yes, it is very painful indeed – they use no painkillers." The singing stopped. I poured the coffee!

Within an hour we were in the bush sitting amongst a fifty-strong herd of elephants. I looked to Eric for reassurance. Although I'd had many elephant encounters through Africa, I was unaccustomed to being in such a large herd with babies and bulls in musth (on heat!). I expected some 'aggro' having seen how agitated they can quickly become. Eric laughed at my worried face as one came inches from my open window and said,

"Don't worry. They are very friendly here as we have no trouble with poachers". I was not convinced,

"But what if he puts his trunk in the car?" Eric's confident reply was simple,

"He won't". And he didn't.

Suddenly there was a chase on. One of the large bulls had picked his female and, in turn, she was running away as fast as she could. She looked tiny compared to him and very young. We gave chase, weaving our way through the other elephants and through the bushes, until the female stopped and the bull took his prize. It lasted all of two minutes and by the time he had finished, some of the other females had caught up. Unbeknownst to me, they too had been chasing the chase and ran straight over to the young female to comfort her as the male swaggered off. It was a sight that confirmed to me all I had read about elephants. They are intelligent and caring animals who look out for each other – especially their own family.

Eric had a way with words. We drove along the 'massage roads' (corrugated!), stopping for more animal encounters along the way. Sometimes we would sit staring at something in the distance trying to decipher whether it was an animal like a big cat, or an inanimate but 'animal-like thing' or 'ALT'. Sometimes it would turn out to be a log.

We visited the wells and met many of the villagers who rely on them. The villages were made up of stick buildings that looked as if they were built for extremely small people. I was invited into one of the huts to look around and had to bend down so far I was almost crawling. It was very small and made up of two bedrooms, with a corridor in between. These days they tend to have a separate hut for cooking, as the smoke has been proved to cause cataracts.

The boys of the Samburu tribe go through three stages in life. First is childhood where they are taught to shepherd the goats and cows. At the age of fifteen they are circumcised. During the circumcision, the boy is not allowed to move at all. He must not even blink while it is being performed. Any movement is a sign of weakness and is considered bad form. The procedure is watched by the other men and boys in the village (but not by the women). This is when they become warriors. The warrior stage lasts until they are thirty and during this fifteen-year period they are not allowed to eat in their village or in front of anyone other than other warriors. They can only take milk from the village. To eat, they must go out and fend for themselves. At the age of thirty they marry and can have as many wives as they want – or can afford.

The girls can be married as young as eleven. On their wedding day their vulvas are cut in the traditional and dreadful way of female genital mutilation (FGM) which is far worse, in terms of long-term pain, damage and ill-health consequences, than male circumcision. Thankfully, this tradition is slowly being eradicated through education). Each wife has a house of her own and will raise her children there. It is common for each woman to have nine or ten children.

Nowadays, many of the children go to school. I spoke to one woman at the well with the help of Eric as translator. She had nine children and seven of them went to school. Three had been kept back to tend to the livestock. When Eric himself was raised with his eight siblings, only three of them went to school and the rest worked on the land. The ones who go to school help with finances once they are educated.

The wells have helped more kids to get educated as there is now a lot less work to do and fewer of them get sick. Before the wells were dug, most women and many children would have to walk for three hours to get to the river to fetch dirty water for their families, then three hours back; in many areas they still do. Now, the whole water-fetching process takes only half an hour and the water is clean. So many diseases have been eradicated thanks to these wells and to education. The tribe were extremely welcoming and very happy to allow me to take their photos, play with the kids, and generally hang out for a while. This was, of course, all thanks to Eric, my wonderful host.

327

Chapter 30
The Hamar Tribe

I felt my phone vibrate just as I was putting my gloves on to leave Archers Post. Removing them again and digging through to the inner pocket of my riding jacket, I pulled out my battered old iPhone to discover a WhatsApp message from Chris and Erin; a husband and wife team from Colorado I had met back at Jungle Junction (outside Nairobi). They had taken the Turkana route in their converted Land Rover, in convoy with another couple, Rob and Emi from Holland, a few days before. The message read that Rob and Emi had lost control of their vehicle (another Land Rover) on the gravel and found themselves rubber side up and a long way from anywhere. Thankfully no one was hurt, Chris had towed them out of trouble, and they had all made it over to Henry's Campsite just south of the Ethiopian border. I checked the map and texted back '*I'm 300kms away. Get the kettle on*'.

That night we discussed our next moves around the campfire over a 'jello' sandwich. We had heard from David, so we knew he'd made it back to the main road from the lake and had actually crossed the border already. Rob and Emi's car would not be going any further North. The roof was almost level with the bonnet and there was no windscreen left, but Rob was confident he could straighten it up enough to drive it slowly back to Nairobi where they would decide what to do. I really wanted to go to Omo Valley, in southern Ethiopia, but I was feeling apprehensive about going alone. We'd heard news reports that one of the tribes had gone on a killing spree and shot thirteen people on the road into the valley after one of their tribe was run over. Thankfully, Chris and Erin felt the same as me, and so we agreed we'd just have to go in together. The killings were the week before and so the likelihood of more trouble was *probably* quite slim.

The Omo valley is known as a cultural crossroads because so many people had migrated there over thousands of years, from all directions. There are now around 200,000 people living along the river, consisting of sixteen different tribes. Each tribe has its own cultures and ceremonies, which can be vastly different from its neighbours'. The Mursi tribe, for example, is known for its lip plate tradition; an unmarried woman's lower lip will be pierced and then progressively stretched over the period of a year. A clay disc, with an outer groove like a pulley wheel, is squeezed into the hole in the lip. As it stretches, ever-larger discs are forced in, until the lip, now a loop, is so long it can sometimes be pulled right over the owner's head. The size of the lip plate determines the bride price, with a large one bringing in fifty head of cattle.

Most of the time the tribes live peacefully side by side, as cattle herders and farmers, but the occasional cattle rustling or land disputes can lead to eruptions of violence. This recent violence though, was aimed at road users.

I still had to go. Where else on the planet was I likely to find so many different faces together telling so many different stories? I hoped I could capture some of those stories through my lens. We had all missed out on the full Turkana experience in the end, and there was no way I was going to miss out on this one. Chris and Erin were easy to spend time with and I felt very lucky to be in their company. As Chris so often liked to say – this was going to be *awesome!*

Our journey across the border and into the valley was uneventful and peaceful, thankfully. No tribal rampages, just a partly tarmac, partly dirt road sweeping through miles of green and golden patchwork countryside; it was flanked either side by acacia trees and clusters of cacti. Tiny thatched huts mingled with domed haystacks, making them difficult to tell apart, while small children with big sticks ran between them chasing goats, cows and chickens. The road itself was quiet, apart from the odd herd of camels, usually shepherded by slightly older children (but not much older). In most cases, all sizes of children would run at the vehicles, arms out, shouting, "Money! Money!" when they saw us coming – no doubt the only words they knew in English.

In one memorable case, we were stopped by some slightly more industrious kids walking on stilts made from long branches. To this day, I cannot tell you how they got up on them – or down again! Decorated in tribal paint, and dressed in little more than a loincloth, they wandered this stretch waiting for a passing car they could entertain for money; and entertain us they did. Their skills were very impressive, and they showed them off, running up and down the road, before they too, put their hand out and repeated, "Money! Money!" We gladly paid up and moved on.

It was nice to have some company behind me, especially as they were also kindly carrying my luggage! Unencumbered by it, Rhonda felt light and playful; but two days into the ride down into the lower section of the Omo Valley, the Land Rover caught fire. The rough roads had broken a bracket holding the battery and rattled a bolt out too. Luckily it was put out safely with little damage caused, but Chris wanted it checked over and repaired before going on, so we turned back and aimed for the nearest small town. This turned out to be a stroke of good luck. While we were there, we got chatting to a local who told us there would be a bull-jumping ceremony at one of the Hamar tribe villages that day and if we were quick, we might just be able to get an invitation to join them. There was no way we were going to miss this opportunity, so we quickly hired a guide with a truck to sort out the invitations and take us the rest of the way. Of course, invitations came at a price and were extended to a small number of outsiders in order to help pay for the ceremony. The chief set the price with the guide before we were driven there at the appropriate time.

Our guide explained en route that a Hamar man comes of age by leaping over a line of cattle – naked. It's the ceremony which qualifies him to marry, own cattle and have children. The timing of the ceremony is up to the man's parents and happens after harvest. The guests, who come from miles around, receive their invitations in the form of a strip of bark with a number of knots – one to cut off for each day that passes in the run up to the ceremony. After the ceremony, they have several days of feasting and drinking sorghum beer.

We found the tribe sitting around on the parched riverbed. Most of the girls were huddled together, wearing a combination of old ripped T-shirts, or goatskin, or a combination of the two. Many had short dreadlocks and okra in their hair, like the tribes in Angola. Some wore nothing but a goatskin around their waist and beaded necklaces around their neck. Many had bells around their calves. The men were busy painting each other or just sitting under the trees talking, having already been prepared. Some carried rifles. Some had what looked like whips in their hands, their legs smeared in mud and decorated in white paint over the top. Many wore feathers on their head, held on by a thin band. Most wore red, checked cloth not dissimilar from that of the Masai tribe in Kenya. There were a couple of obviously western people already present. They stood out a mile with their cameras and clean, modern clothes. I was disappointed to see them. It ruined the picture.

This was not an event put on for tourists. This was a personal and important ceremony. We were interfering, changing things by being there and I did not feel comfortable or even welcome, despite being invited and having paid for the privilege. Not everyone from the tribe saw the benefit, and the hostility was not lost on me. I didn't want to be part of it, so I put my camera down and went to sit quietly on the riverbank near a group of men from the tribe. I smiled at them as I sat quietly leaning against a tree, trying to be as unobtrusive as possible. Within a few minutes I felt something hit my arm. I looked over at the guys sat nearby, who turned out to be the *maza*, or the 'whippers', and found one of them smiling at me, having just thrown a twig to get my attention. The ice had been broken and I felt just a little bit accepted, in a childish, playground kind of way.

I looked over toward Chris who was sitting in the middle of the dry riverbed surrounded by some of the tribe. He had given them his phone and camera and seemed to be giving them a lesson in how to take selfies. He might have looked like a wise old elder sitting with his pupils, had it not been for his dyed bright red hair, white skin and digital gadgets. These guys live miles from any town or infrastructure and although I am sure these were not the first cameras they had seen, some behaved as if they were completely alien to them.

It seemed as if everyone was starting to relax as the ceremony continued throughout the afternoon and I slowly joined, always showing them the photograph I had taken when I was done. They were not used to seeing themselves and it often caused a shriek of delight. After a while I felt welcomed by many but still just tolerated by others. Hardly a surprise.

As I sat quietly amongst the tribe, I heard some shouting and singing coming from further upriver. An old lady stood up and began dancing and chanting, as if in anticipation of the new arrivals. Her skin looked like your favourite leather armchair that you just can't bring yourself to throw out; well worn, but familiar and comfortable. It was hard to tell her age, but I would guess she was in her late seventies or early eighties and yet she kicked up that dust with the energy of a spring chicken. The bells on her ankles joined in her chants and invited others to join her. Then, more women arrived in a cloud of dust. Some were topless. All had beads, red dreadlocks and the usual goatskin skirts. They danced their way over to the men with the whips, who in turn looked vaguely uncomfortable, as if they knew what was coming; knew their role, yet would really rather not.

They were the *mazas*, and their role was to whip these female relatives, who were far from cowering at the prospect. In fact, they were shouting; pushing, shoving, vying for attention; wanting to be the chosen one. Finally, the *maza* made his choice and the other women stood back in defeat as the chosen one stepped forward and stood stock still; one arm raised as if in salute. A statue.

The *maza's* arm went back as far and as high as it could. Still, the girl did not move; not even a blink of anticipation. The whip came down against her skin with a crack so loud, it echoed across the riverbed. A yelp escaped my lips as the blow tore through the flesh of her belly and spilled her crimson blood against her dusty black skin. It was almost beautiful in contrast. Glistening and real, like lava erupting from a volcano. The girl did not move. Not even a murmur escaped her lips, and yet it must have hurt like hell. I felt shame for making a noise, when all that had touched me was the shadow of his strike.

My translator made his way over and joined me under the shade of the tree while the whipping and the bickering and the dancing continued to the sound of horns and chants and laughter all around us.

"How are you doing?", he asked with genuine interest.

"I'm feeling a little more comfortable being here now" I replied, "I wasn't sure at first".

"Not everyone is accepting" he said, "Many find you very strange and this is new to them, but you being here is what will pay for much of the celebrations later, so *most* are happy to have you", he replied honestly.

332

"So, tell me", I asked "What is the whipping all about? Why are they begging to be whipped?"

"One effect of this ritual whipping is to create a strong debt between the young man and his 'sisters'" he explained. "If they face hard times in the future, he'll remember them because of the pain they went through at his initiation. Her scars are a mark of how she suffered for her 'brother'. You see how they sometimes rub dirt in the wounds?"

"Yes of course. What the hell?"

"This is to ensure the scars are prominent"

I thought about this for a minute.

"Well," I concluded, "We have tattoos back home. We put ourselves through that pain, and what for? We don't even get a goat at the end of it!" The translator laughed.

"While you are here" I continued, "See this guy next to me?" I pointed with my eyes, without moving my head at the young man who was crouched not five feet away, in the shadow of the neighbouring tree.

"Yes" he confirmed.

"Will you please ask him if I may take his portrait photograph."

"Just take it."

"I'd rather ask first. I want to get quite close to capture his face, and I would consider it very rude to just wander in uninvited".

The translator shrugged and turned to the man in question. He was a beautiful young man. Wearing nothing but a loincloth and tribal paint on his skin that looked so ridiculously soft, it seemed out of place in such a rugged environment. His eyes were big, brown and wise. His body so toned and athletic, he looked capable of both power and stealth, delivered in equal measures, and at the flick of a switch. He was the finest specimen of a tribal man I had ever seen. Here was a man who could quite possibly get me to leave all my western comforts and ways and live out the rest of my days in this dusty thorn-infested wilderness, herding goats and hunting for my supper. As this image played out in my mind, I saw myself as one of the guys. I wasn't the childbearing, stay-at-home wife, who fetched water and kept the home while my beautiful husband tended to the herd. I was one of the guys, hunting, herding and providing. It was *my* daydream after all.

The translator looked at me and said,

"He says you can take his picture, but only if he can whip you".

I looked into the eyes of this beautiful man, who stared back at me with a deadly straight face. He did not blink or look away as I found myself lost in his eyes actually considering this proposal. After all, how many western men would kill a goat for you these days? Then my thoughts were interrupted by the translator who, in that moment, had conveniently blended out of my vision,

"He's joking" he said with a touch of despair.

A wry smile crossed the thin lips of my goat-herding tribal warrior potential husband and I blushed as I snapped back to reality. *A sense of humour too! Wow! This guy was perfection personified.*

"I knew that" I giggled.

My 'husband' posed for me as I took the picture; his eyes forever captured; his soul shining through. A photograph I would cherish. A priceless moment; a fleeting connection I would recall and smile at, time and time again. No amount of possessions could bring such a richness or depth to my future as these memories.

The noise increased as more people joined in the dancing, and then as one, with everyone decorated and the whipping apparently finished, the procession moved on up the bank and through the bushes. I followed about mid-pack. In the clearing at the top, there were several cows lolling around, mercifully oblivious to the small fact that they were about to become bovine steppingstones. (Yes, some were cows rather than bulls).

The girls huddled together and brought the chanting to an almighty crescendo, jumping in unison, blowing their horns and chanting louder than ever. The *mazas* expertly grabbed the bulls by their horns and wrestled them into a line, with one holding them in place by the head and another by the tail. The little additional rituals seem to go on for an age, but eventually, the naked boy jumped. He jumped onto the first bull and ran across seven others before jumping down, turning around and jumping back over the line. This happened several times, until his brother – my future husband – stopped him. The boy didn't seem to want to stop and tried to protest. He could do more, but there was no need. He had done enough to prove himself a man. Now he could marry the wife chosen for him by his parents and start to build up his own herd.

It was an amazing experience and despite my conflicting emotions about being there, I was very glad I had been given the opportunity. We left as we had come, only now I was desperate to see more, to learn more. There were, after all, fifteen other tribes in this area. But it was time to hit the road again and see some more of Ethiopia. It was a big country and I only had thirty days in total on my visa.

Once back in signal I got an email from David. Unfortunately, I lost the signal before the body of the email was downloaded, and all I could read was the subject line, which said, 'BE CAREFUL'. We knew there had been some political protests throughout the country, with the army being called in and many turning violent. But it was hard to get the latest information as we moved further north. We didn't know where David was and so weren't sure which bit to be careful of! Regardless, Erin and Chris wanted to check out one route, and I had plans for another, so we said our goodbyes and agreed to meet up again in Lalibela, north of Addis Ababa, for Christmas. We hoped to meet David there too, along with his friend who was flying into the capital and who would be travelling with him for a while.

Back on my own again, I travelled through villages and towns along the sometimes dirt roads, sometimes well-tarmacked, Chinese-built roads that twisted through the hills like long black snakes. Each road was only ever covered in cattle, goats or camels, rather than cars. Fuel stations were sparse and most of the time I would have to buy black market fuel from the little villages that dotted the mountain roads.

"Benzina? Gasolina?" I'd shout to the first person I saw, who would point me to one hut. They in turn would point me to another, and eventually I would find someone with fuel-filled barrels, and the bartering would begin. In the meantime, I would be surrounded by the villagers, sometimes thirty at a time, who had spotted the lady on the iron steed rocking up, and wanted to get a closer look.

The most popular language in Ethiopia is Amharic. I had learned from an English-speaking waiter at one of my coffee stops that Amharic for 'thank you' is *'ameseginalehu'*.

"Is there a shorter version?" I had asked hopefully, knowing how terrible my memory for language was. He laughed and shook his head. But I would not be perturbed and remembering then at how good I was at remembering lyrics I continued my ride singing this word in my helmet over and over again.

Riding through a village with my visor up, still singing,

"ameseginalehu oh yeah *ameseginalehu* la la la" I suddenly felt a sharp sting in my eye. It felt like someone had stuffed an amazingly hot chilli right on the eyeball. My eyes watered and naturally closed up. It stung so badly I felt it in the nerves all over my head. I pulled over making appropriate, 'arghhhh!' noises and blindly jumped off the bike. I sucked water out of my Kriega bladder, spat it in my hand and frantically washed out my eye; hair everywhere and dirt creating patterns on my face as I washed out the offending and unidentified article. Eventually I could see again and looked up to survey my surroundings. I found maybe fifty surprised faces staring back at me. Someone from the crowd shouted,

"What happened?".

"ameseginalehu" I said proudly, "I'm OK. I got something in my eye"

As one, they decided I was not crazy and moved in to get a closer look. They turned out to be a friendly bunch, who were clearly glad of the distraction for the day and I waved a fond goodbye as I moved deeper into the countryside. To this day I have absolutely no idea what got in my eye. It was quite scary at first and the pain was intense.

Ethiopia's countryside is a textbook illustration of what pastoral living is supposed to look like: lush, rolling hills, sweet-smelling cypress air, idyllic village scenes. It provides an extraordinary variety of landscapes, with its huge and lofty central plateau area contrasting totally with its hot desert tropics. For the most part, despite popular belief, it is a lush and fruitful land, farmed beautifully by its people. I would stop on the roadside for my usual coffee – now of course from Ethiopian beans, sourced locally – to watch young men threshing barley, or old men working the fields with their ploughing cattle. It felt peaceful here. A simple and honest way of life.

This was the lifestyle that the other half of my brain dreamed of when left to its own devices; one side always contradicting the other. As I began to settle into one way of life, my cravings for the other grew stronger. Was it possible to have both? A foot in both camps? A healthy, crazy-rolling-adventure/simple-sedentary life balance? Perhaps I was just greedy. That was it. A greedy dreamer. Still, it worried me that my cravings for home now, would soon be replaced with a need for more action later. Never settling. Forever ruined by the excitement of my exploration. Then again, perhaps I would never want to leave Wales again; my desires fulfilled. By the time these tyre treads had worn down, I guessed I would have my answer.

I really didn't want to go to the city, but it had to be done. The sprawling, choking capital city of Addis Ababa, that sat on the highlands, was another necessary evil I would have to endure. I would have to tackle its great smog and population of four million to apply for my next visa, into Sudan. I still had no phone signal. I found out later that the government had cut access to mobile phones to try to stop the spread of the protest. I didn't know if Addis had protests, or indeed if my friends were well, but I guessed I would soon find out.

As I fought my way through the city throng, working my way towards the hostel where I hoped I would find David, and his friend Falgony, I found myself on busier and busier roads. We were close to the centre and so I thought nothing of it. This last road though was at a standstill. Again, I thought nothing of it, and began working my way through the cars, with the drivers growing impatient and keeping their hands firmly on their horns. This was not an unusual scene.

I turned and looked to my right – there was a clearing in the buildings. A large green square, maybe a football field. There were hundreds of people on it though. *What was going on?* I saw no match. I saw no game at all; just people gathering. As I squeezed through the line of cars, I noticed many people on the pavement now. Some were running towards the front of the line; others were running away from something. Then up ahead I realised the crowd were in the road, just a few cars ahead. I stopped and asked the guy in the car with the open window. He did not speak English, but he gestured for me to turn around and go back. He was clearly stuck. I wasn't in much of a better position myself. There were too many cars in the way. The kerb of the central reservation next to me was too high to ride over from this angle, and this close. I would need to take a run at it, but how could I, with so many cars in the way? *Should I ride through the crowd ahead?* I looked again. They seemed to be gaining momentum and I wasn't altogether sure I wasn't about to ride straight into the middle of a demonstration. The cars were going nowhere, and the crowd was increasing in size and volume all the time. There must have been two hundred people all shouting with arms waving. I was aware that some people had been killed in some of the other cities. Some by angry mobs, but mostly by the military who had gone in guns blazing. *Was I about to ride straight into the middle of another massacre?*

I didn't hang around to find out. I had seen enough. I gestured for the guy next to me to stay still as the car in front of him inched forward. Just enough for me to squeeze Rhonda in. The gap I had left would soon be filled so I shouted over my shoulder and waved my hand at the driver. Thankfully he stopped. Somehow, I managed to get the front wheel straight enough, and with enough momentum, to get over the foot-high kerb. I didn't stop. I rode over the reservation and onto the other side of the road and away from the trouble. Eventually I found my way to the hostel, where I found David and Fal camped out in the car park.

"Good to see you", David said, as I pulled in.

"You too" I replied, "I think I just had a very close call with a demonstration."

"We had a *really* close call" laughed David, "Did you get my email?".

"I got the bit that said be careful but nothing more."

"Come and meet Fal, and let's grab a beer. I'll tell you all about it".

Fal was a lovely lady of Indian origin who lived in Canada. She and David had worked together as graphic designers in Vancouver. Fal had been jealous of David's upcoming adventure so he had invited her to come along for one leg of it. She had decided on Ethiopia and was certainly getting her money's worth so far, by all accounts! They had headed towards Haridwar – the place I had planned to go next – and had come across crowds in the street. Then the military showed up and things got pretty violent. They turned back and even picked up a couple of the locals who were trying to escape. Several people had died already. Thankfully, they got out unharmed, but not without quite a scare.

"Well that's Haridwar out then" I said, "Bit too much adventure for me!"

Chris and Erin turned up the next day and we all went together to experience the chaos and rugby tackles in the Sudanese embassy for a couple of days. A little man would occasionally come out from behind the toughened glass screens and beat people back from the counters. If anyone got out of line, he would grab them and manhandle them out the door. He kept clear of us though, as if some magical force protected us. I guessed it was the colour of our skin. After the hassle with the Ethiopian embassy in Nairobi I decided to go with my Canadian passport straight away. No one minds the Canadians.

I was right. When we arrived at the desk, they took one look at my passport, then one look at Chris and Erins, and said,

"Canadian. No problem. American. No visa".

It would take a couple more days for them, but eventually we all hit the road together; heading towards Lalibela for Christmas.

Ethiopia does not celebrate Christmas until the 7th January. Based on the ancient Coptic calendar, the Ethiopian Calendar is also seven and a half years behind the Gregorian calendar, owing to alternate calculations in determining the date of the annunciation of the birth of Jesus Christ. There were always differences with each border but never before had I travelled back in time! We were now in 2010 not 2017!

Our Christmas Eve (24th December) was spent around the campfire listening to Christmas songs previously downloaded by Erin until I persuaded the group to let me put a Motown mix on instead; still nice and chilled and of the right vibe. Erin and David cooked two separate meals. David's offering was a duck that he had been saving in his tiny freezer for nine months! It had travelled with him all the way from France. Erin offered a stew with dumplings. Both were delicious and we even had pudding of banana and melted chocolate, cooked on the fire. I certainly hadn't expected a three-course meal and it was lovely to see everyone busying around the campsite, really getting into the Christmas spirit before snuggling up in our sleeping bags, content and glad to be in such good company.

The following morning, I snuck out of my tent early, found my one clean pair of socks and stuffed them with previously bought 'local' chocolate.

It wasn't exactly Green&Black's, but it would do. I then hung them on the other two vehicles with a note saying, *'Merry Christmas! Love, Santa'*. I don't think they believed me when I asked for the socks back later and said that Santa had borrowed them.

Lalibela was a lovely town. The population is almost completely Ethiopian Orthodox Christian. Lying in the Amhara region of Ethiopia, it is famous the world over for its monolithic churches cut out of the solid rock. We spent a few days there wandering around them and enjoying the calm after the city. This was my fourth Christmas on the road and one I would treasure, along with all the others (Australia on the beach, Mexico on the beach and Canada by the lake). I was never a big fan of Christmas back home. The build-up, the adverts and the money spent on things people didn't really want. These last four had been so different, so simple, with focus on all the right things: food, fun and friends. The only thing missing was family, but thanks to modern technology, we even managed that, in a virtual fashion.

From Lalibela we rode up to Mekele in convoy. I was the fastest vehicle because I could ride the dirt roads faster, nip between the potholes on the tarmac, and just about manage the switchbacks with a little more speed than a Land Rover 130 Defender. Only just mind. I never was that fast in corners. It was fun to nip back and forth between them, race off ahead, get pictures as they came along the dirt roads and occasionally, just find some shade to chill in for a while as they caught up. It never took long; David was leading the pack at quite a pace that day.

Mekele was our base to get to the Danakil Depression. This area required an armed escort because if we strayed off the path, we were told, we were likely to get shot. It was unclear by whom, but this was exactly what had happened to a German tourist just two weeks before our arrival. There were many different stories as to who shot him, but it seemed he had a new guide who got them lost. The guide was shot in the leg but got out alive. The German did not.

Danakil is the lowest point in Ethiopia at 125 metres below sea level. The area is located in the Afar Region of the north-east, near the border with Eritrea. The climate there can only be described as cruel. Yet against all odds, people do live there. The Afar people call it their home. It is a geological depression that has resulted from being the junction of three tectonic plates in the Horn of Africa. It is the hottest place on earth. We drove in as far as we could and then walked the rest. There were volcanoes with bubbling lava lakes, multi-coloured hydrothermal fields, and great salt pans that dazzled the eyes. We were warned not to touch the water as it would burn like acid. I didn't test it.

The next day, Mike and Sue turned up in their ambulance. Then another David, a German traveller we had met back at Jungle Junction, in Kenya. This German David was a solo traveller like me, but that's where the similarities ended. Tall, with scruffy blonde hair, he travelled in a big old converted Mercedes 608 van, that we affectionately called 'the hippy bus'. And it was. He had painted lots of pictures on it, and apparently had started the journey with his wife and five-year-old child. His wife had eventually taken their daughter home and was now filing for divorce – something we would hear about many times as we travelled together. Now we were five vehicles. We had ourselves a proper convoy!

From Mekele, we drove together to Axum for the New Year. A busy little tuk-tuk-filled town with a cheap hotel hidden away in the back streets. 'Hippy' (German) David had saved a firework from Mauritania so we let that off from the roof before the two of us went in search of some local music, and alcohol. Our first stop was a rather bizarre club that offered a local dancing performance and blaring speakers. Local dancing involved a lot of shoulder movement and it wasn't long before I was joining in – the beer helping me to forget my injuries – and apparently doing rather well at it. It was Zanzibar all over again, with a mostly young male crowd and a lot of attention on me. Of course, I loved it! We joined a group of locals to the next club, which turned out to be a fairly modern affair on the inside and the dancing really got underway.

I got a little frustrated with David in the end. He had been obsessing over the beautiful waitress most of the night and was now trying to negotiate a price for her with one of the guys who spoke English, confusing her and everyone else in the process. Having a white boy trying to 'buy' a beautiful Ethiopian lady, was sadly not uncommon here, despite me trying to persuade him that she was not for sale. Apparently, she was. Of course she was. David wasn't exactly rich, but by their standards he was a millionaire and was the key to a better life. Now though, he wasn't just trying to sleep with her. He was trying to persuade her to leave her job and come with him to Europe in his big green hippy bus. She was seriously considering it too, but had no passport, so between them they were now trying to figure out how she might get one quickly,

"David, for fuck's sake", I snapped eventually, "She can't speak English for starters. How will you communicate? Bring this guy with you as translator?"

"I don't care" he replied, "I like her". He took her hand and inspected it as if checking out a pedigree horse.

"You don't know her. Secondly, don't make her walk out of her job on your schoolboy whim. You won't get her into Europe, and you'll be sick of her by then anyway. Stop filling her with false hope."

David had met many women during his travels and paid for their services in one way or another, calling them his 'guides', before dropping them off in various towns and picking up another one. He was an amusing guy. A bit dippy, in a very childlike way, and apparently completely sex-crazed, but I enjoyed his slight tinge of craziness at times. But this was going too far. He, like so many others who had come through, no doubt, were taking advantage of the locals' poverty. These women were stunning though, and he didn't see it like that. He was a good guy who wanted to have fun. That was all. He really was. He just didn't know where to draw the line. Sleeping with her was one thing, but promising her a new life? That was just crazy in my opinion. Eventually I pulled him away, without the girl. He didn't thank me for it.

We repeated the drinking and dancing the following night and I thought I was going to have to let the convoy go without me the next morning, having only had six hours sleep in forty-eight hours. Somehow though, with the help of a coffee from Chris out of the Land Rover, I pulled on my riding jacket and got myself on the road again.

Our drive that day was a long one but the route between Axum and Gondar was stunning. The dirt road section that took our convoy through the Simien Mountains was a breathtaking ride, with views for miles. Overhangs covered in dripping, rainforest-like shrubs, big drop-offs from the steep and narrow track, and the ever-present threat of landslides. It was wonderful.

Gondor would see us all leaving at different times. I found out later that German David had managed to get himself arrested before he left Ethiopia after hitting a small boy with his van. He spent three months in an Ethiopian jail before being deported. French David, and Fal had unexpected visa issues and had to go all the way back to Addis to sort them out. The other guys remained just a day or so behind me.

I would later put Ethiopia up there as being one of my favourite countries, but for most of the travellers I met along the way, it was a place that became too much after a while. They grew weary of the dust, the rough roads and the lack of showers. Often there was no power, water or working toilets. In fact, I would later also give Ethiopia the dishonourable title of 'worst toilets in the world', and that was no mean feat. On many occasions, I would open the door to a dark and stinking toilet hut to see a thousand cockroaches scattering in all directions as the light hit them. What was left behind was arguably worse: a squat toilet covered in, and surrounded by, shit.

The constant request for money was far worse than any other equally poor country I had visited. I could understand why fellow-travellers were tired of it. Saying "No" all the time created a kind of negativity within, and if it had been a different time for me, then perhaps my perspective might have been different, but I was on a high. Ethiopia was the perfect place for riding a motorcycle, with great dirt roads and a cooler temperature than most of the countries we had been through in Africa. Every time someone asked me for money, I simply replied with, "You give *me* money". They would look confused as I smiled at them, and then would often just smile back and not ask again.

On one occasion, a boy asked me for a pen instead of money. I replied, "You give *me* a pen", he looked as confused as those before him, but then pulled a pen out of his pocket and gave it to me. I laughed and returned it, wishing I could double it, but I had no pen to give him. It certainly was tiring though. Apart from the nights I was out dancing with the locals until 3am, I was tucked up in my tent or a crumbling hotel room by 9pm and asleep by 9.03pm. The past month had been an epic part of my journey. For me, it had been just the right degree of challenge; a nice cool temperature, and above all, it had been different from anything I had seen before. Of course, there was nothing that could have beaten that bizarre looking convoy rolling on through those mountain roads.

Chapter 31
Sudan Bikers and Egyptian Escorts

The border into Sudan at Metema/Gallabat looked more like a barnyard than an official post. There was a host of livestock from half-plucked chickens, to mangy goats, to wimpy-looking cattle milling around in the dust amongst the locals and the tea vendors – none of whom seemed in any rush to get anywhere. The assumption might have been that it was going to be a less-than-smooth ride through the paperwork, yet this could not have been further from the truth. In stark contrast to the outside, the office was quiet and organised. The only minor delay was when the officer had to down paperwork so he could go and pray. He was very apologetic – and in good English for the area – but it was no bother. I took the opportunity to find some fully-plucked and well-cooked chicken for lunch in one of the side huts – typically furnished with plastic garden chairs positioned around wobbly tables on dirt floors – then bought a sim card from a guy called Daniel who lived in the refugee camp nearby. By the time I went back, with a full belly and a topped-up phone, Allah had been satisfied, the paperwork was complete, and I was granted permission to roll on into Sudan.

It still amazed me the difference a border could make. Neighbouring countries can have such obvious differences in the space of a few hundred metres. There were two that struck me immediately when I left Ethiopia and entered Sudan. The first was the lack of livestock on the roads. There were still camels and goats, but in Sudan the shepherds would keep them just off the road with a practised crowd control that was impressive. No longer did I have to stop and weave my way through the meandering herds that gave no precedence to motor vehicles. Don't get me wrong, I didn't mind meandering herds. The novelty of riding through a dozen grumpy-looking camels never really wore off – though the kamikaze goats *were* mildly annoying – but I loved to watch the art of shepherding performed well. No dogs were used (traditionally dogs are considered haram, or forbidden, in Islam as they are thought of as dirty) but back home one of my greatest pleasures in Wales is to watch the farmers and their dogs working as one to bring a flock of sheep down from the vast expanse of mountainside for the winter. Each country has its own methods and I enjoyed exploring the diversity of tactics. The second difference in Sudan was the lack of people asking for money. We were back to the friendly, no-agenda waving now, and as much as I loved Ethiopia, I found this utterly refreshing. And thirdly, we were now out of Ethiopian Orthodox 2010 and back in The Year Of Our Lord 2017, although by the Islamic Calendar in Sudan it was actually the year 1438AH after the Hijra of Mohammed/Muhammad and his followers in our year 622AD...

It took me two days to get to the capital city of Khartoum, with an overnight stay in Al Qadarif en route. The hotel cost 400 Sudanese pounds a night. This was expensive at the usual conversion rate of seven Sudanese pounds to the US dollar, but I had bought black market at the border at a rate of 23SP to the dollar. Using the black market was standard practice for most things in Sudan. I found I still had to buy black market fuel too. Since the country had been divided into two back in 2011 (Sudan and South Sudan), Sudan's economy had dropped (with a black-market rate four times that of the bank rate) and petrol had become scarce with most stations closed or empty. It cost me twenty Sudanese pounds a litre instead of the six I would later pay at the gas stations. Still, it was cheap enough.

It was a long, straight windy road with little traffic other than military vehicles. At times there were so many of them I wondered if I should stop and listen to the news. *Have I missed something? Is Sudan preparing for war?* There were military camps everywhere, and a lot of movement on the road. I never discovered why, but I guessed it was something to do with the long-standing border dispute between Ethiopia and Sudan which occasionally erupted into fighting and deadly skirmishes.

Once deeper into the Al Taif area of Khartoum I turned left at the large flock of city-dwelling sheep, then right past the pack of scrawny street dogs, and onto a sandy, pot-holed back road where I found the guest house I would be staying at. It didn't look like much from the outside, but what did, in this part of the world? A thick layer of sand covered most things, so why bother? Keep it simple and plain. You never could tell what was inside.

I parked Rhonda outside and walked into a pretty courtyard with a pool in the middle, a few potted plants and some tables placed in the shade of a pergola that had a grapevine successfully conquering it. There was no one around. I walked to what looked like an office door, opened it and found an empty desk with a half-eaten sandwich on it and a TV showing one of the popular Arabic soaps. I'd seen this soap in a few places. "Very hammy acting", I'd thought, and then giggled at the irony of it being 'hammy'. Perhaps in a Muslim country it would be more appropriate to call it 'beefy' or 'goaty'. Anything but ham.

"Hello!" I shouted, and waited. Nothing. Then I heard a noise outside. It sounded like a scraping chair; as if someone had just got up. Odd. There was no one there. Then, as I looked over towards the pergola, the chair moved all by itself. It scraped sideways and then fell over, taking the adjacent chair with it.

"What the…"

Part of my view had been obscured by plant pots, but when I moved around slightly, I found the culprit under the table. A very large, and by the looks of him, very old, tortoise was making his way towards me, and taking the chairs with him.

"Well hello there big man. You gave me quite a fright!" I giggled striding the distance between us to save him time…possibly three months at his rate! As I tickled him under his chin, the door behind me opened and the manager came out to greet me,

"Ah hello. I'm sorry Madam, welcome to our guest house".

That evening, Mohamed, from the Sudan Bikers Club came over to welcome me. I knew nothing about him, apart from what was on his rather sparse Facebook page, which mostly showed pictures of him doing wheelies, and a message he had sent me some weeks back inviting me to meet him and the gang when I arrived in Khartoum. It was a short initial visit, but taking a quick look over my bike Mo said,

"You need a new sprocket".

"I know. Can you get that size?" I asked hopefully, and after a short debate over what size I actually needed, he left, promising to make some enquiries and pick me up for coffee later that evening.

I don't think I spent any money while I was with the Sudan Bikers –

much as I tried,

"This is not our way" they would say every time I got my wallet out to pay for the copious amounts of Turkish coffee and cheesecake we consumed over many a great conversation over the next few days. These guys were into *big* bikes. They were *men*, and as such, the more horsepower the better as far as they were concerned. This was a place where testosterone was king and so, determined to hold my own, we light-heartedly argued over everything from brand, to horsepower and back to sprocket size again.

"The only size we can get you is a 45-tooth" Mo announced, on Day Three.

"That will be like a tractor" I said

"It will get you to Egypt and anyway, you don't ride fast so it's OK"

"How do you know how I ride?" I enquired, knowing damn well what he was going to say. After all, it was a revelation that I was riding at all.

"You are a woman", he confirmed. That did it. I was annoyed now.

"The sprocket will be fine. I'll take my chances with the old one". This was not wise, but I refused to be told what to do with my bike and I was quite prepared to 'cut my nose off to spite my face' in this case.

"Then you must not ride any sand", said Mo, "Stay only on the road."

'We shall see." I said with a wry smile. Mo smiled back with a sigh.

"You are a stubborn woman" he laughed.

They were a great bunch and we got on well, but I did find myself getting more stubborn and possibly argumentative the more time I spent with them. Our cultural differences showed in some cases, yet we shared a similar sense of humour and most of it was taken, if not always said, in jest. The guys showed me the sights of Khartoum and taught me a great deal about Sudanese culture. We visited the wonderful National Museum and Tuti Island; the point where the White Nile and the Blue Nile merge, and on the Friday (the holy day of the week in Islam), I was dropped off at the weekly 'whirling dervish' ceremony. The boys would not stay and seemed to take offence at such proceedings, calling them "crazy people". The dervishes gather an hour before sunset around some of the big mosques in Khartoum and Omdurman. A circle is formed, and the ritual begins. The ceremony starts with the *Madeeh* – chanting words of gratitude to the Prophet Mohammed/Muhammad). The audience interacts with the chanters, dancing to the rhythms of the percussion instruments. This Sufism (a mystical form of Islam) is about the purification of the soul in pursuit of inner peace.

The famous whirling dance came from the Mevlevi Order in Turkey and is just one of the physical methods used to try to reach religious ecstasy. While the better-known Mevlevi dervishes wear white robes, these Sudanese dervishes were dressed in green and red. Then comes the *Zikr*, in which the dervishes repeat the word 'Allah' many times. The dervishes start whirling around inside the circle. With the music, the fragrance of burning frankincense, the endless repetition of religious chants, and the dizziness, they go into a state of trance. It was a wonderful experience and the build-up of energy was electrifying. It reminded me of being in the Green Fields of Glastonbury back in 1994 when I and a load of tripping-hippies chanted the rising of the mid-summer solstice sun, while a guy in a kilt climbed a wooden totem pole and played the bagpipes! It was awesome! That was the year before the television cameras came and ruined it all.

On the Saturday before I left, Ihab, one of the younger members of the club, invited me to a wedding party. He assured me I was fine in my grey *Bench* hoody and jeans, but I don't think I have ever felt so out of place. The men, mainly on one side of the room, were dressed in traditional startling white djellabas, while the women on the other side wore the most exquisite and colourful silk dresses. Everyone sat around luxuriously-decorated tables being served a succession of sugary treats from silver trays. This was a pre-wedding party for the groom and there must have been three hundred people there. The groom entered the room dancing with a line of people behind him. People stood up to hug and kiss him before joining in. All this was filmed from a very professional-looking crane that floated around the room and despite sticking out like a sore thumb, I was made to feel very welcome.

Khartoum was a city that really did welcome guests, though I saw few others. A city that had a heart as big as the tripartite metropolis itself (South, North and Omdurman).

Before I left Khartoum, I was presented with a leather waistcoat with a 'Sudan Bikers' badge proudly stitched on the front. I guess that made me an honorary member; probably their first female. And honoured I was. I made some great friends. In fact, I almost became too close with one of the group; certainly, in *his* mind. He was a married man, and so despite his many advances towards me, I pulled away. Aside from my own moral code, Sudan was not a country that took kindly to sex out of wedlock, never mind adultery. It could have meant arrest for both of us had we indulged; if caught I would have been sentenced to either death by stoning or a flogging and three years in prison. Still – the danger did make it briefly tempting. Illegal sex in Sudan was almost as appealing as sex in Antarctica with a Ukrainian meteorologist! I was a wicked western woman all right, but not *that* wicked! Besides, he really wasn't my type. I'm not sure I could have got over the ingrained misogyny.

It was only a short ride of about 225-kilometres from Khartoum to the Meroë pyramids the next morning. I had planned to camp there but I arrived too early, it was too hot and there were too many people. Local tourists, guys offering camel rides *and* I had to pay to go in. It was a waste of time. Not what I had expected at all, so I continued to Atbara and found a cheap hotel with no water and no shower. I put on some cooler clothes over my sticky body and went in search of some fried chicken!

From Atbara, I rode north to Karima along a perfectly straight black line of tarmac that cut through the Nubian desert like a welcome mat for bikers. Smooth as a racetrack and not a pothole or killer road goat in sight. I switched on the cruise control (a piece of plastic jammed between the brake and the throttle in my case), turned up the in-helmet speakers and enjoyed the uninterrupted beauty of the Sudanese wilderness. The only enemy here was the wind, which never seemed to be going my way, and never stopped. This drastically increased my fuel consumption, which was still a problem in that area. Any fuel stations I did find open were often dealing with queues of irritated drivers. Thankfully, as a woman and a guest in the country, I was always offered a spot at the front of the queue.

The small towns in the desert were a wonderful place to lap up the warm inviting culture of Sudan, although the welcome I received in Karima was completely unexpected and while warm, was not Sudanese *at all*. It was British.

As I rounded the corner into the town square, I spotted an odd-looking vehicle with two guys next to it looking slightly bewildered. They were clearly not local and stood out a mile, despite the layer of dust that had attempted to blend them in. I raced over, jumped off the bike and introduced myself, "Hi, I'm Steph. How you doing guys? You have a problem there?"

"I'm Reece. This is Matt" came the reply, "and we know who you are".

"Oh Wow. You're British. How lovely to hear your accent. How did you know who I was though?" I asked, a little bemused.

"We have been following your journey", replied Matt, "I can't believe we've bumped into you".

Matt and Reece had left home from sunny England three months earlier on a Honda SH300i scooter with a homemade sidecar attached and were on a mission to become the first people to ride a scooter and sidecar around the world in aid of Anti-Slavery International – their chosen cause. They had made it this far with no mechanical skills to speak of and now had a minor problem in the form of a loose exhaust. After a good old chin wag, we went our separate ways. I would see them again when they returned triumphant to the UK twelve months, and 34,000 miles later.

That evening I went in search of food from the street vendors. The one I eventually stopped at remembered seeing me earlier and so proceeded to do motorbike impressions. I did sheep impressions in return as I wanted some lamb for tea. I didn't get it. It seemed every vendor in the square was doing chicken and there was nothing else on offer. I got a half-chicken which he had cooked with a cigarette in one hand and a blow torch in the other while I had sat patiently on my plastic chair next to him. Everyone smiled, stared or said hello as I sat devouring my meal with my less-than-clean hands. The cats scrabbled about as usual for the bits that fell, and I got brought some titbits to try from other vendors. All using *their* less-than-clean hands too of course. If my body wasn't used to alien germs by now, then it never would be. I really didn't care.

After dinner, I went back to my room for an early night, with a plan to get up early to catch the sunrise over the nearby pyramids. The pyramids here, in the ancient Nubian area of Napata, were older, and from my point of view, much better than the ones at Meroë: no-one around, no fence, and no hassle. Perhaps I could attempt to fly my drone before I had to send it back home. I'd discovered I couldn't take it across the border into Egypt as it was illegal, and if found, they would confiscate it. Mo (from the Sudan Bikers Club) had thoughtfully arranged to have it sent back to the UK for me from the border. It would travel back to Khartoum with a passing bus driver and he would send it on from there – probably at great expense to himself.

It was still dark when I got to the pyramids at 5.30am, and beautifully eerie as the sun began to rise. Structures thousands of years old, unprotected and largely unvisited, I found myself alone in the shadow of their magnificence. Just me, my motorcycle, and five thousand years of history, surrounded by soft golden sand. Now *this* was a place for contemplation. After a few minutes of just that, I began setting up.

The drone didn't work at first. In 'track' mode it should have followed my movements, but this was often glitchy. I switched it off and on again, and it was ready to go…briefly. After just a few metres, it stopped. Distracted, I ended up riding far too slowly through some soft sand and getting the back wheel stuck – buried right up to the swinging arm. Of course, I tried to get out and of course I just dug it in deeper. There was no bottom to this sand. With all my strength and bad shoulder, I managed to dig her out, get her on her side and drag. I then managed to heave her upright again with great effort (with luggage still attached) and promptly buried her again! *Arghhhh!* It was over an hour before I decided to go in search of help by walking a hundred metres to a high point and waving frantically at the first car that came past. It turned out to be a policeman. He turned off the road onto the sand, and as he approached, I started doing motorbike falling over impressions. Surprisingly, he understood and laughed at my explanation and predicament. As he reached me he gestured for me to get in, but as we drove across the sand, he too got stuck. Thankfully we *both* found it funny.

Much pushing, panting, shoving and buried police-issue flip-flops later, and we managed to get both vehicles out and I was on my way again shouting,

"Shukraan! Shukraan!" (Thank you! Thank you!) over and over until he could no longer hear me.

Later that day I found myself on a 200-kilometre stretch of straight empty road through the desert to Dongala. The black tar looked so dramatic in contrast to the vast beige nothingness around it, and I stopped about halfway along to do some filming. It was easy on these stretches. There was no one around so I could safely leave my camera and tripod on the side of the road and start rolling while I did a few 'fly-bys' – something I had done so many times over the last four years.

I'd just finished packing everything back into its carefully allocated compartments on the bike, when I noticed a truck in the distance heading my way. It was the first vehicle I had seen on this stretch. As he came nearer, he began to slow down. This was not unusual. The truck drivers often kept an eye out for me and would check I was OK by slowing down and putting up a questioning thumb. I would reply with a thumbs-up and a broad smile to make it clear all was good and I wasn't in need of assistance. They would wave and drive on. I loved them for it. I followed this same routine that I had done in so many countries and so many situations before and looking up I saw a smiling face looking back at me from the cab. He was still slowing down though, and as he opened his window, he ground to a halt next to me,

"*Shukraan.* I'm OK" I said, "All good" smiling and thumb still in the air. The driver spoke back in Arabic before pulling out a bottle of water and leaning down to hand it to me. I didn't need water but sometimes it was easier to accept than not.

"*Shukraan*" I repeated reaching up for the gift.

He then said something else, and I raised my hands in a gesture of, "No idea what you're saying mate." Then he decided to pull over with the truck and get out. As he was doing so I quickly switched on the helmet cam and turned the helmet slightly on the handlebars (where it was perched) to ensure it captured whatever was about to happen. I also took note that the truck was from Saudi Arabia and not local to Sudan. I felt it was important to make a mental note of these details. Something about his expression didn't sit well with me. He was smiling but there was something shady about him. I couldn't say what exactly – more a feeling than anything concrete. My gut instinct had certainly had its fair share of training by now.

Climbing down from his cab, he ran over, his white djellaba flowing loosely around him. I was jealous of his clothing. My black riding gear was getting pretty sweaty right now. I had had no intention of staying still this long with no shade – and every second counted in this heat. Now it seemed we were about to get into a conversation that I suspected would be nothing more than a few hand gestures and smiles and possibly a couple of selfies.

I was right in part. The man began talking at me, completely ignoring the fact that I couldn't speak Arabic. I smiled and nodded before, as expected, he gestured for me to take a selfie with him. Amenably, I took my phone out and obliged. As we posed, his cheek came right up next to mine, till it touched. I took the picture quickly and pulled away, still being sure to keep smiling. He then insisted we do it again. "One more and I'm gone." I thought. This time as he leaned and the camera snapped, he turned and kissed me on the cheek.

"Ah ha ha. No." I said. Gently, but a little more firmly now and shaking my head. "Not cool." He then pointed at me, then at him and then rubbed his fingers together before pointing at the cab. It was clear he wanted us to get it on right there and then.

"No!" I said very firmly, (yes, with *that* voice) and put my phone in my pocket in preparation to leave. I was feeling quite uneasy by now and very much aware of the fact that we were miles from anywhere or anyone. He started walking back towards the truck and gesturing for me to follow. I refused and began doing up my jacket. That's when he walked back over, pulled me in close and tried to kiss me. Again, I pulled away and used *that* voice. But as I turned and reached for my helmet, he grabbed my arm firmly and began pulling me towards the cab. I yanked my arm. He lurched forward and grabbed me again. That's when I reached for my Leatherman. It was always attached to my belt these days. Now, bearing in mind he had hold of one arm and my Leatherman needed two hands to undo, it was pretty useless. That said, the sight of a multi-tool with built-in knife was enough to stop him, and to my surprise, he backed off straight away.

I got on my bike without doing up my helmet or putting my gloves on and rode off shouting obscenities at him as he walked back to his cab. Racing away as fast as I could I felt the adrenaline energy switching from 'flight' to 'fight'. *How dare he make me feel so vulnerable*. If he had continued in that vein, I knew I would have had no chance. *How dare he. Stupid man!* I was angry now, really angry. Instead of speeding up though, I started slowing down. I even contemplated stopping and waiting for him so I could give him a piece of my mind. Maybe even scream at him in Welsh. I wanted to stand my ground and tell him what I thought, even if he didn't understand the words. *He'd understand the tone all right*. Thankfully, my sensible brain wrestled my angry brain to the floor and agreed a compromise. Instead of stopping and screaming, I opted for riding at my own pace, determined to enjoy the ride as I had been before he had arrived. He would *not* ruin my day. I wouldn't let him. No one was going to make me rush. I put Skunk Anansie on in my helmet and rode at a steady pace for the rest of the journey. I didn't stop for any more photos or filming though.

Arriving in Dongala I quickly found another cheap hotel. I was tired now. It had been a long day, so after another chicken dinner with bread – no salad or butter – I read for a bit, before remembering I had filmed the guy dragging me towards his cab. After viewing it, I deleted it. I can't say why, other than the fact that I wanted to forget it ever happened. I had hated feeling his strength against mine. That brief second had made me feel far more vulnerable than I cared to admit, and I wanted no more to do with it. Once it was gone, I fell into a deep sleep.

Riding to the Egyptian border the following morning was a strange experience. Bleary-eyed, having had no coffee or breakfast, I joined the queue at the fuel station in town, got ushered to the front as usual, then headed back out onto the windy desert road. Just out of town, I found the usual police stop checks where I expected to be waved through with a smile – but I got a little more than that. As I was passing the police car on the side of the road, just before the roadblock, the passenger shouted through his loudhailer,

"HEY YOU!" I visibly jumped out of my skin and looked over at him in a mild state of shock. There were two policemen in the car laughing their heads off. "HOW ARE YOU?" he asked, still through the loudhailer; still giggling. Clearly just a prank to wake me up and probably a chance to show off his English. I laughed back at them and waved before being waved through by the other officers who were also having a giggle at my expense. Still, it had woken me up as well as any coffee.

Half an hour later, out on the open road, and fully awake, I began to notice dead cattle along the roadside. Why were they out here in the desert? Why were they only on the roadside? There were loads of them, in varying stages of decay. Some were even tied to the signposts. It was quite bizarre. That, along with the dust devils and the strange guy in a car with dark windows who followed me so closely I thought he was going to run me off the road, made the whole experience a little eerie. It felt as if the only thing missing was a sign saying *'Turn back now. All hope is lost!'*.

I found out later that the cows are dragged out of the cattle trucks when they die (or are dying) en route to Egypt. From what I had seen, they were not treated well and so it was no surprise, but rather sad, pointless and positively wasteful.

That was Sudan. A country where beds were hard, bacon butties were illegal, showers (when available) were cold, and the only place to get a beer was at the British Embassy on a Thursday night. Some of the laws there went against every ounce of my personal beliefs. Some of the opinions on women's rights were very much against my own, but try as I might, I couldn't help but fall in love with the country and its people.

****

I was first onto the ferry that would take me across Lake Nasser and the Egyptian border to Abu Simbel and was quickly hemmed in by the cattle truck full of nervous-looking cows. They had good reason to be nervous. Their fate would be sealed in Egypt (probably with cling film and a nice polystyrene tray). Mine, hopefully, would not! The airbrakes went off and a half-ton bull looked as if he might come right over the top. I edged my way onto the stairs, not wanting our fates to collide, and imagining what that would look like if it did! Given enough time, it would make for a funny ending I guessed, *'She rode around the world by motorbike and was killed by a flying bull!'*. There would be jokes – in time.

The other vehicles soon filled the remaining gaps. There was no room left to walk around without breathing in and shoehorning yourself between the bumpers. There *was* one small space though, and that was next to Rhonda. A tiny triangle between her and the truck, just big enough for a prayer mat – or two at a push. Soon enough, two men and their mats were in there. Surveying the scene from my elevated position, it looked as if they were praying to Rhonda. It was a great image that I would have loved to photograph, but somehow it seemed intrusive to me. Perhaps not to them, but I wasn't going to interrupt to ask permission. Instead, I was accosted by a young Sudanese guy who spoke pretty good English and made *me* the subject. We took many selfies and shared some music (one ear of his headphones each), before we were interrupted by one of the crew with a personal invitation onto the bridge to meet the captain. I was even allowed to steer. It was a great start to Egypt, and all added to my excitement that I was just a few kilometres away from a glass of wine. Something I had been denied the pleasure of in 'dry' Sudan. It was time to put that right.

At the border, I had met Tom and Stephen, a couple of British lads in their late twenties perhaps, driving a Land Rover Defender. They too were heading for Cairo and once off the ferry, we travelled together to the same hotel in the small town of Abu Simbel. As soon as we arrived, we dumped our gear and practically ran to the bar! No shower, no unwind time. It was 'beer o'clock' and we didn't care who knew it! We had some catching up to do and our priorities were well and truly in order. I did take my riding boots off first though, swapping them for my trainers – much easier to run in! That night I enjoyed a soft bed, clean towels, loo roll and even had the luxury of a toilet seat. It was divine.

Egypt didn't really go well for a day or two after that. The initial border 'buzz' soon wore off and left me in Luxor with what seemed like a nation of chancers and thieves. Having spent the last four years 'bigging up' the people of the world and, in my own little way, proving the general consensus was wrong, I really struggled with this. I wanted to believe I could find the best in people, with the usual positive approach and quick-witted response that I had developed during my time on the road. Yet here it seemed I had to lower my expectations, and that put me on a downer for a while. I had been to some of the poorest countries in the world, stayed in some of the dodgiest neighbourhoods and managed to find common ground with some of the sketchiest people, but I had to change my game in Egypt and I couldn't excuse or accept the natives' behaviour. It was clearly an ingrained habit and it annoyed me.

In the first few days I had petrol stations trying to rip me off, hotel staff squirrelling around in my room as soon as I left it, and a waiter charging me for breakfast when it was already included in the price. When challenged, they backed down or suddenly spoke no English, but the waiter was actually quite funny. When he tried to come up with an excuse, I lowered myself to saying something like,

"In *my* country we call this bullshit!".

"Please, what is bullshit?" he replied.

I had to laugh. The timing and delivery, combined with his innocently inquisitive face and strong accent was comedy genius. He was a rip-off merchant if ever I saw one.

Street vendors played every trick in the book to get what they could out of me, even when all I wanted was a bottle of water, and most of the guys I spoke to were 'professional masseurs' who could offer me a good rub down should I need one! When I suggested that it seemed they were *all* masseurs, they would warn me not to trust anyone else. I felt disheartened for a while but soon I found myself adjusting my outlook and seeing the funny side. I had switched off then switched back on again (in best computer fashion) and suitably re-set, I was ready to play the Egyptian game. "You want to play dirty?" I thought, "Bring it on!" I found the best approach, as always, was with a sense of humour. Call them out, but with a smile. People soon gave up when they realised I could not be fooled, and quickly softened when they saw I had not taken offence at them trying.

As I pushed on northward, I found the good in people again, in the little 'non-tourist' villages that dotted the banks of the Nile. After being turned back by the police on the western desert road because it was 'too dangerous' to travel alone, I found myself forced onto the bustling streets heading north on the east side of the Nile. My tyres were too soft, so I stopped in one of the little villages to get some air. I sat listening to the call to prayer while the helpful garage owner pumped up my tyres and sent 'the boy' for tea! We had a lovely twenty minutes of chit-chat (despite both our language skills being sorely lacking) before I was on my way again – no charge for the air or the tea.

At the first police checkpoint on this road I was pulled over and asked to wait for the boss to arrive before they decided what to do with me. It seemed they were concerned for my welfare. As I pulled my helmet off, I was greeted with a gasp and a collective "Wow!", as if I had just stepped out of a L'Oréal advert!

"Beautiful hair", they said, "Beautiful eyes!".

"You should see me after a shower!" I replied with a grin from behind a layer of road grime. I was soon offered a seat, a cup of tea and some marijuana, in that order.

"Yes please on the tea, but I don't smoke marijuana", I replied politely.

"But why not?" said the more flirtatious of the three.

"Because there are too many police around."

We laughed and changed the conversation back to my hair, which was particularly 'helmet chic' that day!

This was the beginning of my police escort through Egypt which took me practically all the way to Cairo. The police were all very polite and stopped regularly to check to see if I needed anything like a pee break, a cigarette, a coffee, or petrol. Each truck would have between four and six armed police as they drove ahead, instructing me to stay close whilst watching my every move. It was quite off-putting as they would jump up and shout at any other vehicle that got too close to me. It was also very funny, and I pottered on behind tapping my foot to my music and enjoying all the attention. Often, they would stop and tell me to keep going alone. I would wave, thinking that was the end of my convoy, only to find a police roadblock set up just for me, ten kilometres up the road. They would pull me over, pull up the cones and explain that I was safe now because they all had guns. One officer spoke good English, so I took the opportunity to ask him why I was getting such special treatment.

"You are a very important person in our country." he replied.

"Really?" I said, with a doubtful tone and a smile. He grinned and said the last police escort had called ahead and told them there was "a beautiful lady with beautiful hair" on her way and that she must be protected. "Well, a girl could get used to this." I giggled.

Halfway to Cairo I was taken to a military base for a stopover. Once past the armed soldiers on the gate I was greeted by Oscar, who took the time to show me around the base. He was twenty-six and had started his compulsory military service a little later than the normal age of twenty. This was because he was studying at university, but he had no choice but to enter once he had finished his leisure and tourism degree (although, if he had married a European lady, he would have got away without having to do it). He hated it but only had six months left of his two years. He told me the first seven months were at barracks where, "You learnt a lot about how to survive, not showering for ten days and going without sleep.". Now he worked at the officers' digs for just five hundred Egyptian pounds per month (£20). He worked 45-days on, then got eight days off.

I was not allowed to leave the next morning until my police escort arrived. They were late but I decided not to argue and waited patiently (under close scrutiny) with as much coffee as I could drink, delivered by a small man with a tray, as I stood next to my bike making roll-ups. It was some nasty instant brand, but it kept me occupied.

On one of my short 'alone' sections that day, I decided to jump off the main drag and try another road. I don't know if there were more police waiting but I never heard anything more. The road I chose was like riding in India all over again: Chaos with a capital C. It took two hours to ride the last thirty kilometres into Cairo, with trucks, potholes, people, cows and horses and carts creating a scene much like the old *Wacky Races* cartoon from my childhood.

Once in the city, I made my way to the overwhelmingly ex-pat neighbourhood of Maadi, where I pottered around trying to find my destination. Suddenly, I spotted a small Egyptian woman with wonderfully big hair jumping around and waving a Welsh flag at me! I laughed and pulled over. This *had* to be for me! Chris (A fellow Welshie), and Rania (the Welsh flag-wielding Egyptian lady with the big hair) had offered me a safe haven in the chaos of Cairo several months previously. Once again, the power of Facebook connecting bikers had proved its worth and there I was in what turned out to be a home-from-home in a strange city. What a difference that made to me and what great hosts/friends they became over the next few weeks; not only offering a place to rest up and prepare for my last leg, but also gave me shepherd's pie *with ketchup*, 'proper' tea, a soft bed, and best of all, the feeling of being with friends from the moment I walked into their apartment.

It was mostly at the Ace Club (an ex-pat bar) that I was introduced to many of their friends and soon felt very much at home there too. A friendly community; mostly Welsh, English and Scottish, with the odd South African thrown in. One particular South African caught my eye and Rania quickly introduced us by sending him one of my YouTube videos. *Was she matchmaking? Hmm!* I wouldn't put it past her, and I didn't mind a bit. He was quite dishy. He was also a stunt coordinator in the film industry. That definitely added value for me. I could overlook the fact that he said 'mulk' instead of 'milk' and 'braai' instead of 'barbi'! He was a guy with a sense of humour and big Afrikaans shoulders. He also had contagious dance moves and a great backing voice for my karaoke rendition of Tina Turner's *Proud Mary,* which I sang in the club upstairs before dancing until around 4am the next morning.

Soon after, I was on a plane to Baltimore, (Maryland, USA), to present at the Timonium Motorcycle Show for the second time in my trip; hung over and all danced out, but smiling from ear to ear! Not only had I had a great night, but I'd sung karaoke on all seven continents, and now I was heading for a job that would earn me the money I needed to get home. Life was good!

Within ten days I was back in Cairo with a pocket full of dollars and some great memories, ready to continue where I'd left off. A few days later, Anton (the dancing South African with the big shoulders), accompanied me to Alexandria, where I was to drop Rhonda off at the shippers', ready to be taken to Italy on the people-free ferry for the last leg of my journey.

Hilariously, his Jeep Cherokee broke down en route (so much for support truck!) and finished the journey on the back of a recovery vehicle. My back was causing some weakness in my legs again. It was also a miserable motorway on a rainy day, so we put Rhonda on the recovery truck too. We then climbed into the Cherokee and rode into Alex together, in a truck on top of a truck. It felt very much like being on a carnival float. Cornering was *very* strange. This was definitely Africa!

The Jeep was soon revived and something hit me as we drove back to Maadi through the chaos of the Cairo traffic. Thankfully, it wasn't one of the many unpredictable cars or bike-blind truck drivers in their daily scrum. It was the realisation that I had arrived at the end of the seventh continent. Rhonda was on her way to Italy. The road was coming to an end, and with it, the journey that had given me some of the happiest, most exciting, and most terrifying days of my life. Soon I would be back in Europe and ending this chapter with no idea what I would find in the next. A surge of conflicting emotions rushed through me. None of them a stranger and all of them welcomed in like an old friend; once again, a reminder that I was alive and – most importantly – I was living!

My often questionable and always unpredictable life choices had led me to many strange places in my forty-two years. We all make some bad decisions and some of them snowball to crush you, but, in my case, some of them turn into motorcycles and lead you all the way around the world.

This journey alone had led me to some of the remotest and most hostile environments on earth. It had also led me to the most beautiful. Sometimes I would struggle to differentiate between the two. There was beauty and beast combined in most places, just as there was in most people. In the weather, the terrain, the people and the cultures. That's why I would struggle to choose a favourite place. I remembered them in moments and friendships. Borders may shape a country, but they could not define my experiences.

Africa was clearly 'not for pussies' as I was told so many times before arriving, but it certainly was a place to remember and had actually given me some of my most treasured memories. I wondered how leaving would affect me. Perhaps my love for the place was partly down to the sense of achievement I had felt in completing what I set out to achieve so many years before. I had often said that my state of mind was more important than where I was – and lately I had been on a roll – but I suspected it was more than that. Africa had given me all the ingredients for a great adventure. Africa was ever-changing and often unpredictable. That had been part of the fun, but add into the mix the diversity of culture, the unique wildlife, and...well, the occasional hunky South African stuntman, and you soon begin to understand how it could leave an indelible mark on one's soul.

Rhonda and Skully (the springbok skull) with the Himba tribe in Angola.

One of the many baobab trees I would see in Angola. The reason Angola had called me in the first place.

Shepherd boy
in Angola with
a traditional
mohawk haircut.

On the roadside
in southern Namibia.
Dusty and tired and
taking a drink from my
Kriega hydration pack.

A long sandy road in
Namibia. These roads
are tiring to ride all
day but the vast open
landscape promotes
a certain playfulness
in me.

A roadside pit stop in the shade of a baobab (Angola) for some much-needed coffee and a reprieve from the midday sun.

A very friendly local Ethiopian lady feeding me beer having just bought one for both of us from her little shop booth.

Buying black market fuel in Ethiopia always draws a crowd. I think the whole village had arrived by the time I left.

Chris and Erin in
their 110 Land Rover
Defender waving
to a local in Ethiopia.

Camel caravans
carrying salt blocks in
the Danakil depression,
Afar region, Ethiopia.

Donkeys and camels
are used in this ancient
and gruelling trade in
one of the hottest and
driest places on earth.
They travel for
a week at a time
with each load.

My warrior 'husband' from the Hamar tribe in Omo Valley, Ethiopia.

Rhonda and the elephants in Botswana.

Rhonda gets buried at the pyramids near Karima, Sudan.

Some of my Egyptian escorts en route to Cairo.

Chapter 32
Home by Seven

The separate journeys of Rhonda and myself from Africa back to Europe went smoothly. While Rhonda re-traced the path of the Allied invasion force in 1943 and landed at Salerno, south of Naples, I flew from Cairo to Rome, (with a lot of my kit) then caught a train down the west coast of Italy to join her. She was waiting for me in the car park, with keys in the ferry company's office. No stress, thankfully. Back on the road though, things were fairly uncomfortable in the wintry early March weather.

Raindrops slipped between my well-used visor and the duct tape that held it in place, dripping onto my nose and forming a thick layer of mist on the inside of the visor that was almost impossible to see through. The clouds obscured the sun, leaving a greyness that I had failed to recall during my 'nostalgia-for-the-rain' moments in the sweltering desert stretches of Iran or the chaotic traffic scrums of an Indian summer.

What was it I always said? Oh, yes, "Riding a motorcycle is the true path to happiness and enlightenment.". I must have said that back in the days of warm summer breezes. The sky was now in full sarcastic mode and beating a tune on my helmet that said, "Welcome back, SUCKER!". Still, the weather did not fully reflect my mood. Despite the 'Beast from the East' flicking its nonchalant wet tail in my face, it was good to be back in Europe. I was finally closing the circle. I was coming home, having ridden all seven continents. It felt good. OK, it kind of felt good. It was soggier than I had imagined, and I wasn't doing so well physically. I looked terrible too. My hair was beyond helmet-hair-chic now. It was so weather-beaten and brittle that I couldn't get a comb through it, but I *was* closing that circle.

Riding the wet and boring *autostrada* between Naples and Rome, I decided to call it a day early, and pulled off towards Cassino. I was betting there was a motel nearby. I'd only been riding a couple of hours, but the closer I got to home, the more my joints seemed to hurt; I put it down to the weather rather than anything psychological.

As I unfettered my complaining body from my faithful old motorcycle outside the last in a chain of uninspiring budget motels in Italy, I wondered when my bags had got so heavy and how they got heavier every day. I had happily carried this luggage with me all over the world, and now, on the final leg, I was almost unable to carry it from my bike for the aches and general lethargy I was feeling with every move. *Talk about limping home!*

After a warm shower, I lay on the bed and checked my emails only to find one from Chris – the Swiss guy that Shane and I had had met in Mexico, way back in December 2015 with the Africa Twin sidecar outfit and the family he had created whilst on the road. It read;

Hi Steph

Just saw on your blog that you are back home soon. Wow, you did all seven continents! Congratulations. Weather is horrible in Italy right now. I would stay in Egypt for a couple of months until it is much warmer in Europe ;-) Don't forget to say hello if you are close to Atina :-) Best wishes and a lot of fun for the last leg of your magnificent journey. Chris, Francesca and Leonardo.

I had forgotten they'd decided to settle in Italy. I guessed I'd missed them but pulled out my map anyway on the off-chance.

"Atina, Atina" I muttered as I scanned the length of the country, "Where are you Atina?". Then I found it, "No way!" I laughed. Atina was only ten miles away. I had taken the turning off the main road in exactly the right place. If I followed this road I would come right to their doorstep! I replied straight away,

"Put the kettle on. I'll see you for breakfast. I'm only in bloody Cassino!".

The road was full of wonderful little surprises like this. Chance meetings, chance moments that lift your spirits just at the right time. The more I travelled, the more I found them, and it seemed the road wasn't quite done with me yet.

It was so good to see them and catch up over coffee. Leonardo was a bit bigger by now of course, but Chris and Francesca were just as I remembered them. I dried my soggy gloves on the fire as we caught up on the last two years since we had last met, and after just a couple of hours and a croissant, I was off again; still transient, for now. As I pulled away, I wondered how many inspiring moments like this I would find back home. After all, you have to be 'in it to win it'. You have to put yourself out there for life and luck to find you in the first place, right?

It was a long ride up to Pete Bog's godfather's place near Grasse, in Provence. Then another long ride up to Motobreaks, between Limoges and Poitiers in central France. But it felt very good to get there.

"Full bloody circle baby! GET IN!" I shouted as I met Chris (the owner) at the Motobreaks HQ. My parents, Dani 'the leg' and 'Smilie Steve the fork yoke' turned up shortly afterwards. Pete 'Bog' met us the following day. They had all ridden their motorcycles through horrible conditions to keep a promise – despite my best efforts to stop them under the circumstances. It was great to see them all and Chris had kindly put on a party for us, including a wonderful helmet-shaped cake!

This, of course, was where I should have met up with Tim – almost exactly one year after his death. The place where I could buy him a beer, give him a big hug and thank him for his friendship. A place where I could finally say, "Thank you for having faith in me". And I did, in a way. Holding back the tears, I bought him a beer, put it on the bar (that was now named after him) and quietly said,

"Thanks mate. We did it.", before taking a deep breath, knocking back a tequila, and partying until the early hours. For me. For him. For all those who helped me along the way. For my parents and friends who had cared enough to brave the freezing temperatures on their bikes. For bikers everywhere who grab life by the handlebars and go for it. For every person who ever took the seed of adversity and nurtured it into something beautiful.

****

As we crossed on the ferry from Dieppe, the weather warnings in the UK were still in full force, and I put a message out to all the riders from across the country who had planned to ride out and meet us at Newhaven not to come. I didn't want anyone risking these conditions. In the end they had no choice but to pull out as snow covered much of Blighty.

Once off the ferry, we made our way to a hotel in Newhaven where we would stay the night before riding over to officially close the loop at the Ace Café in North West London, where I had started four years ago, almost to the day. As we pulled into the hotel car park, I could see a lone figure stood near the doorway smoking a cigar. It was dark, but I could just about make out who it was, and I couldn't believe my eyes! I pulled my helmet off and ran over,

"You bloody nutter. What are you doing out here in the cold?".

"A promise is a promise" he replied in a half-drunken Irish slur (he had been waiting a while) and handed me a small wooden box. I knew what it was immediately.

I hadn't seen James Fitzsimmons since our adventure together back in 2008; ten years previously. He had been one of the Enduro Africa gang (My first motorbike adventure abroad where I believed I would surely die for my lack of off-road ability, and the beginning of so much more). We were all still in touch on social media, and four years previously, he had promised he'd be waiting for me at the port with a bottle of Middleton Extra Rare whiskey – a whiskey we had both agreed was a damn fine drink. And there he was – this retired military man who had ridden his motorcycle all the way from Ireland and refused to let any beastly storm get between him and delivering a promise; a bottle of perfectly blended and oak- matured golden pleasure. What a guy!

Calum, another Enduro Africa team member, and his wife joined us by car along with John, a blog follower whom I'd spoken to many times but had never met, and that night we had a good feed, a great catch-up and a well-earned sleep, ready to brave the weather the next day.

Despite the freezing conditions and snow-covered verges, the roads were clear between Newhaven and the Ace Café in the morning, and so our small posse of die-hards went for it. We were joined en route by a couple of bikers who'd been following my blog, and then by my cousin, a Metropolitan police officer, who escorted us into the Ace Café car park on his police motorcycle, and I can tell you, a police escort is a sure-fire way of making a girl feel special – or in trouble; depending on your conscience. My conscience these days was clear, and I smiled then at the thought of what I had achieved. To most, this was a four-year circle; a circumnavigation of the globe. To me, it was the closure of a circle encompassing over twenty years; half my life.

I loved watching the faces in the cars as we rode around the west side of the M25 ring road. Some slowed automatically, seeing a police bike; whether they were speeding or not. Most stared over at me, probably wondering who this person was that was worthy of a police escort. This rock star feeling was about on par with the time I'd received red carpet treatment when landing at the airport in Sumatra. Actually, this was better. This was my very tall 'little' cousin escorting me in his uniform, and behind me, each on their own bike, were my mum, my dad and some very special friends.

It must have been -3C when we pulled into the Ace Café car park that day. It felt even colder. Fittingly, (and not by coincidence) March 18th was their annual Adventure Travel Day, so there were lots of like-minded world travellers, and would-be world travellers there.

Most of them were inside, huddled around extra-large mugs of tea and fry-ups. Their ears pricked up as they heard us rocking up, and as I pulled in and jumped off my bike, a few ran out to greet us. Paddy Tyson – the editor of Overland Magazine was first. I liked his magazine and had been particularly proud when my article and photographs on Angola had been printed in it. I can't remember what he said as he shook my hand. Probably something along the lines of,

"Welcome home". I do remember what *I* said though, as later, I wished I had come up with something a little more profound. All I managed was,

"Cheers Paddy. Bloody hell it's freezing, isn't it? I could do with a brew!"

It seems the British in me dies hard. The first thing I did was talk about the weather and how I'd like a nice cup of tea! I'd imagined this moment so many times but had never quite considered what to say. That would have been pretentious surely? Either way none of it was as I had imagined. I hadn't quite imagined returning as a conquering hero, but I *had* expected floods of emotions. Perhaps, a little more momentous than it felt right then. And yet, looking back, I realise I was in a bit of a daze. Over the coming hour, people, faces greeted me, and I wasn't quite registering them. I wasn't quite sure where I knew them from.

"Congratulations" came a voice from the crowd. I looked up and took the offered hand. "Thank you", I said, and looked into the eyes of a man I knew I liked. A man I knew I had got on with yet couldn't quite remember where from. Then his wife came up behind him, and as I saw the two of them together, it clicked. It was Wagner and Tatiana, the couple I had met way back in Namibia who had so kindly given me a lift into the petrified trees at Sossusvlei, and had stopped to offer me sweets and water along the long, corrugated three hundred kilometres of sandy track the following day when my energy was waning. Again, I don't remember what we said, or even if perhaps I appeared rude at any point. So many people have told me since then that they were there that day, and yet I don't really remember. I guess I really was overwhelmed. I was certainly not the same person who had wobbled out of that same car park exactly four years previously. It would take time to figure out exactly how I was different, but I knew, without a shadow of a doubt, that I was.

Epilogue

It had felt strange being home at first, awkward; as if I was revisiting a world I had never seen before. As if I'd walked in through my front door, only to find I'd accidentally walked into someone else's house. Still, I enjoyed the rest; buzzing from the little things like a roof over my head and not having to pack my things every morning. I took great delight in going to my wardrobe-sized fridge and pulling out cold milk. I rejoiced in the soft pillows and a dog at my heel once more. It was regularised, dependable; unexceptional.

Then I woke one morning, maybe a couple of weeks later, in tears, and found myself staring down into a big black hole. An enormous void, that felt freakishly similar to a bereavement, and in a way, I guess it was. I was mourning the loss of what had become a lifestyle. Moving daily had become addictive. I was always on my way to paradise. Never frustrated by annoyances of the day because I was riding off to the next. It was a kind of reverse institutionalisation. A disordered daily existence. A transformational journey. I had dropped out from the continuity of life. There had been no walls, no rules, and I'd loved it. Now I was back and it scared the hell out of me. What if I got trapped?

For a few weeks after that I didn't look at Rhonda. I put my faithful camera in the cupboard and took no interest in taking pictures or looking at any of the 25,000 photographs in my library. I struggled to write. Anything that had given me joy over the course of my journey had suddenly become impossible to bear. I focused only on the pain in my joints and took on each day with a distinct lack of energy or motivation. When people asked me what I was doing next – as they all did – I told them that I was 'processing'! If I was a computer, I would have been stuck on that annoying spinning circle, and the more they watched it, the longer it would take. I needed time.

It took several months to get my energy levels back up and for my mind to stop wandering, but slowly I found focus again, and the beauty was that I found it somewhere between home and travel; a foot in both camps.

In April 2019 Rhonda was delivered to her temporary home and is currently being enjoyed by many at The National Motor Museum, Beaulieu, in the New Forest. In June 2019 I decided I wanted to see more of my own country and walked and wild-camped a thousand kilometres across the UK from Scotland down to Wales, climbing several peaks along the way, including the three highest peaks in England, Scotland and Wales.

In September 2019, I went back to Nepal with Dani 'the leg' (as we had promised we would after her crash) and led the first group of all-women bikers from Kathmandu up to the Everest Base Camp in Tibet – with the support of Nomadic Knights.

In December 2019, I travelled through Southern Africa again (this time in a Unimog), and on my way through Zambia I picked up 'Skully' the springbok skull that I had promised to liberate and bring back to Wales. Skully now takes pride of place on my bookshelf.

In February 2020 I went back to India (this time in a tiny car called a Tata Nano) and this time at a cooler time of year, to recce for a car rally. In between, I relished my time at home in my beautiful home country of Wales.

Have I changed much? Adventure certainly means something different to me now. I look for it everywhere; in the nooks and crannies of everyday life. It's close to home as much as it's far away. It's a day trip with an old friend. It's baking a cake. It's walking the dog. It's in two tequilas too many! Adventure comes from within. Adventure is our inner child; curious, playful, energetic.

But I guess the biggest change is that now, whenever I watch the news, I see people who accepted me into their homes without questioning my 'wicked' western ways. People who fed me wine out of a teapot, or bought me a meal as I crawled out of my freezing tent or shaded me from the sun while I was fixing a puncture – or even sent me 'dick pics'! People who took me into their communities and offered not only food, but food for thought. I'm in; I relate; I am fully connected.

I like to think that I still live my life adventurously, weirdly, passionately. I still search for 'life hits' and make a daily effort not to lose my inner child and *never* to take my freedom for granted. It's all too easy to build your own prison; to become so accepting and comfortable with the boundaries you have placed on yourself that you abandon hope of ever giving your dreams a chance to be fulfilled. I still have dreams. I always will. After all, from dreams come new adventures…

Made in the USA
Las Vegas, NV
06 June 2023

73049831R00236